JAAVSO

The Journal of
The American Association
of Variable Star Observers

Volume 49
Number 2
2021

AAVSO
49 Bay State Road
Cambridge, MA 02138
USA

ISSN 0271-9053 (print)
ISSN 2380-3606 (online)

Publication Schedule

The Journal of the American Association of Variable Star Observers is published twice a year, June 15 (Number 1 of the volume) and December 15 (Number 2 of the volume). The submission window for inclusion in the next issue of JAAVSO closes six weeks before the publication date. A manuscript will be added to the table of contents for an issue when it has been fully accepted for publication upon successful completion of the referee process; these articles will be available online prior to the publication date. An author may not specify in which issue of JAAVSO a manuscript is to be published; accepted manuscripts will be published in the next available issue, except under extraordinary circumstances.

Page Charges

Page charges are waived for Members of the AAVSO. Publication of unsolicited manuscripts in JAAVSO requires a page charge of US $100/page for the final printed manuscript. Page charge waivers may be provided under certain circumstances.

Publication in *JAAVSO*

With the exception of abstracts of papers presented at AAVSO meetings, papers submitted to JAAVSO are peer-reviewed by individuals knowledgeable about the topic being discussed. We cannot guarantee that all submissions to JAAVSO will be published, but we encourage authors of all experience levels and in all fields related to variable star astronomy and the AAVSO to submit manuscripts. We especially encourage students and other mentees of researchers affiliated with the AAVSO to submit results of their completed research.

Subscriptions

Institutions and Libraries may subscribe to JAAVSO as part of the Complete Publications Package or as an individual subscription. Individuals may purchase printed copies of recent JAAVSO issues via Kindle Direct Publishing. Paper copies of JAAVSO issues prior to volume 36 are available in limited quantities directly from AAVSO Headquarters; please contact the AAVSO for available issues.

Instructions for Submissions

The *Journal of the AAVSO* welcomes papers from all persons concerned with the study of variable stars and topics specifically related to variability. All manuscripts should be written in a style designed to provide clear expositions of the topic. Contributors are encouraged to submit digitized text in MS WORD, LATEX+POSTSCRIPT, or plain-text format. Manuscripts should be submitted through the JAAVSO submission portal (https://www.aavso.org/apps/jaavso/submit/) or may be mailed electronically to journal@aavso.org or submitted by postal mail to JAAVSO, 49 Bay State Road, Cambridge, MA 02138, USA.

Manuscripts must be submitted according to the following guidelines, or they will be returned to the author for correction:

Manuscripts must be:
1) original, unpublished material;
2) written in English;
3) accompanied by an abstract of no more than 100 words.
4) not more than 2,500–3,000 words in length (10–12 pages double-spaced).

Figures for publication must:
1) be camera-ready or in a high-contrast, high-resolution, standard digitized image format;
2) have all coordinates labeled with division marks on all four sides;
3) be accompanied by a caption that clearly explains all symbols and significance, so that the reader can understand the figure without reference to the text.

Maximum published figure space is 4.5" by 7". When submitting original figures, be sure to allow for reduction in size by making all symbols, letters, and division marks sufficiently large.

Photographs and halftone images will be considered for publication if they directly illustrate the text.

Tables should be:
1) provided separate from the main body of the text;
2) numbered sequentially and referred to by Arabic number in the text, e.g., Table 1.

References:
1) References should relate directly to the text.
2) References should be keyed into the text with the author's last name and the year of publication, e.g., (Smith 1974; Jones 1974) or Smith (1974) and Jones (1974).
3) In the case of three or more joint authors, the text reference should be written as follows: (Smith et al. 1976).
4) All references must be listed at the end of the text in alphabetical order by the author's last name and the year of publication, according to the following format: Brown, J., and Green, E. B. 1974, *Astrophys. J.*, **200**, 765.
Thomas, K. 1982, *Phys. Rep.*, **33**, 96.
5) Abbreviations used in references should be based on recent issues of JAAVSO or the listing provided at the beginning of *Astronomy and Astrophysics Abstracts* (Springer-Verlag).

Miscellaneous:
1) Equations should be written on a separate line and given a sequential Arabic number in parentheses near the right-hand margin. Equations should be referred to in the text as, e.g., equation (1).
2) Magnitude will be assumed to be visual unless otherwise specified.
3) Manuscripts may be submitted to referees for review without obligation of publication.

Online Access

Articles published in JAAVSO, and information for authors and referees may be found online at: https://www.aavso.org/apps/jaavso/

© 2021 The American Association of Variable Star Observers. All rights reserved.

The Journal of the American Association of Variable Star Observers
Volume 49, Number 2, 2021

Editorial

The Range of Content in *JAAVSO*
Nancy D. Morrison — 119

Variable Star Research

V963 Persei as a Contact Binary
Joel A. Eaton, Gary W. Steffens, Andrew P. Odell — 121

Discovery of Romanov V20, an Algol-Type Eclipsing Binary in the Constellation Centaurus, by Means of Data Mining
Filipp Dmitrievich Romanov — 130

The Long-term Period Changes of the Cepheid Variable SV Monocerotis
Pradip Karmakar, Gerard Samolyk — 135

Binaries with Mass Ratios Near Unity: The First BVRI Observations, Analysis and Period Studies of TX Canis Minoris and DW Canis Minoris
Ronald G. Samec, Daniel Caton, Jacob Ray, Riley Waddell, Davis Gentry, Danny Faulkner — 138

An Analysis of X-Ray Hardness Ratios between Asynchronous and Non-Asynchronous Polars
Eric Masington, Thomas J. Maccarone, Liliana Rivera Sandoval, Craig Heinke, Arash Bahramian, Aarran W. Shaw — 149

High Cadence Millimagnitude Photometric Observation of V1112 Persei (Nova Per 2020)
Neil Thomas, Kyle Ziegler, Peter Liu — 151

Spectroscopic and Photometric Study of the Mira Stars SU Camelopardalis and RY Cephei
David Boyd — 157

CCD Photometry, Light Curve Modeling, and Period Study of V573 Serpentis, a Totally Eclipsing Overcontact Binary System
Kevin B. Alton, Edward O. Wiley — 170

Distances for the RR Lyrae Stars UU Ceti, UW Gruis, and W Tucanae
Ross Parker, Liam Parker, Hayden Parker, Faraz Uddin, Timothy Banks — 178

25 New Light Curves and Updated Ephemeris through Analysis of Exoplanet WASP-50 b with EXOTIC
Ramy Mizrachi, Dylan Ly, Leon Bewersdorff, Kalée Tock — 186

Characterization of NGC 5272, NGC 1904, NGC 3201, and Terzan 3
Paul Hamrick, Avni Bansal, Kalée Tock — 192

Retraction of and Re-analysis of the Data from "HD 121620: A Previously Unreported Variable Star with Unusual Properties"
Roy A. Axelsen — 197

CCD Photometry, Light Curve Modeling and Period Study of V1073 Herculis, a Totally Eclipsing Overcontact Binary System
Kevin B. Alton, John C. Downing — 201

Table of Contents continued on next page

Pulsating Red Giants in a Globular Cluster: 47 Tucanae
John R. Percy, Prateek Gupta — 209

CCD Photometry, Light Curve Modeling, and Period Study of GSC 2624-0941, a Totally Eclipsing Overcontact Binary System
Kevin B. Alton, John C. Downing — 214

A Photometric Study of the Eclipsing Binary LO Ursae Majoris
Edward J. Michaels — 221

Variable Star Data

Photometric Observations of the Dwarf Nova AH Herculis
Corrado Spogli, Gianni Rocchi, Stefano Ciprini, Dario Vergari, Jacopo Rosati — 232

Photometric Determination of the Distance to the RR Lyrae Star YZ Capricorni
Jamie Lester, Rowan Joignant, Mariel Meier — 247

Southern Eclipsing Binary Minima and Light Elements in 2020
Tom Richards, Roy A. Axelsen, Mark Blackford, Robert Jenkins, David J. W. Moriarty — 251

The Photometric Period of V1674 Herculis (Nova Her 2021)
Richard E. Schmidt, Sergei Yu. Shugarov, Marina D. Afonina — 257

The Photometric Period of V606 Vulpeculae (Nova Vul 2021)
Richard E. Schmidt — 261

Recent Minima of 218 Eclipsing Binary Stars
Gerard Samolyk — 265

Instruments, Methods, and Techniques

Daylight Photometry of Bright Stars—Observations of Betelgeuse at Solar Conjunction
Otmar Nickel, Tom Calderwood — 269

Errata

Erratum: Four New Variable Stars in the Field of KELT-16
Daniel J. Brossard, Ronald H. Kaitchuck — 276

Index to Volume 49 — 277

Editorial

The Range of Content in *JAAVSO*

Nancy D. Morrison
Editor-in-Chief, *Journal of the AAVSO*

Department of Physics and Astronomy and Ritter Observatory, MS 113, The University of Toledo, 2801 W. Bancroft Street, Toledo, OH 43606; jaavso.editor@aavso.org

Received December 2, 2021

The *JAAVSO* aims to provide the richest possible resource for the variable star community, and one of the ways it can do so is to include a broad range of topics on variable stars. We publish articles on almost any topic—research, data collection, history, and education—as long as it concerns variable stars.

Concerning the research and the data articles, I have been tracking the number published during the past five years on the different types of variable star, grouped into categories. Figure 1 displays those numbers. There, the categories, arranged in approximate order of largest to smallest number of articles, are as follows.

Eclipsing	Eclipsing binaries
Misc	Miscellaneous, which may mean more than one kind of variable star, unknown kind, or belonging to none of the other categories
Pulsators	Cepheids of all types, RR Lyrae stars, δ Scuti stars
Novae, SN	Novae and supernovae
Red giants	Mira variables and other pulsating red giants
R CrB	R Coronae Borealis stars and other stars with dust eclipses
Be, LBV, SG	Hot emission-line stars: Be stars, luminous blue variables, and hot supergiants
Exoplanets	Transiting exoplanets
Methods	Instrumentation, observing and data analysis techniques, and modeling

In each category, articles are of two kinds: "Research," which present research results, and "Data," which present data compilations with minimal analysis.

The individual graphs in Figure 1 show how our published articles in each volume are distributed among these categories. The first thing to notice is that there has been general (albeit not monotonic) growth in the total number of research articles per volume. Less clear is whether there has been growth in the breadth of the distribution. Volume 49 appears to have a broader distribution than the others, since the peaks are of more nearly equal height and only one of the categories has no articles. Another indicator of breadth is that the median number of articles per bin in Volume 49 is 5, while in the other volumes it is 1 or 2. Of course, these details are influenced by my choice of categories. It is too early to tell whether a real trend toward greater diversity of topic is in progress.

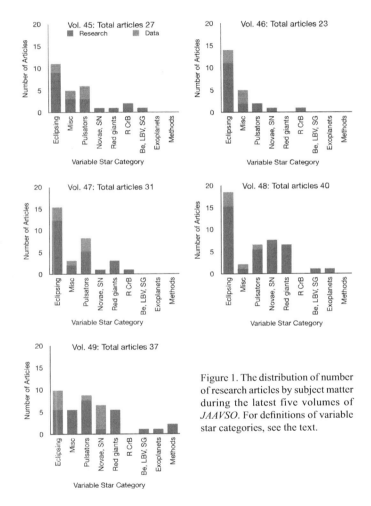

Figure 1. The distribution of number of research articles by subject matter during the latest five volumes of *JAAVSO*. For definitions of variable star categories, see the text.

In Volume 49 and looking ahead to articles in process for Volume 50, two growth areas in particular are evident: pulsating stars—RR Lyrae stars in particular—and transiting exoplanets. During the past two years, both have been the subject of group educational projects: Our Solar Siblings[1] and Exoplanet Research Workshop[2], respectively. They encourage their students to submit papers on their research results to the *JAAVSO*. We are grateful for the submissions!

Growth in transiting exoplanet research is especially to be anticipated because of the activities of NASA projects, such as the *TESS* Follow-Up Observing Program (TFOP)[3] and Exoplanet Watch[4], which provides transit modeling software, provides a facility for uploading results to the AAVSO Exoplanet Database[5], and then siphons those results back to the NASA facility to be incorporated into a global analysis for each exoplanet. AAVSO-affiliated observers are active in both.

Transit timing data are important because most exoplanets have been observed for only a short length of time, and prediction of future transit times is correspondingly uncertain. Continuing observations of transits serve to sharpen the accuracy of the periods and thereby "refresh" the planets' ephemerides. Targets are chosen in order to prepare for future observations with large telescopes such as the *Hubble Space Telescope* and the *James Webb Space Telescope*, which will aim to characterize the planets' atmospheres and other physical properties. Minimizing uncertainties is crucial so that those large telescopes' time can be accurately scheduled and then effectively utilized. We at the AAVSO look forward to contributing to this exciting effort (Zellem *et al.* 2020).

As this year ends, my thanks go out to our authors, our volunteer referees, and our devoted and highly competent editorial staff. The journal would not function without all of you.

References

Zellem, R. T., *et al.* 2020, *Publ. Astron. Soc. Pacific*, **132**, 054401.

---

[1] https://ro.ecu.edu.au/ecuworkspost2013/5236/
[2] https://exoplanetresearch.netlify.app
[3] https://tess.mit.edu/followup/
[4] https://exoplanets.nasa.gov/exoplanet-watch/about-exoplanet-watch/overview/
[5] https://www.aavso.org/databases

V963 Persei as a Contact Binary

Joel A. Eaton
7050 Bakerville Road, Waverly, TN 37185; eatonjoel@yahoo.com

Gary W. Steffens
Morning Star Observatory, Tucson, AZ; gwsteffens@yahoo.com

Andrew P. Odell
Northern Arizona University, Flagstaff, AZ 86011; deceased, May 10, 2019

Received October 1, 2020; revised October 4, 2021; accepted November 3, 2021

Abstract We have reanalyzed V963 Persei, a close binary star which R. G. Samec claimed to have components with very similar masses (q = M_2 / M_1 = 0.87), finding that the mass ratio is actually q ≈ 0.35. The system seems to be marginally in contact with a large temperature difference between the components, similar to a class of binaries analyzed by Kałużny. Primary eclipse is a complete transit, and the peculiarities of the light curve and, more particularly, its changes, are best explained by a cool spot on the more massive component and a hot spot on the less massive one. We classify the spectrum as F9–G1, present radial velocities for both components, and analyze the light curves for various combinations of cool and hot spots. The system overfills its Roche lobe in all our solutions, but the degree depends uncomfortably on assumptions about spottedness. The masses are M_1 = 1.60 ± 0.50 and M_2 = 0.54 ± 0.20 $M_\odot$. Finally, we discuss limitations on our ability to determine properties of contact binaries and the apparent absurdity of some of our results.

1. Introduction

We became interested in the close binary V963 Per (GSC 3355 0394; m_B ≈ 13.2) as an analogue of the star W Crv (Odell 1996; Ruciński and Lu 2000), a close possibly-contact system with components of decidedly different effective temperature, but with a masses uncharacteristically similar to one another for such a system. Samec *et al.* (2010a, b; hereafter SAMEC) had obtained photometry of this faint binary on two nights and analyzed the light curve, finding a transit primary eclipse and a mass ratio q = M_2 / M_1 = 0.8731. In their solution the eclipses were partial. The secondary eclipse (of the cooler, less-massive star), however, seemed to show phases of second and third contact, as though it were total.

Given its period and the shape of its light curve, this star would seem to belong to a class of close binaries with large ellipsoidal variation, transit primary eclipses (larger, more massive star eclipsed), and a temperature difference much larger than in the typical cool contact binary (W UMa binary). See Kałużny (1983, 1986a–d; Kałużny and Pojmański 1983) for a comprehensive discussion of these stars. The large temperature difference is unexpected for a binary in physical contact, for which the first-order theory of structural stability predicts a rather uniform surface temperature (Lucy 1967, 1968a, b). In addition, these stars show unexpected waves in their light variation, brightness increasing from phase zero (primary eclipse) to phase 0.5 (secondary), like the sine-theta phase variation of the well-known reflection effect. And there is usually a difference in brightness between phases 0.25 and 0.75, with phase 0.75 usually fainter and much more variable.

The unexpected sine-theta variation may be explained in a number of rather different ways depending on one's proclivities and the fashion of the day. Kałużny reproduced it in his analyses with an elevated reflection effect, acknowledging that he was using a high albedo merely as a fitting parameter, not claiming that the effect was actual reflection. A number of us have implicitly taken this approach in fitting close binaries. Another way of fitting the queer phase dependence is putting a rather large dark spot on the hemisphere of the primary (more massive) component facing the secondary. This reflects the notion that cool magnetic spots might be expected in these rapidly rotating stars. A third alternative is to use a bright spot on the inner face of the secondary, which might result from mass flowing onto it from the primary (e.g., SAMEC, section 5). Both of these uses of a spot imply some sort of temperature gradient through the neck region of the binary. In fact, postulating a smooth variation of local effective temperature through this region gives a surprisingly good representation of the data. Speculating, we may imagine that it has to do with the energy-transfer mechanism in this sort of star.

We have reobserved V963 Per, obtaining extensive photometry (see Odell *et al.* 2011; hereafter ODELL) and the first spectra suitable for measuring radial velocities. The high quality of the photometry challenges us to analyze the light variation both to fit it definitively at a single epoch and to explore the physical mechanisms for its variation. The light curve is variable, and our data for 2010–2011 define a change that we will use to test ideas about what produces such variability. The spectra measure the velocity amplitude of the primary, clearly detect the secondary, and constrain the mass ratio. They also give a much clearer determination of the spectral type of the primary star than SAMEC could infer from colors.

A second reason for analyzing this star was our concern with the quality of SAMEC, a paper with rather many errors, both careless and substantive. The most egregious of these sins of publication have been discussed by ODELL, but we can now comment on the light curve solution. It turns out that the mass ratio of this binary is not the large value found by SAMEC

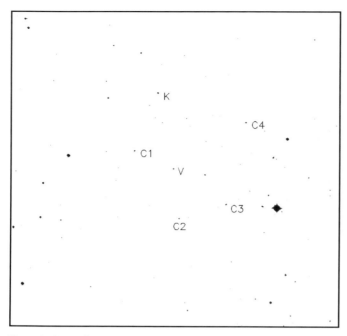

Figure 1. Comparison stars used. This is a 20 × 20 arcmin. field from the red Palomar Sky Survey; N to top, E to left. V963 Per is marked with a V; we used the four numbered stars plus the check star, K, for our five comparison stars. This check star was SAMEC's comparison.

but a much smaller one more consistent with those of similar binaries. The late spectral type found by SAMEC is also wrong, probably as a result of neglecting interstellar reddening.

We have adopted the following ephemeris for our analysis:

$$\text{HJD(Obs)} = 2455563.6833 + 0.462078\varphi, \quad (1)$$

φ being the phase, as determined by ODELL.

2. Observations

Our observations consist of both precise photometry and moderate-dispersion spectra. Odell obtained photometry on four nights in late 2010 (11–13 and 29 Dec. UT) and seven nights in early 2011 (5, 9, 11, 15, 16 Jan. and 9 and 26 Feb. UT). We are dividing these observations into two groups, the first for 11 Dec.–3 Jan, which we call 2011-dec, and the second for 5 Jan.–9 Feb., which we call 2012-jan. We are omitting the night of 26 Feb. from our photometric analyses because it fell noticeably below the data for 2012-jan. These data have been published in ODELL and are some of the most precise measurements ever made for a star of this type. Steffens has recently observed it again (28–30 Oct. and 1 Nov. 2019 UT), getting a light curve for 2019-oct. This photometry consists of differential magnitudes measured with the usual commercially available BVR_cI_c filters; they are not transformed to the standard system via observations of standard stars. Since the variable and comparison stars were all on the same CCD images, we have not corrected them for differential extinction, either. We used the same five comparison and check stars as ODELL (see Figure 1). There are roughly 226, 234, and 306 data in each color for the three epochs 2011-dec, 2012-jan, and

Table 1. Measured Radial Velocities for V963 Per.

RJD	Phase	RV_1	RV_2
2455580.6581	0.7360	35	—
2455580.6722	0.7663	17	—
2455580.6937	0.8128	43	—
2455580.7077	0.8431	24	—
2455580.7336	0.8994	17	—
2455580.7476	0.9295	8	—
2455580.7985	0.0396	–152	—
2455580.8124	0.0697	–100	—
2455580.8372	0.1236	–125	—
2455580.8515	0.1543	–160	—
2455583.6220	0.1500	–157	—
2455583.6359	0.1802	–134	—
2455583.6566	0.2249	–171	—
2455583.6705	0.2550	–170	—
2455583.6908	0.2990	–169	—
2455583.7048	0.3293	–160	—
2455583.7268	0.3769	–131	—
2455583.7407	0.4070	–143	—
2455583.7613	0.4515	–105	—
2455583.7819	0.4961	–71	—
2455583.7959	0.5264	–61	—
2455583.8267	0.5931	–5	—
2455583.8407	0.6234	–1	—
2455583.8618	0.6990	–4	—
2455583.8758	0.6993	18	—
2455931.6137	0.2517	–147	173
2455931.6277	0.2820	–147	178
2455931.6761	0.3868	–120	113
2455931.6901	0.4171	–115	69
2455931.7128	0.4662	–87	—
2455931.7269	0.4967	–75	—
2455931.7516	0.5501	–49	—
2455931.8448	0.7518	19	–213
2455931.8589	0.7824	18	–247
2455931.8812	0.8306	1	—
2455931.8953	0.8611	2	—
2455931.9183	0.9109	–11	—
2455931.9323	0.9412	–9	—
2455931.9645	0.0109	–165	–11
2455937.8126	0.6670	18	–200
2455937.8267	0.6975	17	–243
2455937.8657	0.7819	3	–255
2455940.8276	0.1919	–136	172
2455940.8487	0.2375	–151	142
2455940.8698	0.2832	–159	161
2455940.8904	0.3278	–157	96
2455940.9108	0.3719	–138	66
2455966.6535	0.0826	–126	—
2455966.6759	0.1311	–136	—
2455966.6939	0.1701	–155	116
2455966.7145	0.2146	–160	149
2455967.5918	0.1132	–135	—

2019-oct, respectively. The data for 2019 are available from the AAVSO ftp archive as the ASCII file Eaton-492-V963Per.txt at ftp://ftp.aavso.org/public/datasets/. Listed are the Reduced Julian Date (RJD = HJD-2400000) of observation, and differential magnitudes of the variable and check stars for the four passbands. This dataset is identified by the symbol 2019-oct at the end of each line. Entries with missing data are identified with magnitudes equal to 99.999.

The spectra come from the Steward Observatory 90-inch telescope and Meinel spectrograph. We took 25 spectra of V963 Per covering the range 4750–5300 Å at a resolution of roughly R=5000, on 2 nights (19 and 22 Jan. UT) in 2011. All of the exposures were 1200 s, 0.03 phase. We then bagged 27 more spectra in 2012 (5, 11, and 14 Jan. and 9 and 10 Feb. UT) covering the wavelength range 4050–4950 Å with a new CCD, which gave much better signal-to-noise and has allowed us to isolate the spectrum of the faint secondary component. See Table 1 for dates and measured velocities.

Spectral Type SAMEC inferred a spectral type around K2 from the color of the system. This is much later (cooler) than expected for a binary with the light curve and period of V963 Per (e.g., Qian *et al.* 2017, Figure 10). In fact, the relative strengths of the Hβ and Mg I b lines in our spectra for 2011 are inconsistent with such a cool star. Instead, they imply a type near G0. The newer spectra for 2012 lead directly to a similar classification, namely F8–F9, but certainly no later than G0.

Radial Velocities of the Components The spectra for 2012 have high enough S/N (and resolution, 43 km/s/pixel) that we could isolate profiles of both components in cross-correlation functions derived from them. These are based on the metallic lines between Hγ and Hβ, and exclude the H lines and G band. Figure 2 shows an example of an IRAF session (splot) in which Odell has fit a Gaussian to the profile of the primary component. It shows the averaged line profile (cross-correlation function with the G0 V star HD 50692 as the template) for one of these spectra, showing the relative strengths of the lines in the two components of the system. There Odell was fitting Gaussians to the blended profile with IRAF to get the velocity shifts of the stars.

Errors of the velocities deduced from fitting the profiles with IRAF are 10 km s^{-1} for the primary and 40 km s^{-1} for the secondary. Sine curves fit to the velocities give semiamplitudes of $K_1 = 88.7 \pm 2.6$ km s^{-1} and $K_2 = 199 \pm 6.7$ km s^{-1} for the components (see Figure 3). For 2011, $K_1 = 104 \pm 4.7$, so that a mean amplitude for both years is $K_1 = 92.3 \pm 2.4$ km s^{-1}. These values show that the mass ratio is likely no larger than $q = M_2 / M_1 = 0.46$. However, the systemic (γ) velocities of the stars differ by 20 km s^{-1} in the sense of secondary's velocities after phase 0.5 being too positive. If we require both stars to have the same γ velocity and assume the secondary's velocities after secondary eclipse are discordant, we get $K_2 = 219$ km s^{-1} and $q < 0.42$, the inequality reflecting the effect that the expected hot spot on the inner face of the secondary would have on its measured velocity. Fitting light curves using the Wilson-Devinney code gives an even smaller photometric mass ratio, roughly 0.34 (section 4.2 below), which allows for the asymmetrical surface-brightness distribution on the secondary. This small q is consistent with the relative strengths of the line components (Figure 2). Indeed, the line profiles require it, both because of the relative strengths of the components' profiles and because the hot spot on the inner face of the secondary biases the velocity-curve solution to smaller K_2. A large mass ratio, such as the 0.87 from SAMEC, is not consistent. Thus q is <0.4, and the system is roughly as we argued in the Introduction.

3. Ephemeris

We now have enough times of minimum for this sparsely observed system to begin to define its period and look for

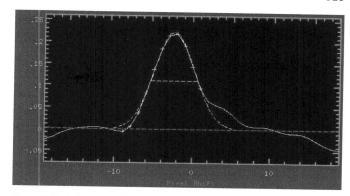

Figure 2. Part of a screen shot of an IRAF session showing the decomposition of the profile at quadrature, phase 0.25 (first entry in Table 1). Notice the difference in the strength of the profiles of the two stars.

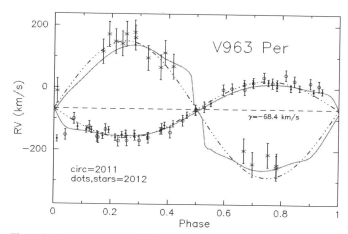

Figure 3. Velocity curves for V963 Per. Dots and stars are data from 2012 for the primary and secondary, respectively. Circles are data from 2011 for the primary. Dash-dotted (black) curves are sine curves fit to the data as described in section 2, $K_1 = 92.3$, $K_2 = 219$, and γ = –68.4 km s^{-1}. Solid (magenta) curves are fits from WD mentioned in section 4.1.4.

changes in it. Table 2 lists all the times of minimum we have found and derived. In addition to the times measured by ODELL, we have added five that Odell derived from archival SWASP (Butters *et al.* 2010) data, the one from SAMEC, four more from the literature, and four measured by Steffens. See Table 2; uncertainties listed represent measurement only, not the potentially much larger ones caused by distortions of the light curve.

The times from Odell and Steffens were determined graphically. They entered the data for a minimum in a spread sheet, plotted them for an assumed time of minimum, replotted them reflected about that time, and adjusted that assumed time of minimum until the direct and reflected data lined up to the eye. This procedure relies on the averaging qualities of human perception. However, it does not give an uncertainty; this we estimate from our experience in fitting times of minimum of GSC 3208 1986 with the Wilson-Devinney code (Eaton *et al.* 2019).

The point from SAMEC deserves a further comment. That paper listed four times of primary minimum, which we could not identify with any of the times of observation given. We think it likely that the times of observations listed are bogus as

Table 2. Times of Minimum for V963 Per.

RJD	σ	Epoch	(O–C)	Source
2454363.6680	0.003	–2597.0	0.0013	S-Wasp
2454381.6950	0.003	–2558.0	0.0072	S-Wasp
2454407.5700	0.003	–2502.0	0.0059	S-Wasp
2454438.5200	0.003	–2435.0	–0.0034	S-Wasp
2454439.4500	0.003	–2433.0	0.0025	S-Wasp
2454828.9913	0.0025	–1590.0	0.0120	Samec (2010a,b)
2455563.6834	0.0004	0.0	0.0001	Odell (2011)
2455564.6077	0.0004	2.0	0.0002	Odell (2011)
2455576.6211	0.0004	28.0	–0.0004	Odell (2011)
2455601.5749	0.0004	82.0	0.0012	Odell (2011)
2455618.6717	0.0004	119.0	0.0011	Odell (2011)
2455922.2549	0.0003	776.0	–0.0009	Banfi (2012)
2455923.6407	0.0005	779.0	–0.0014	Banfi (2012)
2455947.6703	0.0005	831.0	0.0002	Diethelm (2012)
2456312.7041	0.0005	1621.0	–0.0076	Diethelm (2013)
2457060.8021	0.0004	3240.0	–0.0139	Steffens
2458784.7950	0.0004	6971.0	–0.0340	Steffens
2458785.7197	0.0004	6973.0	–0.0335	Steffens
2458788.9544	0.0004	6980.0	–0.0333	Steffens

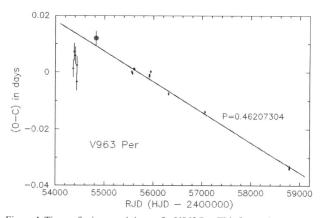

Figure 4. Times of primary minimum for V963 Per. This figure shows deviations of measured times of primary minimum from the linear elements of Equation 1. The large red symbol is the point from SAMEC discussed in the text, and the fitted line gives the revised period of Equation 2.

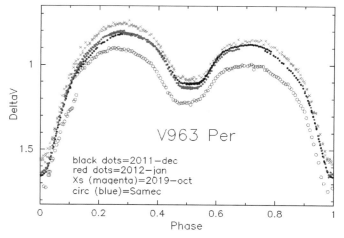

Figure 5. Visual light curves of V963 Per for four epochs. The difference between 2011 and 2012 probably results from a mismatch of the photometric bands; the lower general level for SAMEC (2008-dec), from variation of the comp star.

described in ODELL and section 4.1 below. The time listed in Table 2 is the one likely legitimate one.

The deviations of these measured times from Equation 1 (Table 2, column 4; Figure 4) do not show any trends implying a period change. Fitting a line to them gives the slightly improved elements of Equation 2.

$$\text{HJD(Obs)} = 2455563.6847(1) + 0.46207304(4)\varphi. \quad (2)$$

4. Analysis of the light curve

We have analyzed the light curves with the Wilson-Devinney (WD) code (2015 version; see Wilson and Devinney 1971; Wilson 1990, 1994; Wilson and Van Hamme 2015) because it allows spots at arbitrary positions on the components of a contact binary. This is important because the light curves of V963 Per and similar stars tend to be asymmetric, as though parts of the surface are hotter or cooler than expected in the standard picture of a binary system. In particular, phases either side of phase 0.5 are brighter, when the side of the smaller secondary star facing the larger primary is most exposed. This implies the secondary has either a bright spot on its neck facing the primary, or a dark (cool) spot on its rump facing away. Alternatively, there could also be a dark spot on the neck of the primary facing the secondary, although one does not show up in Doppler profiles of W Crv (Eaton et al. 2021). The system also shows a pronounced O'Connell Effect, being fainter by roughly 0.10 mag at phase 0.75 than at phase 0.25; that feature was roughly the same at all four epochs (Figure 5).

Practically all modern solutions of light curves are based on a standard Roche model in which the stars' surfaces are represented as gravitational equipotentials, Ω, in a system of two synchronously rotating centrally condensed masses. Surface brightnesses in this model are determined by theoretical limb- and gravity-darkening laws (parameters x_i and g_i) and some average or reference temperature for each component, T_i, with the mutual irradiation of the stars (reflection effect; bolometric albedos A_i) handled with schemes of varying sophistication.

Stars like V963 Per do not fit this model, in two ways. First, observationally, they show variations in brightness that the model cannot produce (see Figure 5). Second, theoretically, if they are contact binaries transferring luminosity through flows in a common envelope, they are not strictly in hydrostatic equilibrium and must have gravitational heads or other pressure gradients to drive and regulate these flows. So one must modify the standard model by using physical intuition to figure out how the star differs from our normal assumptions. As we explained in our Introduction, there are various ways to account for both the unexpected sine-theta wave in the light variation and the difference in brightness between phases 0.25 and 0.75. We could represent the former directly as a gradient in effective temperature through the neck, the details of the gradient based on some theory energy transfer in a contact binary. This works to first order; we have coded such a gradient into a program of Eaton's—see Figure 6. However this approach is not coded into the rather opaque WD program, so we would have to simulate it with a combination of dark and bright spots.

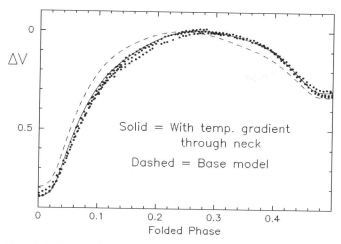

Figure 6. Light curve for a contact binary with a smooth temperature gradient through the neck between the components. Surface temperature assumed varies linearly with distance along the line between the stars between the centers of mass, between the reference temperatures for the two stars.

The difference in level between phases 0.25 and 0.75 likewise requires guesses about what produces it. It probably reflects spots of some sort on one of the stars. Furthermore, since the phenomenon seems to be common to this class of stars (see the papers by Kałużny), such spots must be a common property of the class. Are they dark spots on the trailing hemisphere of the larger, more massive primary star or bright spots on the trailing hemisphere of the secondary? Or, perhaps, dark spots on the leading hemisphere of the secondary? We suspect dark spots on the primary for two reasons: the light around phase 0.25 is relatively constant and the primary contributes most of the system's light, as well as the light variation during secondary eclipse when the smaller star is covered.

Furthermore, we have the challenge of explaining changes in light curves on surprisingly short timescales. This has become particularly acute with the very precise observations for this star by ODELL, reduced with meticulous care to minimize systematic errors between nights, in which changes cannot be dismissed as easily as those in other, less extensive or carefully handled data of this and similar stars. Observations around secondary eclipse over three consecutive nights in December 2010 showed variation of no more than 0.5 to 1%, yet data for a mere month later were consistently about 3 to 4% fainter in mid-eclipse, when only the larger primary component was visible (see Figure 5 and Figure 2 of ODELL). That change cannot be caused simply by a variable comparison star, since ODELL used an ensemble of *five* comparison stars whose relative brightnesses did not change materially.

What parameters, then, can one expect to change on timescales short enough to matter? Some properties, such as the masses, period, semi-major axis, and inclination must be constant on any timescales of interest in fitting seasonal light curves. Except in a few rather special circumstances, if these properties must change to fit two light curves of the same star, we know the physical model of the system is simply wrong. Other properties might change on various thermal or even dynamical time scales. Changes with timescales of order 10^7 years, such as the putative long-term period changes of many close binaries, correspond to the thermal timescale of a solar-type star. Thermal timescales of the outer layers of such stars, however, can be much shorter. Dynamical timescales might be of the order of a day for gas in a star's atmosphere ($R_\odot$ divided by the 10 km s^{-1} sound speed); less for flows in any free space around the binary components. The levels and timescales of variation in these stars imply that their atmospheres are quite dynamic, as Ruciński (2015, 2020) found in high-dispersion spectra of AW UMa and epsilon CrA, W UMa systems with rather deep common envelopes.

Tests of Program To get a better idea of possible systematic errors in the mass ratio derived with the Wilson-Devinney program, we calculated a light curve for W UMa with the Eaton code (Eaton 1986b, 1991; Eaton *et al.* 1993), with a point at every 0.01 phase, and fit that theoretical light curve with the WD code. The main difference between the two programs seems to be how they handle the reflection effect. Both programs gave the same light variation to within 0.001 mag but with slightly different albedos ($A_1 = A_2 = 0.5$ for Eaton, $A_1 = 0.33$ and $A_2 = 0.47$ for WD). However, when we adjusted the properties by differential corrections, the WD code found a mass ratio 0.425 vs. 0.448 in the input data. This would seem to caution against generally accepting the formal errors derived by WD as true errors of the elements. However, in a second test, we calculated another theoretical light curve with properties more like the star in question here ($q = 0.35$, $f = 5\%$, $i = 85$, $g = 0.32$, $T_1 = 5850$ K, $T_2 = 4853$ K, $x_V = 0.53$, and $A = 0.01$). In this case an adjustment of salient parameters (q, f, i, T_2, and L_1) with WD found a mass ratio of 0.3508, an inclination of 84.8, and a filling factor of 5.2% with residuals less than 1 mmag. The WD program found the assumed mass ratio to within 0.0008 when approaching from both higher and lower assumed starting points.

4.1. Application to V963 Persei

We have made several classes of solutions to test various ways of explaining the deviations of the light curves from predictions of the standard binary model. The possible combinations of complications remind us of Polonius' fatuous classification of various types of drama (*Hamlet*; Act 2, Scene 2). Suffice it to say there is a bewildering range of both cool and hot spots on both components, as well the use of a large albedo for the secondary. To limit this range, we will consider only two to three spots divided between the two component stars. Table 3 identifies the combinations of spots assumed and the properties derived.

We must fix some of the parameters of the model to theoretical values. Specifically, we adopted a temperature of the primary consistent with its spectral class, convective gravity darkening (Lucy 1967), convective reflection effect (Ruciński 1969), the Kurucz-atmospheres option in the WD code, and linear limb-darkening coefficients from van Hamme (1993) and al-Naimy (1978). Of course, those values of the limb-darkening coefficients might not apply to a star like V963 Per with likely spots on the inner, eclipsed face of at least one component.

Finally, we are concentrating on the two epochs for ODELL because they are on the same photometric system, somewhat less on our data for 2019-oct, and will attempt to fit the data from SAMEC for 2008-dec. These latter data are problematic

Table 3. V963 Per: Light-Curve Solutions.

Parameter	2011-dec	2012-jan	2019-oct	2008-dec	§4.1.2 Spots	Big A_2
$i(°)$	82.80(8)	82.96(13)	85.05(21)	83.12 (fixed)	83.43(19)	83.07(11)
$q(M_2/M_1)$	0.3353(8)	0.3397(5)	0.3165(20)	0.3351 (fixed)	0.3497(11)	0.3578(11)
Ω	2.522(2)	2.543(2)	2.464(5)	2.530(15)	2.569(3)	2.539(2)
fillout	9.9%	4.7%	19.5%	6.1%	2.2%	23.4%
T_1 (K, fixed)	6000	6000	6000	6000	6000	6000
T_2 (K)	3941(15)	4070(28)	4284(38)	3387(609)	3638(38)	4113(47)
A_2	0.5	0.5	0.5	0.5	0.5	4.15(5)
$<\sigma_{fit}>$	0.010	0.009	0.012	0.014	0.013	0.008
Spots on the More Massive Component						
long(°)	44(1)	47(2)	53(3)	45(4) & 180(4)	none	76.7(9)
rspot(°)	13.0(1)	7.9(6)	12.2(5)	15.2(7) & 13.7(7)		13.1(3)
T_{spot}/T_1	0.80	0.80	0.80	0.8 & 0.8		0.44(12)
Spots on the Less Massive Component						
long(°)	14.9(4)	38(1)	30(1)	17(4)	74(6) & 4(4)	none
rspot(°)	79(1)	90(1)	79(2)	78(7)	25(6) & 70(3)	
T_{spot}/T_2	1.303(4)	1.282(7)	1.235(9)	1.36(5)	1.27(90) & 1.36(9)	

Note: Numbers in parentheses are the errors of the last digits. All spots are assumed to be on the equator. Limb-darkening coefficients: $x_{1,2}(B) = 0.709, 0.866$, $x_{1,2}(V) = 0.573, 0.723$, $x_{1,2}(R) = 0.491, 0.623$, and $x_{1,2}(I) = 0.411, 0.523$, $<\sigma_{fit}>$ is the directly calculated average standard deviation, in mags, for all the data, weighted equally, not some arcane number calculated by WD.

for several reasons. It is not clear exactly when SAMEC even took them (their text says 2007–2008, but the dates in their data table correspond to 2008–2009), and we have identified a likely 1-day error in the times listed in their Table 1 (ODELL). This is all consistent with the allocations of observing time to various authors of SAMEC. On further reflection, the data for B and V seem to have a curious shift in phase between their first and second night which cannot be explained by another 1-day error or by using the wrong orbital period. We suspect the data for the first night do not have a heliocentric correction applied to them. Applying one (0.0046 d.) tightens up the phasing considerably. This effect is not apparent in their Figures 2, 4, and 5 because of these plots' small scale and use of rather large symbols. Furthermore, their published data for R and I do not agree with those for B and V, since they seem to be intensities, not magnitude differences as advertised. We suspect the B and V data came from an incomplete earlier reduction of the data than the R and I data. This is all judicious speculation; however, correspondence with the authors in 2010–2011 failed to obtain a coherent data set. Consequently, we have decided to use only the B and V data, reducing the published Julian Dates by one day and adding 0.0046 d. to the times for their first night. These are the light curves in Figure 8.

4.1.1. Cool primary spot/hot secondary spot

This approach represents the present canonical model for such stars. We placed a cool spot on the primary component to account for the O'Connell effect and a hot spot on the inner face of the secondary to account for the sine-theta variation. See columns 2 to 5 of Table 3 and Figures 7 and 8. The solutions fit as well as expected given the variability of the light curve and the limitations of our knowledge about these stars. The measured values for q and i agree quite well for 2011-dec and 2012-jan, somewhat less so for 2019-oct. The fit for 2008-dec (SAMEC) is much worse, and we could not begin to fit the upward slope of the apparently total phases of secondary minimum.

4.1.2. Two hot spots on secondary

The rationale for this combination, that both the sine-theta wave and the O'Connell effect are caused by heating of the cooler secondary component, comes from our finding that the secondary of W Crv has hot material on both its leading and trailing sides and that this shows up as lots of extra light throughout the orbit of that star (van Hamme and Cohen 2008). Our solution for 2011-dec is given in Table 3, column 6 and plotted in Figure 9. The fit around secondary eclipse, of the spotted star, is noticeably worse than for the more symmetrical secondary component in section 4.1.1. The fit for 2012-jan is somewhat better ($<\sigma_{fit}>= 0.009$). The change in T_2 between 2011-dec and 2012-jan is roughly +400 K, likely reflecting a drop in the brightness of the primary. Such a temperature change seems unlikely on such a short timescale, making a hot O'Connell spot unlikely.

4.1.3. Cool primary spot/big A_2 for secondary

We are looking at this approach because it worked for Kałużny and because it gives us a way to see how a spot with a different temperature distribution than assumed by WD might improve the solutions. Our solution for 2011-dec is given in Table 3, column 7 and plotted in Figure 10. This situation here is similar to section 4.1.1 but with a hot spot whose brightness is more centrally peaked. This concentration of intensity gives the noticeably steeper partial branches of secondary eclipse seen around phase 0.4 in Figure 10. Our tentative conclusion is that the spots are not likely to be so centrally bright as in this approach, a situation that might be expected of a flow away from the neck between the stars, cooling as it goes.

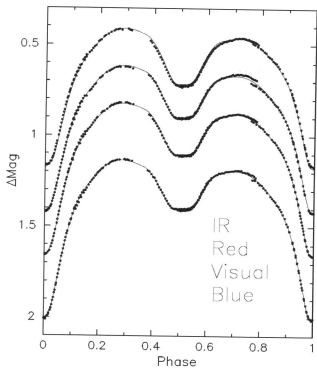

Figure 7. Light curve fit for 2011-dec with a hot spot on the secondary and a dark spot on the primary. This is the solution from column 2 of Table 3.

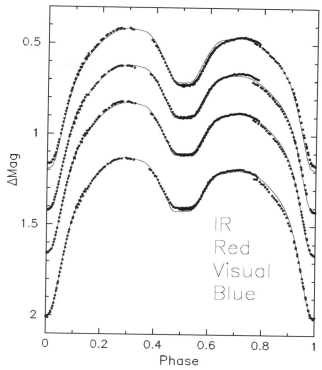

Figure 9. Light curve fit for 2011-dec with two hot spots on the secondary. This is the solution from column 6 of Table 3.

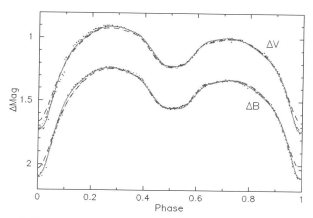

Figure 8. Three "solutions" for 2008-dec. The dots are the photometry from SAMEC, massaged as described in section 4. The magenta solid line is our solution (column 5 of Table 3), the blue dashed line, our representation of SAMEC's solution as a transit, and the dotted-dashed red line, our representation of their solution as an occultation.

4.2. Properties of the stars

We can estimate the masses of the components from a simultaneous solution of the light curves (2011-dec) with the reasonably well determined velocity curve of Star 1 for 2012 ($a = 3.23 \pm 0.36\,R_\odot$ for $q = 0.336$, $M_1 = 1.60 \pm 0.50$, $M_2 = 0.54 \pm 0.20\,M_\odot$, and $R_1 = 1.56\,R_\odot$). The luminosity of the primary star would be $2.8 \pm 0.6\,L_\odot$. These values are roughly consistent with calculations for a 1.2–1.3 $M_\odot$ star in the main sequence to give the observed luminosity and radius (Girardi et al. 2000). Models in this range with the right luminosity and radius are all younger than the Sun, which suggests, weakly, that the primary may have grown through mass exchange (cf. Ruciński and Lu 2000).

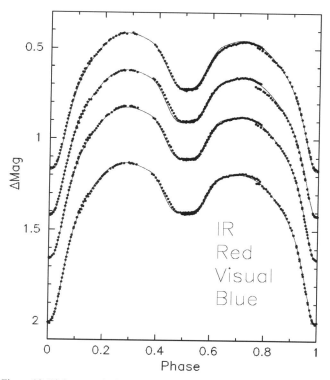

Figure 10. Light curve fit for 2011-dec for a secondary with a largeA2 and a cool spot on the primary. This is the solution from column 7 of Table 3.

5. Implications for contact binaries

Binaries with surfaces enclosed in a common envelope present a problem, both observationally and theoretically. To understand V963 Per we must consider just how it fits into the context of contact binaries as a class. First-order theory holds that these stars would be enclosed in a common gravitational equipotential surface whose local temperature is determined by gravity darkening and a weak reflection effect (Lucy 1967), a surface of roughly constant surface brightness. Warmer contact binaries with possibly radiative envelopes seem to fit this expectation, but the cooler ones with convective envelopes do not. In this latter group, Binnendijk's (1970) W types, the more massive components have markedly lower surface brightness than expected from the gravity-darkening relation. This poorly understood flux deficit corresponds to a surface temperature roughly 4–5% (200–250 K) lower than expected. What causes this deficit? No one really knows. Spots on the primary could be the culprit in these rapidly rotating convective stars (e.g., Eaton 1986a). However, the flux deficit could actually be a signature of the envelope circulation that somehow transfers luminosity from the more massive to the less massive component.

It is clear that these stars must have some sort of circulation in their common envelopes to transfer luminosity from a more massive component to a less massive one that is radiating more luminosity than it can produce. Because their surfaces are in motion, such binaries cannot be in strict hydrostatic equilibrium but must have gravitational heads or other pressure gradients to drive and regulate the flows. In the W-type systems (cooler, convective) a surface flow from the more massive to the less massive would require the primary's surface to be higher and possibly cooler.

Can we actually know if V963 Per is a contact binary? Such binaries are characterized by how much they overfill their Roche lobes. Binnendijk's possibly radiative A-type systems tend to overfill the lobe by several tens of percent, his presumably convective W-type systems by ≤10% (e.g., Smith 1984). Solutions for stars like V963 Per tend to overfill their lobes even less, if at all. However, how much the solutions overfill the lobe depends on assumptions about limb and gravity darkening, photometric elements that might not have their theoretical values in these stars. In addition, light curve programs, such as WD, do not move seamlessly through the transition from very close detached systems, through semi-detached systems, to contact systems. Instead, they have different modes for detached, semi-detached, and contact systems.

It's gratifying that i and q did not change materially between 2011-dec and 2012-jan in spite of marked change in the light curve, nor did the temperature difference between the components. What's problematic, however, is the change in filling factor over our four epochs. Naively, the fillout might change on short timescales given the likely small thermal timescale of the common envelope, as we have argued in section 4. However, such a change in envelope thickness implies a somewhat different distribution of mass beween the components, leading to a change in the orbital period on a rather short timescale. We don't think this effect is observed in these contact binaries. So what would cause the fillout to appear to change?

The results in columns 2, 6, and 7 of Table 3 suggest that spots could change the fillout derived. A cool O'Connell spot at phase 0.75 on the primary flattens out that branch of the light curve, requiring a larger distortion (more overfilling) to give the observed ellipsoidal variation at both maxima. A hot spot on the secondary at phase 0.25 (Table 3 column 6), on the other hand, makes that maximum more peaked, requiring a smaller fillout to fit the distortion around phase 0.75. This effect actually shows up in the solutions, as you can see by comparing columns 2 and 6 of Table 3.

At this point, V963 Per seems to be a genuine contact binary, overfilling its Roche lobe by 5 to 10%. But that is absurd if we believe the theoretical predictions of Lucy (1968a) and others that to be stable, two stars in contact must have a common envelope in which physical properties must be roughly uniform on equipotential surfaces to preserve (pseudo) hydrostatic equilibrium. The temperature difference $T_1 - T_2$, should also remain relatively constant, inasmuch as it represents the thermal state of the gas, for the same theoretical reason and to avoid unobserved consequences of short-term mass redistribution. To conclude, we hope this paper stimulates others of a more theoretical orientation to think about the structure and evolutionary state of these peculiar contact binaries.

6. Acknowledgements

Joel Eaton thanks Jonna Peterson and Brian Skiff for their help in locating various data and analyses Odell had done before he died in May 2019. Andy Odell would acknowledge the gracious amounts of observing time allotted to his research programs over the years by the University of Arizona and Lowell Observatory. He would also thank Patrick Wils for advice on figuring out just what SAMEC had done and Elizabeth Green for help with obtaining, reducing, and analyzing the spectra from Steward. This research used the SIMBAD database, operated at CDS, Strasbourg, France, and the SAO/NASA Astronomical Data Service.

References

Al-Naimy, H. M. 1978, *Astrophys. Space Sci.*, **53**, 181.
Banfi, M., et al. 2012, *Inf. Bull. Var. Stars*, No. 6033, 1.
Binnendijk, L. 1970, *Vistas Astron.*, **12**, 217.
Butters, O. W., et al. 2010, *Astron. Astrophys.*, **520**, L10.
Diethelm, R. 2012, *Inf. Bull. Var. Stars*, No. 6029, 1.
Diethelm, R. 2013, *Inf. Bull. Var. Stars*, No. 6063, 1.
Eaton, J. A. 1986a, *Acta Astron.*, **36**, 79.
Eaton, J. A. 1986b, *Acta Astron.*, **36**, 275.
Eaton, J. A. 1991, *Astrophys. Space Sci.*, **186**, 7.
Eaton, J. A., Henry, G. W., Bell, C., and Okorogu, A. 1993, *Astron. J.*, **106**, 1181.
Eaton, J. A., Odell, A. P., and Nitschelm, C. A. 2021, *Mon. Not. Roy. Astron. Soc.*, **500**, 145.
Eaton, J. A., Odell, A. P., and Polakis, T. A. 2019, *Inf. Bull. Var. Stars*, No. 6263, 1.
Girardi, L., Bressan, A., Bertelli, G., and Chiosi, C. 2000, *Astron. Astrophys., Suppl. Ser.*, **141**, 371.
Kałużny, J. 1983, *Acta. Astron.*, **33**, 345.

Kałużny, J. 1986a, *Acta Astron.*, **36**, 105.
Kałużny, J. 1986b, *Acta Astron.*, **36**, 113.
Kałużny, J. 1986c, *Acta Astron.*, **36**, 121.
Kałużny, J. 1986d, *Publ. Astron. Soc. Pacific*, **98**, 662.
Kałużny, J., and Pojmański, G. 1983, *Acta Astron.*, **33**, 277.
Lucy, L. B. 1967, *Z. Astrophys.*, **65**, 89.
Lucy, L. B. 1968a, *Astrophys. J.*, **151**, 1123.
Lucy, L. B. 1968b, *Astrophys. J.*, **153**, 877.
Odell, A. P. 1996, *Mon. Not. Roy. Astron. Soc.*, **282**, 373.
Odell, A. P., Wils, P., Dirks, C., Guvenen, B., O'Malley, C. J., Villarreal, A. S., and Weinzettle, R. M. 2011, *Inf. Bull. Var. Stars*, No. 6001, 1.
Qian, S.-B., He, J.-J., Zhang, J., Zhu, L.-Y., Shi, X.-D., Zhao, E.-G., and Zhou, X. 2017, *Res. Astron. Astrophys.*, **17**, 87.
Ruciński, S. M. 1969, *Acta Astron.*, **19**, 245.
Ruciński, S. M. 2015, *Astron. J.*, **149**, 49.
Ruciński, S. M. 2020, *Astron. J.*, **160**, 104.
Ruciński, S. M., and Lu, W. 2000, *Mon. Not. Roy. Astron. Soc.*, **315**, 587.
Samec, R. G., Melton, R. A., Figg, E. R., Labadorf, C. M., Martin, K. P., Chamberlain, H. A., Faulkner, D. R., and van Hamme, W. 2010a, *Astron. J.*, **140**, 1150.
Samec, R. G., Melton, R. A., Figg, E. R., Labadorf, C. M., Martin, K. P., Chamberlain, H. A., Faulkner, D. R., and van Hamme, W. 2010b, *Astron. J.*, **140**, 2145.
Smith, R. C. 1984, *Q. Jour. Roy. Astron. Soc.*, **25**, 405.
van Hamme, W. 1993, *Astron. J.*, **106**, 2096.
van Hamme, W., and Cohen, R. E., 2008, in *Short-Period Binary Stars: Observations, Analyses, and Results*, ed. E. F. Milone, D. A. Leahy, D. W. Hobill, Springer, Berlin, 215.
Wilson, R. E. 1990, *Astrophys. J.*, **356**, 613.
Wilson, R. E. 1994, *Publ. Astron. Soc. Pacific*, **106**, 921.
Wilson, R. E., and Devinney, E. J. 1971, *Astrophys. J.*, **166**, 605.
Wilson, R. E., and van Hamme, W. 2015, "Computing Binary Star Observables" (https://faculty.fiu.edu/~vanhamme/binary-stars/).

Discovery of Romanov V20, an Algol-Type Eclipsing Binary in the Constellation Centaurus, by Means of Data Mining

Filipp Dmitrievich Romanov
ORCID: 0000-0002-5268-7735; Moscow, Russian Federation; filipp.romanov.27.04.1997@gmail.com

Received November 2, 2020; revised May 4, August 28, September 7, October 5, 2021; accepted October 7, 2021

Abstract I report my discovery of the large-amplitude Algol-type eclipsing binary system which was initially added to the AAVSO International Variable Star Index (VSX) under the designation of Romanov V20. I describe selection criteria for searching for variability among other stars, the search of photometric data from several sky surveys, and my observations using remote telescopes, and the analysis of the data in the VSTAR software. I find the orbital period, eclipse duration, and magnitude range in Johnson B, V and Sloan g, r, i bands for primary and secondary eclipses.

1. Introduction

During the course of analysis of the AllWISE catalog (Cutri *et al.* 2014), I discovered a new variable star. I checked that the star was not previously known as a variable in the AAVSO VSX, in the VizieR catalogues, or in the SIMBAD Astronomical Database. Using the publicly available All-Sky Automated Survey for Supernovae (ASAS-SN) Sky Patrol (Shappee *et al.* 2014; Kochanek *et al.* 2017) data, I found the period of eclipses and magnitude range in Johnson V band (including magnitude of secondary eclipse) and duration of primary eclipse. The star was added to the AAVSO VSX (Watson *et al.* 2006) on 03 December 2018 under the designation of Romanov V20. Later its variability was confirmed in the *ASAS-SN catalogue of variable stars – V. Variables in the Southern hemisphere* (Jayasinghe *et al.* 2020) under the name of ASASSN-V J112124.71-522143.6.

In this paper, I describe my method of search which led to the discovery of this star and my refinement of the initially determined parameters of the system, in part on the basis of multicolor photometry from remote telescopes.

2. Information about Romanov V20

Its position (epoch J2000.0) according to Gaia DR2 (Gaia Collaboration *et al.* 2018) is: R.A. $11^h 21^m 24.68^s$, Dec. $-52° 21' 43.7''$; galactic coordinates: $289.2744°, +8.1061°$ (Centaurus).

Other names of this star include:
2MASS J11212468-5221437 =
WISEA J112124.66-522143.7 =
GALEX J112124.6-522143 =
GSC 08225-00671 =
UCAC4 189-058773 =
USNO-B1.0 0376-0339496.

Table 1 presents the data about the star: magnitudes and colors from several catalogs and the mean (calculated by the author) of maximum B and V magnitudes from the APASS DR10 (Henden *et al.* 2018) Epoch Photometry Database. The rounded value of the geometric and photogeometric distance posteriors (Bailer-Jones *et al.* 2021) for this star is 1550 pc. The effective temperature (from Gaia Data Release 2) is 8020 K.

3. Selection criteria for searching for variability

This star was found in the AllWISE catalog using the TAP VizieR service (http://tapvizier.u-strasbg.fr/adql/). In this catalog, objects have a variability flag value (from 0 to 9) which indicates the probability that the source flux measured on the individual WISE exposures is variable; values >7 have the highest probability of being true variables. But, because the catalog does not contain information about the detected variable stars, such as classifications or periods, these objects are not known as variables on this basis alone. The text of my request using Astronomical Data Query Language (Osuna *et al.* 2008) is attached in Appendix A. It was designed to exclude (by color indices, with a large margin) red variable stars, such as semiregular or Mira-type, while the magnitude was limited to W1 < 13 to exclude stars that would be too faint to observe with an amateur telescope. The search area was chosen within several degrees of the Galactic plane. From the obtained table, I selected those stars that have the variability flag 9 for the W1 and W2 bands, and checked them for variability in the data of the ASAS-SN Sky Patrol.

4. Observations

After adding this star to the AAVSO VSX, I used the ephemeris given in VSX to calculate times of primary minimum, and I observed the primary eclipse using the remote

Table 1. Magnitudes and colors of Romanov V20.

Source	Magnitude	Color Index
2MASS (Two Micron All-Sky Survey)	J = 12.86; H = 12.65; K = 12.60	J–K = 0.26
AllWISE (Wide-field Infrared Survey Explorer)	W1 = 12.63; W2 = 12.69	W1–W2 = -0.06
APASS DR10 (AAVSO Photometric All-Sky Survey)	V = 13.46; B = 13.66	B–V = 0.20
Gaia DR2	G = 13.43; BP = 13.54; RP = 13.19	BP–RP = 0.35
GALEX GR6 (Galaxy Evolution Explorer)	FUV = 19.57; NUV = 17.09	FUV–NUV = 2.48

telescope T32 (0.43-m f/6.8 reflector + Charge-Coupled Device) of iTelescope.Net at Siding Spring Observatory, Australia. Seventeen photos with 60 seconds exposure time and Johnson V filter were obtained on 26 February 2020, but only part of the duration of the primary eclipse was recorded due to weather conditions. For photometric measurements I used MaxIm DL Pro Version 6.23 Demo software (Diffraction Limited 2020) and the AAVSO star chart. These values are presented in Table 2. Figure 1 shows a finding chart for this variable star, created from images taken at this remote telescope during this observing run.

The magnitudes (from APASS DR10) and positions (from Gaia DR2) of the comparison stars marked in Figure 1 are shown in Table 3.

On 18 January 2021, I observed another primary eclipse on T32. I obtained 10 V-band images, but I still did not detect the moment of minimum brightness. Therefore, I requested observations of eclipses of this variable star on the remote telescopes of AAVSOnet (Simonsen 2011). From 12 March to 27 March 2021, images of Romanov V20 were taken with the Johnson V, B, and Sloan r, i filters on AAVSOnet telescopes OC61 (Optical Craftsman 0.61-m telescope located at Mount John University Observatory, New Zealand) and BSM Berry2 (72-mm refractor of Bright Star Monitor Station located in Perth, Australia). Exposures were 60 seconds for imaging with all filters except B (120 seconds).

I used AAVSO VPhot (online tool for photometric analysis; AAVSO 2021) for my photometric measurements from these images. The magnitudes of the comparison stars were the same as shown in Table 3 (I used r' and i' magnitudes of stars for comparison for images taken with r and i filters, because I ignored minor differences between these magnitude systems). As a result, I obtained 1964 values of brightness of Romanov V20. The purpose of obtaining images from the AAVSOnet remote telescopes was not to clarify the time of eclipses, but to measure the depths of the eclipses in different filters.

I uploaded all my photometric measurements to the AAVSO International Database (AID).

5. Data analysis

In the beginning, I analyzed the first data from iTelescope T32 and the sky surveys data. I made a heliocentric correction for the times of my photometric values and downloaded the photometric data of this star from the following sky surveys: ASAS-SN Johnson V (from 04 February 2016 to 03 August 2018) and Sloan g (from 13 June 2018 to 10 August 2020) bands, APASS V and g' bands (from 19 February 2011 to 01 June 2014), and All Sky Automated Survey: ASAS-3 (Pojmański 2002) V band (from 07 December 2000 to 30 July 2009). ASAS-3 observations are assigned four quality flags, from A to D (in order of decreasing quality). I used epoch photometric data (HJD-2450000) from the MAG_0 column with the quality grades A and B for the analysis.

I analyzed all data in the VStar software (Benn 2012). As a result, I improved the information previously added by me to VSX: period, magnitude range, and duration of eclipses. The fact that the secondary eclipse is at phase 0.5 suggests that the orbit is circular. These results are shown in Table 4. The resulting combined phased light curve of primary eclipse is

Table 2. The first results of photometric measurements of magnitude of Romanov V20.

Time (JD)	Magnitude (V)	Error
2458906.249468	14.571	0.010
2458906.250637	14.571	0.010
2458906.251667	14.510	0.010
2458906.252778	14.531	0.009
2458906.253854	14.455	0.009
2458906.254919	14.443	0.009
2458906.256019	14.427	0.009
2458906.257072	14.385	0.009
2458906.258137	14.339	0.009
2458906.259225	14.318	0.009
2458906.260266	14.303	0.008
2458906.261331	14.265	0.008
2458906.262407	14.182	0.008
2458906.263495	14.220	0.008
2458906.264618	14.173	0.008
2458906.265683	14.135	0.008
2458906.266759	14.148	0.008

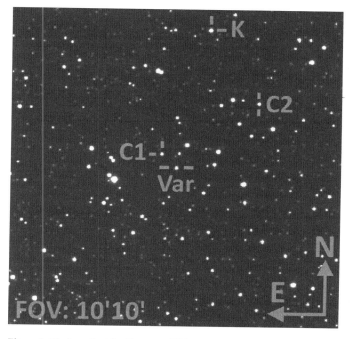

Figure 1. Finding chart for Romanov V20. Var is Romanov V20, C1 and C2 are comparison stars, K is check star.

Table 3. Comparison stars.

Ident.	Name	R.A. (J2000.0) h m s	Dec. (J2000.0) ° ′ ″	B	V	g'	r'	i'	B–V
C1	UCAC4 189-058786	11 21 27.64	–52 21 18.07	15.32 ± 0.046	14.57 ± 0.14	14.87 ± 0.023	14.33 ± 0.131	14.18 ± 0.225	0.75
C2	UCAC4 189-058714	11 21 08.59	–52 19 38.75	14.60 ± 0.039	13.96 ± 0.131	14.22 ± 0.033	13.74 ± 0.119	13.59 ± 0.208	0.64
K	UCAC4 189-058753	11 21 18.49	–52 17 21.02	13.50 ± 0.03	12.98 ± 0.118	13.19 ± 0.029	12.77 ± 0.114	12.69 ± 0.212	0.52

Table 4. Parameters of Romanov V20.

Period: 1.267496 days
Epoch of primary eclipse: 2457934.052 HJD
Duration of primary eclipse: 15% (4.56 hours).

Band	Max	Min I	Min II	Source
B	13.65 ± 0.03	16.7 ± 0.2	13.73 ± 0.03	AAVSOnet
V	13.45 ± 0.02	16.45 ± 0.2	13.54 ± 0.02	ASAS-SN; AAVSOnet
g	13.52 ± 0.02	16.6 ± 0.1	13.59 ± 0.02	ASAS-SN
r	13.37 ± 0.03	15.95 ± 0.15	13.49 ± 0.02	AAVSOnet
i	13.45 ± 0.04	15.65 ± 0.15	13.61 ± 0.02	AAVSOnet
W1	12.50 ± 0.02	13.6 ± 0.1	12.97 ± 0.05	NEOWISE-R
W2	12.55 ± 0.02	13.85 ± 0.15	13.05 ± 0.1	NEOWISE-R

shown in Figure 2. Figure 3 shows the combined phase plot for secondary eclipse in the ASAS-SN V and g data. Figure 4 shows phased light curve from the AAVSOnet data (V, B, r, i bands).

After finding these parameters of the system, I also created the phase plot from the NEOWISE-R data (Mainzer *et al.* 2011): I downloaded data (from 12 January 2014 to 21 June 2019) in the W1 and W2 bands from the NASA/IPAC Infrared Science Archive (NASA/IPAC 2020) from NEOWISE-R Single Exposure (L1b) Source Table. Figure 5 shows the HJD phased light curve plotted with the VSTAR software from these data. Table 4 shows the range of variability in all the observed bands.

The fact that the eclipse depths change with passband, with the primary eclipse being deeper in the B (AAVSOnet data) and g bands (ASAS-SN data) and the secondary eclipse being much deeper in the longer-wavelength bands (AAVSOnet Sloan r and i, and NEOWISE-R W1 and W2), may be explained if the hotter component of the system (judging by the color indices, this is a star of spectral class A or F) is eclipsed by the cooler one (which has relatively brighter magnitudes in the infrared range than in optical) and vice versa.

6. Conclusions

I used both my own photometric measurements in various filters and data from sky surveys to determine the basic parameters of a newly discovered eclipsing binary, Romanov V20 = ASASSN-V J112124.71-522143.6. I conclude that the data from sky surveys are quite enough to determine the period, duration, and epoch of eclipses for bright eclipsing stars.

I showed that an amateur astronomer, who does not have astronomical equipment, but only has a personal computer and access to the Internet, can both search and discover variable stars based on open photometric data from sky surveys, and can research variable stars using observations with remote telescopes. This is a valuable contribution to the science of astronomy; moreover, in the future, such variable stars may become objects of professional astronomical research.

7. Acknowledgements

I am grateful to the AAVSO for granting me a complimentary membership for 2021 (thus giving me access to AAVSOnet,

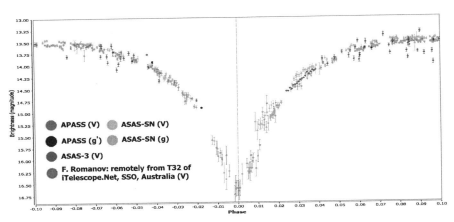

Figure 2. Phase plot for Romanov V20 (primary eclipse).

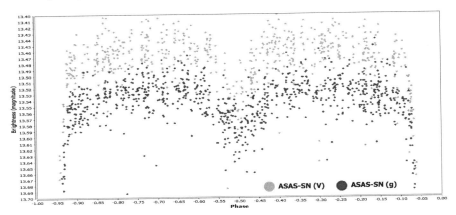

Figure 3. Phase plot for Romanov V20 (secondary eclipse).

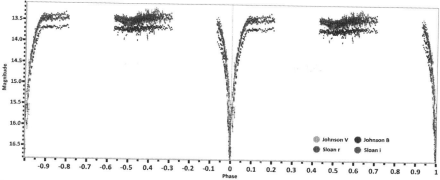

Figure 4. Combined phase plot from the AAVSOnet data.

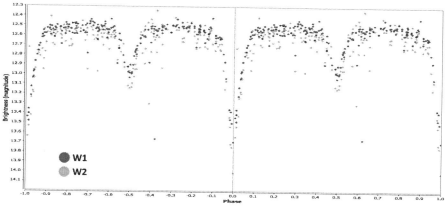

Figure 5. Phase plot for Romanov V20 (based on the NEOWISE-R data).

VPHOT, and APASS Epoch Photometry data) and for imaging Romanov V20 with AAVSOnet remote telescopes after the approval of my proposal #184.

Also, I am grateful to iTelescope.Net for giving me some complimentary points for observing (including for imaging Romanov V20) with their remote telescopes.

This publication makes use of data products from the Two Micron All Sky Survey, which is a joint project of the University of Massachusetts and the Infrared Processing and Analysis Center/California Institute of Technology, funded by the National Aeronautics and Space Administration and the National Science Foundation.

This research has made use of the VizieR catalogue access tool, CDS, Strasbourg, France (DOI: 10.26093/cds/vizier). The original description of the VizieR service was published in *Astron. Astrophys.*, **143**, 23.

This paper is based in part on observations made with the Galaxy Evolution Explorer (GALEX). GALEX is a NASA Small Explorer, whose mission was developed in cooperation with the Centre National d'Etudes Spatiales (CNES) of France and the Korean Ministry of Science and Technology. GALEX is operated for NASA by the California Institute of Technology under NASA contract NAS5-98034.

This research has made use of the NASA/IPAC Infrared Science Archive, which is funded by the National Aeronautics and Space Administration and operated by the California Institute of Technology.

References

AAVSO. 2021, VPHOT AAVSO photometric software (https://www.aavso.org/vphot).
Bailer-Jones, C. A. L., Rybizki, J., Fouesneau, M., Demleitner, M., and Andrae, R. 2021, *Astron. J.*, **161**, 147.
Benn, D. 2012, *J. Amer. Assoc. Var. Star Obs.*, **40**, 852.
Cutri, R. M., et al. 2014, *VizieR On-line Data Catalog: II/328*.
Diffraction Limited. 2020, MAXIM DL image processing software (https://diffractionlimited.com/product/maxim-dl/).
Gaia Collaboration, et al. 2018, *Astron. Astrophys.*, **616A**, 1.
Henden, A. A., Levine, S., Terrell, D., Welch, D. L., Munari, U., and Kloppenborg, B. K. 2018, AAS Meeting #232, id. 223.06.
Jayasinghe, T., et al. 2020, *Mon. Not. Roy. Astron. Soc.*, **491**, 13.
Kochanek, C. S., et al. 2017, *Publ. Astron. Soc. Pacific*, **129**, 104502.
Mainzer, A., et al. 2011, *Astrophys. J.*, **731**, 53.
NASA/IPAC. 2020, NASA/IPAC Infrared Science Archive (https://irsa.ipac.caltech.edu/applications/Gator/).
Osuna, P., and Ortiz, I., eds. 2008, International Virtual Observatory Alliance, IVOA Astronomical Data Query Language, Version 2.0, IVOA Recommendation 30 Oct 2008.
Pojmański, G. 2002, *Acta Astron.*, **52**, 397.
Shappee, B. J., et al. 2014, *Astrophys. J.*, **788**, 48.
Simonsen, M. 2011, AAS Meeting #218, id. 126.02 (AAVSOnet; https://www.aavso.org/aavsonet).
Watson, C. L., Henden, A. A., and Price, A. 2006, in *The Society for Astronomical Sciences 25th Annual Symposium on Telescope Science*, Society for Astronomical Sciences, Rancho Cucamonga, CA, 47.

Appendix A: Query using Astronomical Data Query Language to extract WISE data for this research.

```
-- output format : text
SELECT "II/328/allwise".AllWISE, "II/328/allwise".RAJ2000,
    "II/328/allwise".DEJ2000,  "II/328/allwise".W1mag,
    "II/328/allwise".W2mag,  "II/328/allwise".var
FROM "II/328/allwise"
WHERE 1=CONTAINS(POINT('ICRS',"II/328/allwise".
    RAJ2000,"II/328/allwise".DEJ2000), BOX('ICRS',
    170.00000, -52.00000, 5., 5.))
AND W1mag-W2mag<0 AND Jmag-Kmag<0.3 AND W1mag
    <13
```

The Long-term Period Changes of the Cepheid Variable SV Monocerotis

Pradip Karmakar
Department of Mathematics, Madhyamgram High School (H.S.), Madhyamgram, Sodepur Road, Kolkata 700129, India; pradipkarmakar39@gmail.com

Gerard Samolyk
P.O. Box 20677, Greenfield, WI 53220; gsamolyk@wi.rr.com

Received December 15, 2020; revised August 17, October 14, 2021; accepted October 20, 2021,

Abstract We present the long-term period changes of the Cepheid variable SV Mon, as determined by a parabolic fit to archived observations made from 1911 through mid-2021. We found that its period is increasing at a rate of 4.1 s/yr, showing a long-term evolutionary trend.

1. Introduction

The pulsation period changes of Cepheids have the potential to provide information about their evolutionary trend through the instability strip (Neilson *et al.* 2016). The O–C diagram of a star that has a constant but slow rate of period change is well represented by a parabola. For a Cepheid, such a quadratic fit allows the rate of observed period change to be compared with the theoretical rate of period change. Experience has shown that when the time interval spanned by the O–C diagram approaches a century, the observations are sufficient to reveal evolutionary changes of a Cepheid (Turner *et al.* 2006). Therefore, when the variability of Cepheid periods is studied, the longest possible time interval should be covered by observations.

Observations of SV Mon under the AAVSO Classical Cepheid Program headed by Thomas A. Cragg (Cragg 1972) began in the period JD 2440000–2441000 (23 May 1968–17 February 1971).
The primary purpose of that program was to investigate the slow period changes of the classical Cepheids of period greater than 10.0 days. In this first study 134 observations of SV Mon were obtained which could be well described by a period of 15.2321 days. The value of O–C was nearly zero. The visual magnitude range was 8.5–9.

In their second study, they observed long-period classical Cepheids for the next 1,000 days (JD 2441000–2442000, 17 February 1971–13 November 1973) (Cragg 1975). In this series Landis, one of the observers of this program, collected mostly photoelectric measurements and k-factors were used to adjust individual observers' measurements to the mean light curve. Little meaningful change had been observed in O–C value. In this session, the time of maximum light's date in JD (M) minus time of minimum light's date in JD (m) i.e. M–m, was 0.33 P (P = 15.2 days (0.35P according to GCVS, Kukarkin *et al.*(1969)). In this session a little hip in the light curve of SV Mon was observed.

In their third phase study, JD 2442000–2443000 (13 November 1973–09 August 1976) (Cragg 1983), they observed SV Mon and obtained 147 observations of it. The O–C value was +1.5 days. M–m was 6 days = 0.39 P (0.35 P in GCVS).

In an another study of period change of this Cepheid, Szabados (1981) considered the visual (vis), photographic (pg), and photoelectric (pe) data sets and showed that there was no significant change in period. In this study the O–C residuals were calculated using the formula:

$$C = 2443794.338 + 15.232780 \, d * E \qquad (1)$$

In his next study, Szabados (1991) found that the new pulsation period of SV Mon, determined only from photoelectric data, was somewhat shorter than that determined in his previous study (Szabados 1981). In the 1991 study the O–C residuals were calculated using the elements:

$$C = 2443794.249 \, (\pm 0.019) + 15.232582 \, d \, (\pm 0.000073) * E \qquad (2)$$

In the current paper we study the behavior of the pulsations of the low-amplitude Cepheid SV Mon; the period of its brightness variation (P) used is 15.23278 days, given in the AAVSO's International Variable Star Index (VSX; Watson *et al.* 2014).

2. Techniques and observational data

For this new period change study of SV Mon, initially we considered visual observational data downloaded from the AAVSO International Database (Kafka 2021) from 1967 to mid-2020, i.e. for a time span of nearly 50 years; we excluded the photoelectric measurements observed by Landis in JD 2441000–2442000 as these observational records are not available in the AAVSO International Database.

Next, we considered the V filter (PEP and CCD) AAVSO magnitudes (1954–mid 2021; Kafka 2021) and ASAS (2002–2009; Pojmański 1997).

We also have considered the visual, photographic, and photoelectric data sets collected from Szabados' two papers (1981, 1991).

We divided the total data set yearly. Then we needed to determine the time of maximum light (TOMax) of SV Mon. For this purpose we used the AAVSO's VSTAR package (Benn 2012, 2013). To determine the O–C values, we considered T_0 = 2443794.338 and the initial period, P = 15.23278 days. Both of these values were obtained from VSX.

3. Period change analysis of SV Mon

Figures 1 and 2 show the phased visual and non-visual (CCD) light curves of SV Mon. Figure 3 shows an O–C diagram for SV Mon, using visual and non-visual photometric data, calculated according to the ephemeris (light elements) T_{max} = 2443794.338 + 15.23278E, where T_{max} is the predicted time of maximum light and E is the number of elapsed cycles.

Table 1 lists the times of maxima (JD), cycles, O–C residuals, and type of observation. We subtracted 2400000 from the Julian Date in order to reduce the number of significant digits and so increase the accuracy of the calculation.

Visual inspection shows that the actual period is longer than the VSX period. We used the data set from 1911 to mid-2021 and constructed an O–C diagram, then added a parabolic fitted curve to the O–C diagram (Figure 3).

It is clear from Figure 3 that the period is increasing.

$$JD_{max} = a + bE + cE^2 \qquad (3)$$

We fit Equation (3) to the Times of maximum (JD) and E (Cycle) given in Table 1, and derived the following ephemeris (light elements) (Equation 4):

$$JD_{max} = 9.9 \times 10^{-7} (\pm 1.4 \times 10^{-7}) E^2 \\ + 15.234 (\pm 0.00011) E + 2443795 (\pm 0.091) \qquad (4)$$

The period rate of increase since 1911 can be deduced by Equation (5):

$$\frac{dP}{dt} = 2 \times (9.9 \times 10^{-7}) \times \left(\frac{1}{15.234}\right) \times (86400) \times 365.25 = 4.1\,\text{s/yr} \qquad (5)$$

$$O-C = a + bE + cE^2 \qquad (6)$$

We fit Equation (6) to the O–C and E (Cycle) values in Table 1, and derived the following fitted parabola:

$$O-C = 9.94 \times 10^{-7} (\pm 1.4 \times 10^{-7}) E^2 \\ + 0.0012 (\pm 0.00011) E + 0.684 (\pm 0.091) \qquad (7)$$

From Equation 7, it is clear that the value of c (9.94×10^{-7}) is greater than 0 (zero), so the fitted parabola is in the upward direction. Hence we can suggest that the period of SV Mon is increasing.

4. Conclusions

As presented in section 3, the photometric observations of SV Mon covering about 111 years (1911–mid-2021) indicate the increasing nature of its period, as demonstrated by the parabolic interpretation of the O–C diagram in Figure 3. From our analysis we conclude that SV Mon has a trend of period increase of 4.1 s/yr for the years 1911 to mid-2021.

5. Acknowledgements

We thank the AAVSO and ASAS for the use of the data from their archives and L. Szabados for his two papers (Szabados, 1981, 1991), from where we have collected all the data sets of SV Mon for our paper. We also thank the referee for their help and advice in the improvement of our paper. We are very grateful to Kevin B. Alton and David Benn for their help about using VSTAR to determine the time of maximum light of SV Mon. We are also grateful to Horace A. Smith for his fruitful suggestions for the improvement of our paper.

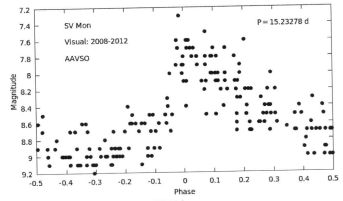

Figure 1. The visual light curve of SV Mon.

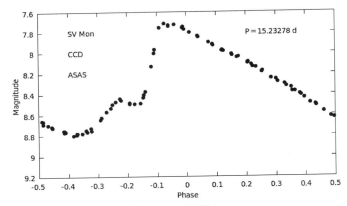

Figure 2. The CCD V-band light curve of SV Mon.

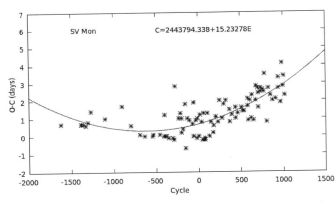

Figure 3. Period change of SV Mon. The O–C versus Cycle diagram of SV Mon, drawn from the TOMax and epochs (cycles) listed in Table 1. The solid curve is a parabolic fit to the data. All O–C residuals (shown as x) were computed with the ephemeris C = 2443794.338 + 15.23278E (via VSX, from Szabados 1981).

Table 1. O–C residuals of the Times of Maxima (TOMax) in Julian days (JD) for SV Mon from 1911 to mid-2021 of visual (AAVSO International Database (Kafka 2021); Szabados (1981)) and non-visual (AAVSO, ASAS (Pojmański 1997), and Szabados (1981, 1991)) photometry.

TOMax (JD2400000+)	Cycle (E)	O–C	Type	TOMax (JD2400000+)	Cycle (E)	O–C	Type
19041.759	–1625	0.7	pg	45683.183	124	0	pe
22621.464	–1390	0.7	vis	46049.58	148	0.8	vis
22728.089	–1383	0.7	vis	46414.6	172	0.2	vis
23047.978	–1362	0.7	vis	46521.61	179	0.6	vis
23459.184	–1335	0.6	vis	46872.5	202	1.1	vis
23809.69	–1312	0.8	vis	47222.5	225	0.8	vis
24419.65	–1272	1.4	vis	47543.6	246	2	vis
27008.827	–1102	1	vis	47999.54	276	1	vis
29964.707	–908	1.7	vis	48274.53	294	1.8	vis
31608.67	–800	0.6	vis	48655.56	319	2	vis
33192.42	–696	0.1	pg	49340.6	364	1.5	vis
34045.326	–640	0	pg	49385.6	367	0.8	vis
35477.219	–546	0	pe	49751.58	391	1.2	vis
35538.224	–542	0.1	pg	50421.63	435	1	vis
36848.276	–456	0.1	pg	50452.68	437	1.6	vis
37564.109	–409	0	pe	50863.63	464	1.3	vis
37656.731	–403	1.2	ccd	51488.46	505	1.6	vis
37899.291	–387	0	pe	51640.64	515	1.4	vis
38097.263	–374	0	pg	51960.56	536	1.4	vis
38386.627	–355	–0.1	pg	52615.8	579	1.7	vis
39209.191	–301	–0.1	pe	52646.697	581	2.1	ccd
39346.203	–292	–0.2	pe	52706.38	585	0.9	vis
39577.7	–277	2.8	vis	53042.62	607	2	ccd
40215.6	–235	1	vis	53315.8	625	1	vis
40474.9	–218	1.3	vis	53408.564	631	2.3	ccd
40672.6	–205	1	vis	53711.8	651	0.9	vis
40732.705	–201	0.2	pe	54093.66	676	2	vis
40976.7	–185	0.4	vis	54170.65	681	2.8	vis
41343.64	–161	1.8	vis	54444.737	699	2.7	ccd
41493.5	–151	–0.7	vis	54520.5972	704	2.4	ccd
41784.6	–132	1	vis	54565.48	707	1.6	vis
42119.63	–110	0.9	vis	54886.29	728	2.5	vis
42728.74	–70	0.7	vis	54947.48	732	2.7	ccd
42836.54	–63	1.9	vis	55297.65828	755	2.6	ccd
42865.108	–61	0	pe	55510.82	769	2.5	vis
43231.63	–37	0.9	vis	55664.15	779	3.5	vis
43489.586	–20	–0.1	pe	55996.56	801	0.8	vis
43551.57	–16	1	vis	56333.59	823	2.7	vis
43794.342	0	0	pe	56668.35	845	2.3	vis
43993.59	13	1.2	vis	57094.56	873	2	vis
44282.56	32	0.8	vis	57658.29	910	2.1	vis
44449.175	43	–0.2	pe	58101.25	939	3.3	vis
44525.339	48	–0.2	pe	58131.04	941	2.7	vis
44648.64	56	1.3	vis	58541.6	968	1.9	vis
44890.941	72	–0.2	pe	58802.74746	985	4.1	ccd
44967.09	77	–0.2	pe	58938.55	994	2.8	vis
44997.62	79	–0.1	vis	59121.88933	1006	3.4	ccd
45364.58	103	1.3	vis	59303.59368	1018	2.3	ccd

References

Benn, D. 2012, *J. Amer. Assoc. Var. Star Obs.*, **40**, 852.

Benn, D. 2013, VStar data analysis software (https://www.aavso.org/vstar-overview).

Cragg, T. A. 1972, *J. Amer. Assoc. Var. Star Obs.*, **1**, 9.

Cragg, T. A. 1975, *J. Amer. Assoc. Var. Star Obs.*, **4**, 68.

Cragg, T. A. 1983, *J. Amer. Assoc. Var. Star Obs.*, **12**, 20.

Kafka, S. 2021, Observations from the AAVSO International Database (https://www.aavso.org/data-download).

Kukarkin, B. V., et al. 1969, *General Catalogue of Variable Stars, Volume 1*, Moscow.

Neilson, H. R., Percy, J. R., and Smith, H. A. 2016, *J. Amer. Assoc. Var. Star Obs.*, **44**, 179.

Pojmański, G. 1997, *Acta Astron.*, **47**, 467 (http://www.astrouw.edu.pl/asas).

Szabados, L. 1981, *Commun. Konkoly Obs.*, **77**, 1.

Szabados, L. 1991, *Commun. Konkoly Obs.*, **96**, 123.

Turner, D. G., Abdel-Sabour Abdel-Latif, M., and Berdnikov, L. N. 2006, *Publ. Astron. Soc. Pacific*, **118**, 410.

Watson, C., Henden, A. A., and Price, C. A. 2014, AAVSO International Variable Star Index VSX (Watson+, 2006–2014; https://www.aavso.org/vsx).

Binaries with Mass Ratios Near Unity: The First BVRI Observations, Analysis and Period Studies of TX Canis Minoris and DW Canis Minoris

Ronald G. Samec
Pisgah Astronomical Research Institute, 112 Idlewood Acres, Hartwell, GA 30643; ronaldsamec@gmail.com

Daniel Caton
Jacob Ray
Riley Waddell
Davis Gentry
Dark Sky Observatory, Physics and Astronomy Department, Appalachian State University, 525 Rivers Street, Boone, NC 28608-2106; catondb@appstate.edu

Danny Faulkner
Johnson Observatory, 1414 Bur Oak Court, Hebron, KY 41048; dfaulkner@answersingenesis.org

Received February 25, 2021; revised July 13, 14, 2021; accepted July 14, 2021

Abstract CCD BVRI light curves of TX CMi and DW CMi were taken in 2020 on 20, 21 January, 22, 23 February, and 4 April with the 0.81-m reflector of Appalachian State University by Daniel Caton, Ronald Samec and Danny Faulkner. Six times of minimum light were determined from our present observations of TX CMi. Fifty-five total times of minimum light were included in the 61-year period study. From these we determined that the period for TX CMi is increasing. Eight times of minimum light were determined for DW CMi and thirty-five total times of minimum light were included in the 19.3-year period study. The period is weakly decreasing with a quadratic term of -1.9×10^{-11}. This could be due to mass transfer to M_1 ($q = M2/M_1$) for DW CMi. A Wilson-Devinney (W-D) analysis of TX CMi reveals that the system is a W UMa binary with a mass ratio near unity, $q \sim 1.00$. Its Roche Lobe fill-out is $\sim 10\%$. One spot was needed in the modeling. The temperature difference of the components is only ~ 90 K, so the stars are nearly twins, with the secondary component the slightly cooler one. The inclination is high, $86.9 \pm 0.1°$. A W-D analysis reveals that DW CMi is a W-type W UMa binary a mass ratio near unity, $q \sim 1.1$. Its Roche Lobe fill-out is $\sim 10\%$. One weak spot was needed in the modeling. The temperature difference of the components is $T_2 - T_1 \sim 260$ K, making the binary of W-type.

1. History and observations

1.1. TX CMi

Even though the TX CMi has been known for some 90 years, very little information is known about the binary. We summarize it here. The variability of TX CMi ([GGM 2006] 12715955) was discovered by Hoffmeister (1929). He found it to be a short period variable and little more. It was classified as an EB (β Lyrae) system with a magnitude of V = 13.461 by Gessner (1966). Paschke (1994), using minima from BBSAG bulletins, improved his earlier period (0.3892 d, Paschke 1992) with elements of TX CMi:

$$\text{Min I HJD} = 2436598.611 + 0.3892173 \, d \times E. \quad (1)$$

Otherwise, a number of times of minima and one low light exist which are noted in the period study.

The system was classified as an EW-type by the All Sky Automated Survey (Shappee *et al.* 2014; Kochanek *et al.* 2017) as ASASSN-V J074019.94+044239.5 (Pojmański 2002). This catalog gives key information: a V_{mean} = 13.58, an amplitude of 0.8, and EW designation, J–K = 0.396, B–V = 0.635, E(B–V) = 0.05, and GAIA distance = 794 ± 10 pc. Their ephemeris is:

$$\text{Min I HJD} = 2458023.09677 + 0.3892165 \, Ed \times E. \quad (2)$$

The ASAS light curve is given in Figure 1.

1.2. DW CMi

DW CMi (GSC2.2 N22123134124) was discovered in 2005 by the SkyDOT team (Polster, Zejda, and Safar 2005) of Copernicus Observatory in Brno, Czech Republic. They gave an R magnitude of 14.3, an EW type, and an ephemeris of:

$$\text{Min I HJD} = 2451965.2876 \pm 0.0009 + 0.30755 \, d \times E. \quad (3)$$

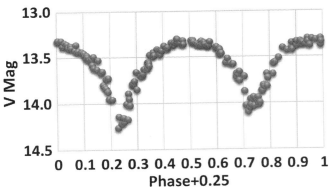

Figure 1. ASAS light curves of TX CMi (Pojmański 2002).

Table 1. Photometric targets.

Star	Name	R.A. (2000) h m s	Dec. (2000) ° ' "	V (mag)	J–K (mag)
TX CMi	AN 146.1929 ASAS J074020+0442.7 NSVS 127159566868894	07 40 20.1[1]	+04 42 39.5[1]	14.23	0.396 ± 0.054[1]
DW CMi	GSC2N22123134124 2MASS J07403307+0442200 USNO-A2.0 0900-05269593	07 40 33.0[1]	+04 42 20.1[1]	14.75	0.429 ± 0.062[1]
C (comparison)	GSC 0187 1415 3UC190-079026	07 40 33.0[2]	+04 43 55.90[2]	14.75	0.38
K (check)	GSC 0187 1966	07 40 43.3[1]	+04 42 36.6[1]	10.46	0.256 ± 0.045[1]

[1]SMBAD. [2]UCAC3: The USNO CCD Astrograph Catalog (Zacharias, N., et al. 2010).

They also gave a position of R.A. (2000) = 07^h 40^m 33^s, Dec. (2000) = +04° 42' 17". The only other information published has been times of minimum light (see Table 3, period study table of DW CMi and plots, Figures 10 and 11). The system was observed by the All Sky Automated Survey as ASASSN-V J074032.97+044219.9 (Pojmański 2002). This catalog gives key information: a V_{mean} = 14.65, an amplitude of 0.59, EW designation, J–K = 0.429, B–V = 0.741, E(B–V) = 0.047, and GAIA distance = 904 ± 18 pc. Their ephemeris is:

$$\text{Min I HJD} = 2457322.13402 + 0.3075535\,d \times E. \quad (4)$$

The ASAS light curve is given in Figure 2.

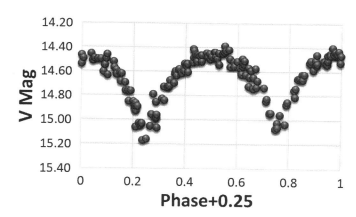

Figure 2. ASAS light curves of DW CMi (Pojmański 2002).

These systems were observed as a part of our professional collaborative studies of interacting binaries at Pisgah Astronomical Research Institute using data taken from DSO observations.

The observations were taken by D. Caton, R. Samec, and D. Faulkner. Reduction and analysis were done by R. Samec.

Our BVR_cI_c light curves were taken at Dark Sky Observatory, on 20, 21, January, 22, 23 February, and 4 April 2020, with a thermoelectrically cooled (−40°C) 2K × 2K Apogee camera and Johnson-Cousins BVR_cI_c filters. Individual observations included 163 in B, 213 in V, 240 in R, and 225 in I. The probable error of a single observation was 9 mmag in B, 11 mmag in V, 16 mmag in R, and 18 mmag in I. The nightly C–K values stayed constant throughout the observing run with a precision of about 1.0–1.5% in V. Exposure times varied from 150s in B, 75–100s in V, and 40s in R and I. To produce these images, nightly images were calibrated with 25 bias frames, at least five flat frames in each filter, and ten 300-second dark frames. The early results of this study were presented at the American Astronomical Society meeting #237, 11–15 January (Caton et al. 2021; Samec et al. 2021).

2. Photometric targets

Table 1 gives basic information on the two variables and the comparison (C) and check (K) stars, including designations, positions, magnitudes, and colors. The finding chart for DW CMi and TX CMi and the comparison and check stars is given in Figure 3.

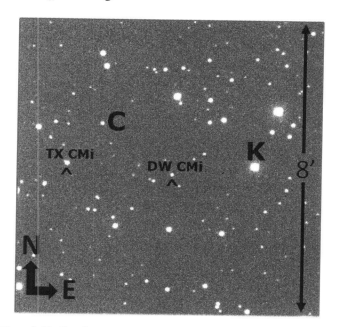

Figure 3. Finding chart (V) for variables TX CMi and DW CMi, comparison (C), and check (K), 4 April 2020.

3. Sample nightly light curves

Two nightly light curves of TX CMi are given as Figures 4 and 5 for 20 and 22 January 2020. Also, nightly light curves of DW CMi are given as Figures 6 and 7 for 20 January and 22 February.

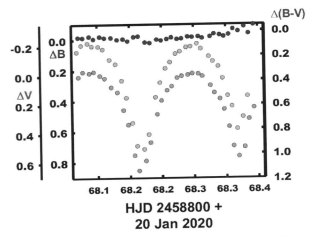

Figure 4. B, V light curves and B–V color curves of TX CMi for 20 January 2020.

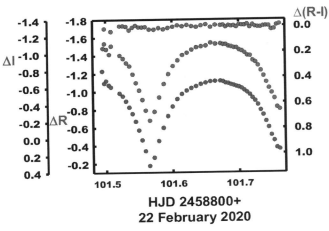

Figure 5. R, I light curves and R–I color curves of TX CMi for 22 February 2020.

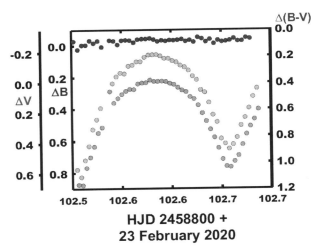

Figure 6. B, V light curves and B–V color curves of DW CMi for 23 February 2020.

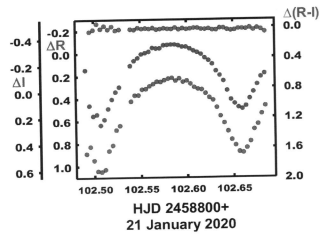

Figure 7. R, I light curves and R–I color curves of DW CMi for 21 January 2020.

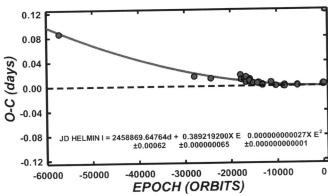

Figure 8. A plot of the quadratic term overlying the linear residuals for TX CMi (showing an increasing period).

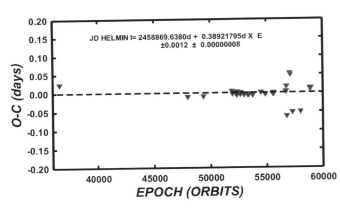

Figure 9. A plot of the linear residuals for TX CMi.

Table 2. Period study TX CMi.

	Epoch	Cycle	Initial Residual	Linear Residual	Quadratic Residual	Wt	Reference
1	36598.6110	−57220.0	0.0378	0.0244	−0.0010	1.0	Paschke (2012)
2	47992.3490	−27946.5	−0.0090	−0.0096	−0.0050	0.1	Paschke (1990)
3	49383.4150	−24372.5	−0.0096	−0.0086	−0.0035	1.0	Paschke (1990)
4	51899.5281	−17908.0	0.0011	0.0049	0.0094	1.0	Brát et al. (2007)
5	51968.4125	−17731.0	−0.0061	−0.0023	0.0021	1.0	Brát et al. (2007)
6	52230.5521	−17057.0	−0.0051	−0.0009	0.0033	1.0	Brát et al. (2007)
7	52692.3610	−15871.0	−0.0038	0.0009	0.0046	1.0	Paschke (2003)
8	53320.9432	−14256.0	−0.0094	−0.0040	−0.0009	1.0	Krajci (2006)
9	53353.8320	−14171.5	−0.0095	−0.0041	−0.0011	1.0	Krajci (2006)
10	53410.2691	−14026.5	−0.0091	−0.0036	−0.0007	1.0	Zejda et al. (2006)
11	53410.4645	−14026.0	−0.0083	−0.0028	0.0001	1.0	Zejda et al. (2006)
12	53464.3712	−13887.5	−0.0083	−0.0028	0.0001	1.0	Zejda et al. (2006)
13	53768.3487	−13106.5	−0.0104	−0.0045	−0.0021	1.0	Hubscher et al. (2006)
14	53768.5416	−13106.0	−0.0121	−0.0062	−0.0038	1.0	Hubscher et al. (2006)
15	52362.3002	−16719.0	−0.0074	−0.0031	0.0009	1.0	Zejda (2004)
16	52367.3610	−16706.0	−0.0065	−0.0021	0.0019	1.0	Zejda (2004)
17	52367.3614	−16706.0	−0.0061	−0.0017	0.0023	1.0	Zejda (2004)
18	52369.3125	−16701.0	−0.0011	0.0033	0.0073	1.0	Zejda (2004)
19	52369.3036	−16701.0	−0.0100	−0.0056	−0.0016	1.0	Zejda (2004)
20	52668.4196	−15932.5	−0.0083	−0.0036	0.0001	1.0	Zejda (2004)
21	52668.4205	−15932.5	−0.0074	−0.0027	0.0010	1.0	Zejda (2004)
22	52668.4219	−15932.5	−0.0060	−0.0013	0.0024	1.0	Zejda (2004)
23	52668.2266	−15933.0	−0.0067	−0.0020	0.0017	1.0	Zejda (2004)
24	52668.2263	−15933.0	−0.0070	−0.0023	0.0014	1.0	Zejda (2004)
25	52668.2264	−15933.0	−0.0069	−0.0022	0.0015	1.0	Zejda (2004)
26	53000.4215	−15079.5	−0.0097	−0.0047	−0.0013	1.0	Zejda (2004)
27	53000.4210	−15079.5	−0.0102	−0.0052	−0.0018	1.0	Zejda (2004)
28	53318.8024	−14261.5	−0.0095	−0.0041	−0.0010	1.0	Dvorak (2005)
29	54491.7160	−11248.0	−0.0055	0.0012	0.0026	1.0	GCVS (Samus et al. 2017)
30	54890.4652	−10223.5	−0.0106	−0.0034	−0.0027	1.0	Samolyk (2010)
31	55553.8889	−8519.0	−0.0097	−0.0017	−0.0023	1.0	Zejda (2004)
32	55554.0835	−8518.5	−0.0097	−0.0017	−0.0023	1.0	Zejda (2004)
33	55554.2781	−8518.0	−0.0097	−0.0017	−0.0023	1.0	Zejda (2004)
34	55554.4727	−8517.5	−0.0097	−0.0017	−0.0023	1.0	Zejda (2004)
35	55554.6673	−8517.0	−0.0097	−0.0017	−0.0023	1.0	Zejda (2004)
36	55554.8619	−8516.5	−0.0097	−0.0017	−0.0023	1.0	Zejda (2004)
37	55555.0565	−8516.0	−0.0097	−0.0017	−0.0023	1.0	Zejda (2004)
38	55555.2511	−8515.5	−0.0097	−0.0017	−0.0023	1.0	Zejda (2004)
39	55555.4457	−8515.0	−0.0097	−0.0017	−0.0023	1.0	Zejda (2004)
40	55555.6404	−8514.5	−0.0097	−0.0017	−0.0023	1.0	Zejda (2004)
41	55555.8350	−8514.0	−0.0097	−0.0017	−0.0023	1.0	Zejda (2004)
42	55556.0296	−8513.5	−0.0097	−0.0017	−0.0023	1.0	Zejda (2004)
43	55556.2242	−8513.0	−0.0097	−0.0017	−0.0023	1.0	Zejda (2004)
44	55621.4197	−8345.5	−0.0082	−0.0002	−0.0010	1.0	Hubscher et al. (2012)
45	55625.3126	−8335.5	−0.0075	0.0005	−0.0002	1.0	Hubscher et al. (2012)
46	55625.5047	−8335.0	−0.0100	−0.0020	−0.0027	1.0	Hubscher et al. (2012)
47	56713.3747	−5540.0	−0.0054	0.0038	0.0006	1.0	Hubscher et al. (2015)
48	56726.4123	−5506.5	−0.0067	0.0026	−0.0006	1.0	Hubscher et al. (2015)
49	54890.4652	−10223.5	−0.0106	−0.0034	−0.0027	1.0	Samolyk (2010)
50	58868.6769	−2.5	−0.0002	0.0115	0.0023	1.0	Present Observations
51	58869.6501	0.0	0.0000	0.0117	0.0025	1.0	Present Observations
52	58869.8443	0.5	−0.0004	0.0113	0.0021	1.0	Present Observations
53	58901.5674	82.0	0.0014	0.0131	0.0038	1.0	Present Observations
54	58902.5385	84.5	−0.0005	0.0112	0.0019	1.0	Present Observations
55	58943.6028	190.0	0.0012	0.0129	0.0035	1.0	Present Observations

4. Period determination, TX CMi

Six mean times (from BVRI data) of minimum light were calculated from our present observations, three primary and three secondary eclipses:

HJD I = 2458869.65009 ± 0.00030,
2458901.56743 ± 0.00060. 2458943.60276 ± 0.00069

HJD II = 2458868.67688 ± 0.00030, 2458869.84431 ± 0.00060, 2458902.53853 ± 0.000038.

All minima were weighted as 1.0 in the period study except for one time of low light which was weighted 0.1. In total, 55 times of minimum light (References listed in Table 2) were included in this study. This gave us an interval of 61 years.

Table 3. Times of minimum light, DW CMi.

	Epoch	Cycle	Linear Residual	Quadratic Residual	Wt	Reference
1	51876.5672	–22734.5	0.0054	0.0075	0.5	Polster et al. (2006)
2	51899.4704	–22660.0	–0.0043	–0.0022	0.5	Polster et al. (2006)
3	51965.2876	–22446.0	–0.0039	–0.0019	0.5	Polster et al. (2006)
4	52002.3517	–22325.5	–0.0001	0.0018	0.5	Polster et al. (2006)
5	52213.6433	–21638.5	0.0011	0.0027	0.5	Polster et al. (2006)
6	53410.3376	–17747.5	–0.0018	–0.0016	1.0	Zejda et al. (2006)
7	53410.4915	–17747.0	–0.0016	–0.0015	1.0	Zejda et al. (2006)
8	53464.3143	–17572.0	–0.0010	–0.0009	1.0	Zejda et al. (2006)
9	53768.3327	–16583.5	–0.0009	–0.0010	1.0	Hubscher and Walter (2007)
10	53768.4862	–16583.0	–0.0012	–0.0013	1.0	Hubscher and Walter (2007)
11	53813.3886	–16437.0	–0.0018	–0.0020	1.0	Hubscher and Walter (2007)
12	54149.3924	–15344.5	–0.0020	–0.0025	1.0	Hubscher and Joachim (2007)
13	55621.3567	–10558.5	0.0033	0.0023	1.0	Hubscher and Lehmann (2015)
14	55621.5096	–10558.0	0.0024	0.0014	1.0	Hubscher and Lehmann (2015)
15	55625.3539	–10545.5	0.0023	0.0012	1.0	Hubscher and Lehmann (2015)
16	55625.5033	–10545.0	–0.0021	–0.0031	1.0	Hubscher and Lehmann (2015)
17	55987.6523	–9367.5	0.0007	–0.0003	1.0	GCVS (Samus et al. 2017)
18	56354.4143	–8175.0	0.0032	0.0022	1.0	Hubscher (2015)
19	56713.3313	–7008.0	0.0033	0.0024	1.0	Hubscher and Lehmann (2015)
20	56713.4843	–7007.5	0.0025	0.0016	1.0	Hubscher and Lehmann (2015)
21	56726.4008	–6965.5	0.0017	0.0008	1.0	Hubscher and Lehmann (2015)
22	57039.0275	–5949.0	–0.0014	–0.0022	0.1	ASAS–SN (Shappee et al. 2014; Kochanek et al. 2017)
23	57100.8501	–5748.0	0.0026	0.0019	0.1	ASAS–SN (Shappee et al. 2014; Kochanek et al. 2017)
24	57131.7636	–5647.5	0.0068	0.0061	0.1	ASAS–SN (Shappee et al. 2014; Kochanek et al. 2017)
25	57441.7745	–4639.5	0.0021	0.0016	0.1	ASAS–SN (Shappee et al. 2014; Kochanek et al. 2017)
26	57757.0166	–3614.5	0.0002	–0.0002	0.1	ASAS–SN (Shappee et al. 2014; Kochanek et al. 2017)
27	58407.0343	–1501.0	0.0001	0.0003	0.2	ASAS–SN (Shappee et al. 2014; Kochanek et al. 2017)
28	58868.6736	0.0	–0.0010	–0.0002	1.0	Present Observations
29	58868.8245	0.5	–0.0038	–0.0031	1.0	Present Observations
30	58869.7496	3.5	–0.0014	–0.0007	0.5	Present Observations
31	58901.5827	107.0	–0.0002	0.0005	1.0	Present Observations
32	58901.7360	107.5	–0.0007	0.0001	1.0	Present Observations
33	58902.5052	110.0	–0.0004	0.0003	1.0	Present Observations
34	58902.6583	110.5	–0.0011	–0.0003	1.0	Present Observations
35	58943.5632	243.5	–0.0011	–0.0002	1.0	Present Observations

From these timings, two ephemerides have been calculated, a linear one and a quadratic one:

$$\text{JD Hel Min I} = 24558869.63843 \pm 0.00098\, d + 0.3892179629 \pm 0.0000000667 \times E. \quad (5)$$

$$\text{JD Hel Min I} = 2458869.64764 \pm 0.00062\, d + 0.3892192001 \pm 0.0000000658 \times E + 0.000000000027 \pm 0.000000000001 \times E^2. \quad (6)$$

Figure 8 shows the quadratic term overlying the linear residuals and Figure 9 gives the linear residuals. Table 2 gives the minima and the residuals of the quadratic and the linear ephemerides.

This TX CMi period study covers an interval of 61 years. It shows an orbital period that is increasing. It might be due to mass transfer to the more massive, component (probably our primary component) making the mass ratio more extreme. Table 2 give the residuals of the linear and quadratic ephemerides. The initial ephemeris was HJD Min I = 2458869.650088 + 0.3892184012 × E.

5. Period determination, DW CMi

Eight mean times (from BVRI data) of minimum light were calculated from our present observations, three primary and five secondary eclipses:

HJD I = 2458868.67357 ± 0.00090, 2458901.5827 ± 0.0004, 2458902.5052 ± 0.0019

HJD II = 2458868.82446 ± 0.00030, 2458869.74956 ± 0.00017, 2458901.7360 ± 0.0008, 2458902.65829 ± 0.00022, 2458943.56315 ± 0.00022

All minima were weighted as 1.0 in the period study except for the ASAS-SN times of low light which was weighted 0.1. In total, 35 times of minimum light (References listed in Table 3) were included in this study. This gave us an interval of 19.3 years.

From these timings, two ephemerides have been calculated, a linear one and a quadratic one:

$$\text{JD Hel Min I} = 2458868.67452 \pm 0.00041\, d + 0.307555157 \pm 0.000000034 \times E. \quad (7)$$

Table 4. Light curve characteristics, TX CMi.

Filter	Phase Min I	Mag ±σ	Phase Max I	Mag ±σ
	0.00		0.25	
B	−0.093	0.021	−1.091	0.002
V	−0.167	0.024	−1.100	0.005
R	−0.173	0.003	−1.105	0.008
I	−0.274	0.061	−1.090	0.024

Filter	Phase Min II	Mag ±σ	Phase Max II	Mag ±σ
	0.5		0.75	
B	−0.316	0.012	−1.155	0.017
V	−0.326	0.007	−1.158	0.005
R	−0.361	0.034	−1.160	0.006
I	−0.341	0.001	−1.146	0.027

Filter	Min I − Max I	±σ ±σ	Max I − Max II	±σ ±σ	Min I − Min II	±σ ±σ
B	0.998	0.023	0.064	0.019	0.223	0.033
V	0.933	0.029	0.058	0.010	0.159	0.031
R	0.932	0.011	0.055	0.014	0.188	0.037
I	0.816	0.085	0.056	0.051	0.067	0.062

Filter	Min II − Max I	±σ ±σ	Min I − Max II	±σ ±σ	Min II − Max I	±σ ±σ
B	0.775	0.013	1.062	0.038	0.775	0.013
V	0.774	0.012	0.991	0.029	0.774	0.012
R	0.744	0.042	0.987	0.010	0.744	0.042
I	0.749	0.025	0.872	0.088	0.749	0.025

Table 5. Light curve characteristics, DW CMi.

Filter	Phase Min I	Mag ±σ	Phase Max I	Mag ±σ
	0.00		0.25	
B	0.787	0.037	0.058	0.008
V	0.665	0.037	−0.013	0.009
R	0.585	0.006	−0.067	0.011
I	0.518	0.008	−0.099	0.022

Filter	Phase Min II	Mag ±σ	Phase Max II	Mag ±σ
	0.5		0.75	
B	0.622	0.022	0.026	0.008
V	0.553	0.008	−0.032	0.022
R	0.487	0.009	−0.087	0.010
I	0.432	0.040	−0.116	0.036

Filter	Min I − Max I	±σ ±σ	Max I − Max II	±σ ±σ	Min I − Min II	±σ ±σ
B	0.729	0.045	0.032	0.016	0.165	0.059
V	0.678	0.046	0.019	0.030	0.112	0.045
R	0.652	0.016	0.020	0.021	0.098	0.014
I	0.617	0.030	0.017	0.058	0.086	0.047

Filter	Min II − Max I	±σ ±σ	Min I − Max II	±σ ±σ	Min II − Max II	±σ ±σ
B	0.564	0.030	0.761	0.045	0.596	0.03
V	0.566	0.017	0.697	0.059	0.585	0.03
R	0.554	0.019	0.672	0.015	0.574	0.018
I	0.531	0.062	0.634	0.043	0.548	0.076

JD Hel Min I = 2458868.67380 ± 0.00042 d + 0.30755479
± 0.00000009 × E − 0.000000000019 ± 0.000000000005 × E2. (8)

The residuals of the quadratic and linear ephemerides are given in Table 3.

The phased BVRI light curves and B–V and R–I color curves of TX CMi and DW CMi are given in Figures 12, 13, 14, and 15.

6. Light curve characteristics

6.1. TX CMi

The curves are of fair accuracy, averaging better than 2% photometric precision. The amplitude of the light curves varies from 0.87–1.1 mags from B to I filters. The O'Connell effect, an indicator of spot activity, averages 0.06 mag, which indicates the existence of spots. The differences in minima are appreciable, from 0.07 to 0.22 mag, I to B, respectively, indicating a fair difference in component temperatures. The secondary amplitude averages 0.75 mag, large for a contact binary. The light curve characteristics of TX CMi are given in Table 4.

6.2. DW CMi

Again, the DW CMi curves are of good accuracy, averaging about 2% photometric precision. The amplitude of the light curve varies from 0.76–0.63 mag from B to I filters. The O'Connell effect averages 0.03 mag in B and 0.02 mag in VRI, which indicates the existence of weak spots. The differences in minima are appreciable, from, 0.09 to 0.17 mag, I to B, respectively, indicating a small difference in component temperatures. The secondary amplitude is 0.60, 0.585, 0.58, 0.55 mag, B to I, which is fairly large for a contact binary. The light curve characteristics of DW CMi are given in Table 5.

7. Temperatures

The 2MASS J–K = 0.396 ± 0.054 for TX CMi and B–V = 0.635, E(B–V) ~ 0.05. These correspond to ~G5V ± 2, which yields a temperature of 5750 ± 200 K. Fast rotating binary stars of this type are noted for being magnetic in nature, so the binary is of solar-type with a convective atmosphere.

The J–K for DW CMi = 0.429 ± 0.062 and B–V = 0.741, E(B–V) = 0.047. These correspond to ~G8V ± 2, which yields a temperature of 5500 ± 200 K.

8. Light curve solutions, TX CMi

The B, V, R_c and I_c curves of TX CMi were pre-modeled with BINARY MAKER 3.0 (Bradstreet and Steelman 2002). Fits were determined in all filter bands and the results were tabulated. The solutions were that of a shallow contact eclipsing binary. The parameters were then averaged (q or mass ratio = 0.9, fill-out = 0.05, i or inclination = 88, T_2 = 5562.5, and one cool spot with t-fact ($T_{spot}/T_{photosphere}$) 0.87) and input into a 4-color simultaneous light curve calculation

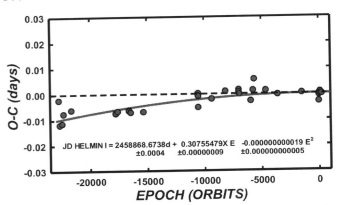

Figure 10. A plot of the quadratic term overlying the linear residuals for DW CMi (showing weakly decreasing period).

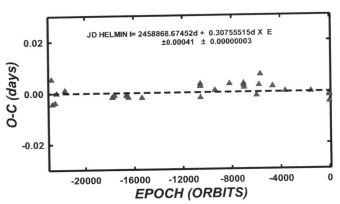

Figure 11. A plot of linear residuals of DW CMi.

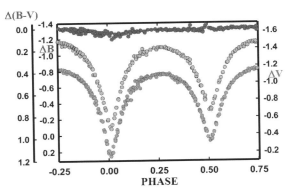

Figure 12. B, V light curves and B–V color curves of TX CMi.

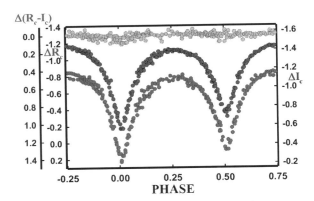

Figure 13. R, I light curves and R–I color curves of TX CMi.

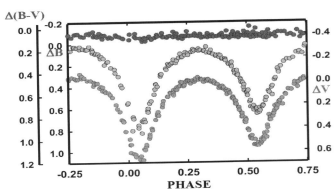

Figure 14. B, V light curves and B–V color curves of DW CMi.

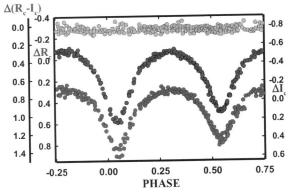

Figure 15. R, I light curves and R–I color curves of DW CMi.

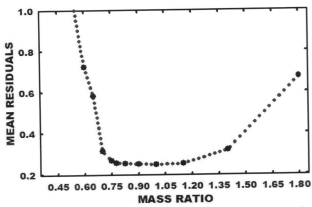

Figure 16. q-search for TX CMi, solutions with fixed mass ratio vs. solution residual.

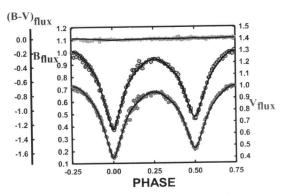

Figure 17. TX CMi B,V normalized flux curves and B–V color curves overlain by B,V solutions.

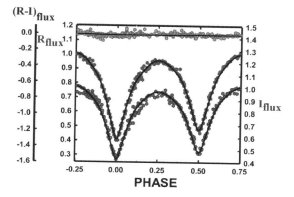

Figure 18. TX CMi R_c, I_c normalized flux curves and R_c–I_c color curves overlain by R_c, I_c solutions.

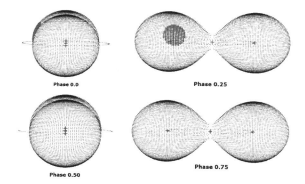

Figure 19. Geometrical representations of TX CMi at quadratures.

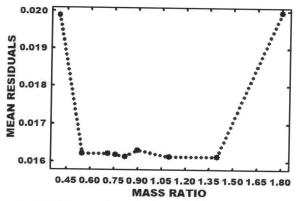

Figure 20. DW CMi, q-search with fixed mass ratio vs. the solution residual.

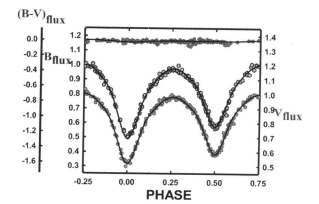

Figure 21. DW CMi B, V normalized flux curves and B–V color curves overlain with B, V solutions.

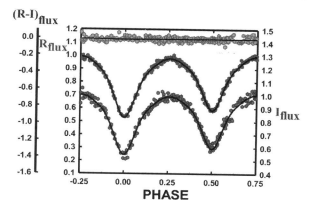

Figure 22. DW CMi R, I normalized flux curves and R–I color curves overlain with R, I solutions.

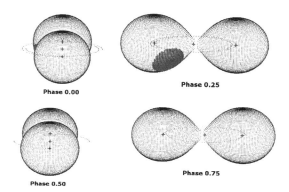

Figure 23. DW CMi geometrical representations at quadratures.

using the Wilson-Devinney Program (W-D; Wilson and Devinney 1971; Wilson 1990, 1994; van Hamme and Wilson 1998). The initial computation was in Mode 3 and converged to a solution (q~0.7). Because there was no total eclipse, a q-search was instigated. In a q-search, a solution is produced for fixed mass ratios. The sum of square residuals is listed for each. The smallest residual is considered to belong to the best estimate of the mass ratio. The minima was fairly broad, with similar residuals between q~0.8 and q~1.15 and the very best goodness of fit residual was at a minima of q=1.0. The q-search is shown in Figure 16. Convective parameters, g = 0.32, A = 0.5 were used. The q=1.0 solution is given in Table 6. Figures 17 and 18 show the BVR$_c$I$_c$ flux overlaid by the light curve solutions. Geometric representations of the surface of the binary at quadratures are given in Figure 19. Table 7 gives the system dimensions and Table 8 gives absolute parameters.

9. Light curve solutions, DW CMi

As with TX CMi, the B, V, R$_c$ and I$_c$ curves were pre-modeled with BINARY MAKER 3.0. Fits were determined in all filter bands and the results were tabulated. The results were, again, that of a shallow contact eclipsing binary. The averaged parameters were q = 0.7, fill-out = 0.05, i = 78.5°, T$_2$ = 5300 K, and one cool spot with t-fact = 0.91 and was followed by analysis by the W-D program. The initial computation was in Mode 3 and converged to a solution

Table 6. B,V,R$_c$,I$_c$ Wilson-Devinney program solution parameters, TX CMi.

Parameter	Value
$\lambda_B, \lambda_V, \lambda_R, \lambda_I$ (nm)	440, 550, 640, 790
g_1, g_2	0.32
A_1, A_2	0.5
Inclination (°)	86.91 ± 0.13°
T_1, T_2 (K)	5750, 5559 ± 1
Ω	3.6953 ± 0.0028
$q(m_2/m_1)$	1.002 ± 0.002
Fill-outs: $F_1 = F_2$ (%)	0.10 ± 0.01
$L_1/(L_1+L_2+L_3)_I$	0.5277 ± 0.0009
$L_1/(L_1+L_2+L_3)_R$	0.5321 ± 0.0011
$L_1/(L_1+L_2+L_3)_V$	0.5381 ± 0.0007
$L_1/(L_1+L_2+L_3)_B$	0.5514 ± 0.0007
JD$_0$ (days)	2458868.8713 ± 0.0001
Period (days)	0.38921992 ± 0.0000009
$r_1/a, r_2/a$ (pole)	0.363 ± 0.002, 0.363 ± 0.0002
$r_1/a, r_2/a$ (side)	0.382 ± 0.002, 0.382 ± 0.002
$r_1/a, r_2/a$ (back)	0.417 ± 0.003, 0.417 ± 0.003
Spots, Star 1	
Colatitude (°)	74.6 ± 2.4°
Longitude (°)	282.3 ± 0.7°
Radius (°)	20.2 ± 0.4°
T-factor	0.781 ± 0.009

Table 7. TX CMi system dimensions.

R_1, R_2 (pole, $R_\odot$)[1]	1.005 ± 0.0006	1.005 ± 0.0006
R_1, R_2 (side, $R_\odot$)[1]	1.058 ± 0.0006	1.058 ± 0.0006
R_1, R_2 (back, $R_\odot$)[1]	1.154 ± 0.0007	1.154 ± 0.0007

[1]Using $a = 2.76778\ R_\odot$.

Table 8. Estimated TX CMi absolute parameters.[1]

Parameter	Star 1	Star 2
Mean radius ($R_\odot$)	1.071	1.071
Mean density	1.078	1.078
Mass ($M_\odot$)	0.94	0.94
Log g	4.42	4.42

[1]Density units are gm/cm^3. a = semi-major axis.

Table 9. DW CMi Solution synthetic light curve parameters.

Parameter	Value
$\lambda_B, \lambda_V, \lambda_R, \lambda_I$ (nm)	440, 550, 640, 790
g_1, g_2	0.32
A_1, A_2	0.5
Inclination (°)	78.36 ± 0.08
T_1, T_2 (K)	5500, 5244.4 ± 1.8
Ω	3.8864 ± 0.0034
$q(m_2/m_1)$	1.100 ± 0.002
Fill-outs: $F_1 = F_2$ (%)	4.5
$L_1/(L_1+L_2+L_3)_I$	0.5202 ± 0.0006
L1/(L1+L2)R	0.5270 ± 0.0057
$L_1/(L_1+L_2+L_3)_V$	0.5369 ± 0.0016
$L_1/(L_1+L_2+L_3)_B$	0.5553 ± 0.0011
JD$_0$ (days)	2458868.67318 ± 0.00011
Period (days)	0.3075569 ± 0.000001
$r_1/a, r_2/a$ (pole)	0.351 ± 0.002, 0.367 ± 0.002
$r_1/a, r_2/a$ (side)	0.369 ± 0.003, 0.386 ± 0.003
$r_1/a, r_2/a$ (back)	0.402 ± 0.005, 0.418 ± 0.004
Spots, Star 1	
Colatitude (°)	119.8 ± 1.5
Longitude (°)	306.6 ± 4.3
Spot radius (°)	30.85 ± 1.01
T-factor	0.939 ± 0.003

Table 10. DW CMi system dimensions.

R_1, R_2 (pole, $R_\odot$)	0.844 ± 0.006	0.882 ± 0.006
R_1, R_2 (side, $R_\odot$)	0.887 ± 0.007	0.929 ± 0.007
R_1, R_2 (back, $R_\odot$)	0.965 ± 0.011	1.010 ± 0.010

Table 11. DW CMi estimated absolute parameters.[1]

Parameter	Star 1	Star 2
Mean Radius ($R_\odot$)	0.899	0.939
Mean Density	1.755	1.819
Mass ($M_\odot$)	0.906	0.997
Log g	4.48	3.49

[1]a = semi-major axis. Density units are gm/cm^3.

(q ~ 0.7). As with TX CMi, a q-search was instigated. The minima of the curve was very broad, with similar residuals between q ~ 0.4 and q ~ 1.4 and the very best goodness of fit residual was at a minima of q = 1.1. The q-search is shown in Figure 20. Again, convective parameters, g = 0.32, A = 0.5 were used. The q = 1.1 solution follows in Table 9. The system dimensions are given in Table 10 and absolute parameters in Table 11. The B, V, R, I normalized flux curves and B–V and R, I color curves overlain with the solutions are given in Figures 21 and 22. The geometric system representations at quadrature's are given in Figure 23.

10. Discussion

TX CMi is shallow contact W UMa binary. As stated earlier, the q-search minimized at 1.0. The system's fill-out is 10%, and a component temperature difference is a ~ 190 K, so the stars are very similar in spectral type. One spot was needed in the solution, a Northern, 15° latitude, 20° radius spot with a t-fact of 0.78. The inclination of ~86.9 degrees did not result in a time of constant light in due to the similar sizes of the components. Its photometric spectral type indicates a surface temperature of ~5750 K for the primary component, making it a solar-type binary. Such a main sequence star would have a mass of ~0.98 $M_\odot$ and the secondary (from the mass ratio) would have a mass of ~0.965 $M_\odot$, making the stars nearly twins.

DW CMi is a shallow contact (4.5%) W-type W UMa binary (if the q = 1.1 (0.002) solution is correct). The component temperature difference was about 255 K, which is reasonable for the shallow contact. The spot was a Southern cool spot with a t-fact of 0.938, –30° latitude, 31° radius spot off the side of the L_1 point. The inclination of ~78.4 degrees was not steep enough to allow total eclipses. Its photometric spectral type indicates a surface temperature of 5500 ± 200 K for the

Table 12. Sample of first ten TX CMi B, V, R_c, I_c observations.

ΔB	HJD 2458800+	ΔV	HJD 2458800+	ΔR	HJD 2458800+	ΔI	HJD 2458800+
−1.063	68.574	−1.100	68.577	−1.120	68.571	−1.064	68.572
−1.091	68.581	−1.101	68.584	−1.108	68.578	−1.073	68.579
−1.094	68.591	−1.094	68.594	−1.104	68.588	−1.082	68.589
−1.068	68.598	−1.080	68.601	−1.103	68.595	−1.099	68.596
−1.045	68.608	−1.050	68.610	−1.077	68.605	−1.072	68.606
−1.034	68.615	−1.040	68.618	−1.062	68.612	−1.090	68.613
−0.992	68.625	−0.993	68.628	−1.047	68.622	−1.013	68.623
−0.950	68.635	−0.927	68.637	−0.976	68.632	−0.969	68.632
−0.836	68.644	−0.819	68.647	−0.914	68.641	−0.894	68.642
−0.754	68.651	−0.736	68.654	−0.839	68.648	−0.815	68.649

Note: First ten data points of TX CMi B, V, R_c, I_c observations. The full table is available through the AAVSO ftp site at ftp://ftp.aavso.org/public/datasets/samec492-txcmi.txt (if necessary, copy and paste link into the address bar of a web browser).

Table 13. Sample of first ten DW CMi B, V, R_c, I_c observations.

ΔB	HJD 2458800+	ΔV	HJD 2458800+	ΔR	HJD 2458800+	ΔI	HJD 2458800+
0.078	68.574	0.008	68.577	−0.046	68.571	−0.027	68.572
0.035	68.581	−0.026	68.584	−0.043	68.578	−0.035	68.579
0.022	68.591	−0.018	68.594	−0.068	68.588	−0.071	68.589
0.041	68.598	−0.027	68.601	−0.071	68.595	−0.097	68.596
0.043	68.608	0.018	68.618	−0.067	68.605	−0.093	68.606
0.061	68.615	0.079	68.628	−0.040	68.612	−0.082	68.613
0.129	68.625	0.114	68.637	−0.017	68.622	−0.014	68.623
0.158	68.635	0.237	68.647	0.043	68.632	0.022	68.632
0.266	68.644	0.338	68.654	0.108	68.641	0.091	68.642
0.378	68.651	0.547	68.663	0.186	68.648	0.200	68.649

Note: First ten data points of TX CMi B, V, R_c, I_c observations. The full table is available through the AAVSO ftp site at ftp://ftp.aavso.org/public/datasets/samec492-dwcmi.txt (if necessary, copy and paste link into the address bar of a web browser).

primary component, making it a solar-type binary. Such a main sequence star would have a mass of ~ 0.94 $M_\odot$ and the secondary (from the mass ratio) would have a mass of ~ 1.03 $M_\odot$ making the secondary star over massive for its type and very similar to the primary component.

The period and epoch were used as iterating parameters for both of the solutions. One can see from the solution plots (Figures 17, 18, 21, and 22) that the data (phased with the linear ephemerides) fits th Wilson-Devinney phased solution plots very well.

11. Conclusions

The period is increasing for TX CMi with the mass ratio departing from unity so that the mass ratio becomes more extreme. We expect that this solar-type binary is undergoing magnetic braking and the binary will ultimately coalesce into a fast rotating late A-type single star. A spectroscopic radial velocity curve is needed to determine the actual mass ratio of the binary, but it is fairly assured that it is between $m_1/m_2 = 0.8$ and 1.2.

The period of DW CMi is decreasing. The occurrence of a spot and the period change does lend us to believe that the star is undergoing magnetic braking so we expect the future scenario to be much like that stated for TX CMi. The mass ratio is less determinable, however, and could range as much as 0.55 to 1.40. If q is > 1.0, as the q-search gives, the system is a W-type W UMa binary (more massive star is the cooler one.) Radial velocity curves are very much needed to obtain the actual mass ratio and absolute (not relative) system parameters. Tables 8 and 12 give estimated parameters. Observations for DW CMi and TX CMi are given in Tables 13 and 14.

12. Acknowledgement

I wish to think the physics students from Appalachian State University that have helped with the observations.

References

Bradstreet, D. H., and Steelman, D. P. 2002, *Bull. Amer. Astron. Soc.*, **34**, 1224.

Brát, L., Zejda, L. M., and Svoboda, P. 2007, *B.R.N.O. Contrib.*, No. 34, 1.

Caton, D., Samec, R., and Faulkner, D. 2021, *Bull. Amer. Astron. Soc.*, **53**, e-id 2021n1i339p24.

Dvorak, S. W. 2005, *Inf. Bull. Var. Stars*, No. 5603, 1.

Gessner, H. 1966, *Veröff. Sternw. Sonneberg*, **7**, 61.

Hoffmeister, C. 1929, *Mitt. Sternw. Sonneberg*, **16**, 1.

Hubscher, J. 2007, *Inf. Bull. Var. Stars*, No. 5802, 1.

Hubscher, J. 2015, *Inf. Bull. Var. Stars*, No. 6152, 1.

Hubscher, J., and Lehmann, P. B. 2015, *Inf. Bull. Var. Stars*, No. 6149, 1.

Hubscher, J., Lehmann, P. B., and Walter, F. 2012, *Inf. Bull. Var. Stars*, No. 6010, 1.

Hubscher, J., Paschke, A., and Walter, F. 2006, *Inf. Bull. Var. Stars*, No. 5731, 1.

Hubscher, J., and Walter, F. 2007, *Inf. Bull. Var. Stars*, No. 5761, 1.

Kochanek, C. S., *et al.*, G. 2017, *Publ. Astron. Soc. Pacific*, 129, 104502.

Krajci, T. 2006, *Inf. Bull. Var. Stars*, No. 5690, 1.

Paschke, A. 1990, *BBSAG Bull.*, No. 95, 6.

Paschke, A. 1992, *BBSAG Bull.*, No. 102, 9.

Paschke, A. 1994, *BBSAG Bull.*, No. 106, 7.

Paschke, A. 2003, in Diethelm, R., *Inf. Bull. Var. Stars*, No. 5438, 3.

Paschke, A. 2012, *Open Eur. J. Var. Stars*, **147**, 1.

Pojmański, G. 2002, *Acta. Astron.*, **52**, 397.

Polster, J., Zejda, M., and Safar, J. 2005, *Inf. Bull. Var. Stars*, No. 5700, 10.

Polster, J., Zejda, M., and Safar, J. 2006, *Inf. Bull. Var. Stars*, No. 5700, 4.

Samec, R., Caton, D., Ray, J., Waddell, R., and Gentry, D. 2021, *Bull. Amer. Astron. Soc.*, 53, e-id 2021n1i128p01.

Samolyk, G. 2010, *J. Amer. Assoc. Var. Star Obs.*, **38**, 85.

Samus N. N., Kazarovets E. V., Durlevich O. V., Kireeva N. N., and Pastukhova E. N. 2017, *Astron. Rep.*, **61**, 80 (*General Catalogue of Variable Stars: Version GCVS 5.1*, http://www.sai.msu.su/groups/cluster/gcvs/gcvs).

Shappee, B. J., *et al.* 2014, *Astrophys. J.*, **788**, 48.

van Hamme, W. V., and Wilson, R. E. 1998, *Bull. Amer. Astron. Soc.*, **30**, 1402.

Wilson, R. E. 1990, *Astrophys. J.*, **356**, 613.

Wilson, R. E. 1994, *Publ. Astron. Soc. Pacific*, **106**, 921.

Wilson, R. E., and Devinney, E. J. 1971, *Astrophys. J.*, **166**, 605.

Zacharias, N., *et al.* 2010, *Astron. J.*, **139**, 2184.

Zejda, M. 2004, *Inf. Bull. Var. Stars*, No. 5583, 1.

Zejda, M., Mikulasek, Z., and Wolf, M. 2006, *Inf. Bull. Var. Stars*, No. 5741, 1.

An Analysis of X-Ray Hardness Ratios between Asynchronous and Non-Asynchronous Polars

Eric Masington
Department of Physics and Astronomy, Texas Tech University, Lubbock, TX 79409; eric.masington@ttu.edu

Thomas J. Maccarone
Department of Physics and Astronomy, Texas Tech University, Lubbock, TX 79409; thomas.maccarone@ttu.edu

Liliana Rivera Sandoval
Department of Physics and Astronomy, Texas Tech University, Lubbock, TX 79409; and Department of Physics, University of Alberta, Edmonton, Alberta T6G 2R3, Canada; lriveras@ualberta.ca

Craig Heinke
Department of Physics, University of Alberta, Edmonton, Alberta T6G 2R3, Canada; heinke@ualberta.ca

Arash Bahramian
Department of Physics and Astronomy, Curtin University, Perth, Western Australia 6845, Australia; arash.bahramian@curtin.edu.au

Aarran W. Shaw
Department of Physics, University of Nevada, Reno, NV 89557; aarrans@unr.edu

Received March 5, 2021; revised June 28, July 5, 2021; accepted July 5, 2021

Abstract The subclass of magnetic Cataclysmic Variables (CV) known as asynchronous polars is still relatively poorly understood. An asynchronous polar is a polar in which the spin period of the white dwarf is either shorter or longer than the binary orbital period (typically within a few percent). The asynchronous polars have been disproportionately detected in soft gamma-ray observations, leading us to consider the possibility that they have intrinsically harder x-ray spectra. We compared standard and asynchronous polars in order to examine the relationship between a CV's synchronization status and its spectral shape. Using the entire sample of asynchronous polars, we find that the asynchronous polars may, indeed, have harder spectra, but that the result is not statistically significant.

1. Introduction

One of the first results on accreting white dwarfs with the International Gamma-Ray Astrophysical Laboratory (INTEGRAL) was that the asynchronous polars represented a disproportionate fraction of its detected cataclysmic variables (Barlow *et al.* 2006). Asynchronous polars are magnetically accreting white dwarfs with deviations between the orbital and spin period (unlike standard polars) and with streamlike accretion rather than accretion disks (unlike the intermediate polars). The difference between the spin and orbital period in the asynchronous polars is typically about 5% or less. It is unclear whether INTEGRAL preferentially detected these objects because the x-ray and gamma-ray spectra of the asynchronous polars are different from those of the standard polars, or merely because they tend to be more luminous, hence at higher fluxes within the well-understood samples. Here, we test whether the spectral indices of these sources in the soft gamma-ray band alone, and between x-ray and gamma-ray, are systematically harder for the asynchronous polars than for the standard polars.

The asynchronous polars are often suggested to have been driven out of synchronization by classical novae, which can affect both the orbital and spin periods of cataclysmic variables, motivated by the association between one of the asynchronous polars, V1500 Cyg, with a classical nova in 1975 (Campbell and Schwope 1999). Searches for additional nova shells around other asynchronous polars have not yielded any new evidence for the nova hypothesis (Pagnotta and Zurek 2016), but it remains a viable one, as nova shells may have lifetimes shorter than the synchronization timescales of the asynchronous polars. Because the sample sizes of the asynchronous polars are quite small, and rather long, well-sampled light curves are needed to identify that there are two separate, but similar, periods in the light curves, it is worth exploring new methods that might work to find new members of the class, and the INTEGRAL discoveries of these objects suggest that gamma-ray surveys might be an interesting approach. With this in mind, we undertake an exploration of whether the high energy spectra of asynchronous polars are fundamentally different from those of the standard polars.

2. Data used

We obtained a set of cataclysmic variables from the Ritter and Kolb catalog, update 7.24 (Ritter and Kolb 2003), hard x-ray data from the 2018 Swift-BAT 105-month All-Sky Survey catalogue (Oh *et al.* 2018) and soft x-ray data from the ROSAT All-Sky Survey (Boller *et al.* 2016). Matching all

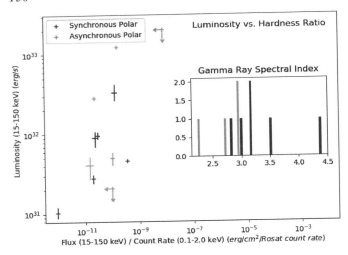

Figure 1. Shown here is the ratio between the flux (Swift-BAT 15–150 keV band) and count rate (ROSAT 0.1–2.0 keV band) plotted against luminosity (calculated using Gaia distances and Swift flux values). As can be seen, no significant relationship can be established between whether or not a polar is asynchronous and the hardness of its flux ratio. It must be considered that the ROSAT and Swift data are taken non-simultaneously, so because the ROSAT count numbers are generally too small for spectral fitting, they cannot be reliably converted into fluxes. Thus, this means that only very strong trends could have been detected using this combination of data. Such trends are not present, but the method would not be particularly sensitive to subtle systematic variations. Also shown are the gamma-ray spectral indices. We can see that the asynchronous polars seem to be harder (lower spectral indices), but more data are needed to confirm this (Gaia Collaboration *et al.* 2016, 2018).

three catalogues with a maximum separation of 3 arc minutes (typical for Swift-BAT for faint sources) yields 10 objects, including 4 asynchronous polars and 6 standard polars. We define an asynchronous polar to be a system with a spin period within 5% of the orbital period. The asynchronous polars are BY Cam, CD Ind, V1432 Aql, and the recently identified IGR J19552+0044 (Tovmassian *et al.* 2017). (We consider a 5-sigma upper limit for V1500 Cyg and RX J0838.7-2827 based on the survey depth, but do not consider upper limits for the much larger class of synchronous polars.) V1500 Cyg and RX J0838.7-2827 were added to the analysis to provide for a more complete sample. The synchronous polars included are AM Her, Swift J231930.4+261517, 1RXS 145341.1-552146, IW Eri, V2301 Oph, and V834 Cen.

3. Analysis and conclusions

First, we looked at the ratio of hard x-rays from Swift-BAT (15–150 keV) to soft x-rays from ROSAT (0.1–2.0 keV). This comparison is done between a count rate for ROSAT and a flux for Swift BAT because the standard ROSAT data include only a count rate, and the standard BAT data include only a flux. The fluxes do, thus, show some model dependence, but since the spectra are all steep power laws, with photon index greater than 2.0, in all cases, the BAT flux is dominated by counts near the lower end of the band, and this comparison is nearly equivalent to a count rate-to-count rate comparison. However, the ROSAT and Swift data are taken non-simultaneously, and because the ROSAT count numbers are generally too small for spectral fitting, they cannot be reliably converted into fluxes. This thus means that only very strong trends could have been detected using this combination of data. Such trends are not present, but the method would not be particularly sensitive to subtle systematic variations. No trend is found in this ratio between the two classes of polars, as can be seen in Figure 1. Also, there exists much uncertainty in the soft x-ray flux values for V1500 Cyg and RX J0838.7-2827, given that they are not in the Swift catalog. Because the plot shows strong scatter between the Swift and ROSAT data, and the ROSAT data in most cases are insufficient for detailed spectral analysis, we simply leave this plot as a ratio of a flux to a count rate.

Next, we consider the spectra within the gamma-ray band alone. Within the Swift band, the mean spectral index is 2.73 for the asynchronous polars and 3.34 for the standard polars. We apply the Anderson-Darling test to the distributions of spectral indices. This is a cumulative statistic, similar to the Kolomogorov-Smirnov test, but with greater diagnostic power in cases where the differences are strongest near the edges of the distributions, and at least equal power in all cases. The Anderson-Darling (AD) test statistic here is 2.28 (computed using https://www.real-statistics.com/non-parametric-tests/goodness-of-fit-tests/two-sample-anderson-darling-test/). For this sample size, the critical value of the AD test statistic is 3.38 for a 99% confidence level detection of a difference. The asynchronous polars do show a different gamma-ray spectral index at the 95% confidence level. Since this is a marginally significant difference, we expect that a larger sample of objects would have a reasonable probability of establishing a difference. Unfortunately, doing so will require finding new asynchronous polars, as we have already investigated the properties of the whole sample, with only V1500 Cyg and RX J0838.7-2827 undetected in the Swift-BAT data. If more asynchronous polars can be found from optical or soft x-ray searches, NuSTAR would easily be capable of measuring gamma-ray spectral indices for objects much fainter than the BAT survey can, so searches for more asynchronous polars would be well-motivated.

References

Barlow, E. J., Knigge, C., Bird, A. J., J Dean, A., Clark, D. J., Hill, A. B., Molina, M., and Sguera, V. 2006, *Mon. Not. Roy. Astron. Soc.*, **372**, 224.
Boller Th., Freyberg, M. J., Trumper, J., Haberl, F., Voges, W., and Nandra, K. 2016, *Astron. Astrophys.*, **588A**, 103.
Campbell, C. G., and Schwope, A. D. 1999, *Astron. Astrophys.*, **343**, 132.
Gaia Collaboration, *et al.* 2016, *Astron. Astrophys.*, **595A**, 1.
Gaia Collaboration, *et al.* 2018, *Astron. Astrophys.*, **616A**, 1.
Oh, K., *et al.* 2018, *Astrophys. J., Suppl. Ser.*, **235**, 4.
Pagnotta, A., and Zurek, D. 2016, *Mon. Not. Roy. Astron. Soc.*, **458**, 1833 (doi:10.1093/mnras/stw424).
Ritter H., and Kolb U. 2003, *Astron. Astrophys.*, **404**, 301.
Tovmassian, G., *et al.* 2017, *Astron. Astrophys.*, **608A**, 36.

High Cadence Millimagnitude Photometric Observation of V1112 Persei (Nova Per 2020)

Neil Thomas
Kyle Ziegler
Peter Liu
Department of Astronautical Engineering, United States Air Force Academy, CO 80840; neil.thomas@afacademy.af.edu

Received April 1, 2021; revised June 11, 2021; accepted June 14, 2021

Abstract The private Lookout Observatory (LO) monitored the classical nova V1112 Persei on 37 nights spanning over 80 days, beginning shortly after its discovery by Seiji Ueda on 25 November 2020. Images were captured at a high cadence, with exposure lengths of initially less than 2 seconds and with some sessions lasting more than ten hours. The standard error of the photometry was typically better than 5 thousandths of a magnitude (5 mmag). This cadence and precision allowed for not only the observation of the expected dimming of the nova, but also variability having a period of 0.608 ± 0.005 day. These data complement the publicly available photometry from the American Association of Variable Star Observers (AAVSO) and the resultant data are combined to perform this photometric analysis. This paper does not attempt an in-depth physical analysis of the nova from an astrophysical perspective.

1. Introduction

Classical novae are close-proximity binary stars comprised of a white dwarf primary and a (typically) late-type main sequence star (Warner 1995). The large mass of the white dwarf combined with their close pairing results in a deformed secondary that consistently loses material to the primary (Bode and Evans 2008). This material either forms an accretion disk or simply flows directly to the surface of the primary, depending on whether the primary contains a magnetic field. Eventually a critical mass of material accumulates on the surface of the white dwarf, resulting in a runaway hydrogen fusion reaction, or nova.

V1112 Persei was discovered on 25 November 2020 (Ueda 2020) at an unfiltered magnitude of 10.6. Spectroscopy obtained by Munari *et al.* (2020) on 26 November identified the nova as a galactic nova near maximum at the time of first observation. A campaign was developed to obtain photometry on this nova starting three days after its discovery. The goals were to both record the long-term evolution of its brightness and leverage the high cadence and mmag precision of the LO configuration to search for short-term variations.

Section 2 details the observations. Section 3 discusses the LO instrumentation and data reduction methods. Section 4 describes the photometric analysis of this nova, while section 5 is the conclusion.

2. Observing campaign

V1112 Persei was observed over 37 nights between 25 Nov 2020 and 16 Feb 2021 and a total of nearly 190 hours of target photometry was collected. Exposures ranged from 2 seconds in November up to 90 seconds by the end of the survey. The dates, duration, and mean nightly magnitudes of the observations are listed in Table 1. Individual LO measurements can be obtained from the AAVSO website (Observer Code TNBA).

Table 1. Observation details.

Date	Duration (hours)	Mag. (Gaia G)	Mag. Err.
28 Nov 2020	5.80	8.252	0.0119
29 Nov 2020	10.44	8.689	0.0186
30 Nov 2020	2.77	8.341	0.0202
1 Dec 2020	2.98	8.271	0.0196
4 Dec 2020	1.42	8.125	0.0109
5 Dec 2020	10.17	8.639	0.0218
6 Dec 2020	9.60	8.525	0.0257
7 Dec 2020	10.18	8.294	0.0275
8 Dec 2020	10.18	8.503	0.0314
9 Dec 2020	9.40	8.874	0.0159
10 Dec 2020	8.50	9.020	0.0178
14 Dec 2020	2.12	8.878	0.0598
16 Dec 2020	9.90	9.049	0.0324
17 Dec 2020	9.80	9.268	0.0172
20 Dec 2020	1.24	9.244	0.0103
22 Dec 2020	9.59	9.573	0.0141
23 Dec 2020	1.24	9.601	0.0070
24 Dec 2020	1.00	9.686	0.0146
25 Dec 2020	6.23	9.778	0.0219
26 Dec 2020	10.19	9.747	0.0090
27 Dec 2020	3.02	9.854	0.0120
30 Dec 2020	8.90	9.800	0.0115
31 Dec 2020	7.80	10.056	0.0173
2 Jan 2021	4.59	10.242	0.0233
5 Jan 2021	3.03	10.638	0.0151
7 Jan 2021	2.23	10.024	0.0306
11 Jan 2021	3.80	12.424	0.0176
15 Jan 2021	5.06	13.083	0.0178
20 Jan 2021	0.90	13.612	0.0530
21 Jan 2021	0.64	13.380	0.0255
23 Jan 2021	2.10	13.865	0.0580
24 Jan 2021	2.57	13.955	0.0997
28 Jan 2021	3.32	14.243	0.0630
29 Jan 2021	4.03	14.367	0.0660
30 Jan 2021	0.52	14.471	0.0830
8 Feb 2021	3.16	14.951	0.0507
16 Feb 2021	2.30	15.133	0.0587

3. Instrumentation and methods

The LO is primarily optimized to maintain the photometric precision necessary to observe exoplanet transits. It consists of an 11-inch Celestron telescope modified to f/1.9 with a HyperStar. A ZWO ASI 1600 CMOS camera performs the imaging, and optical filters are not typically used. Dawn and dusk flats are captured and applied in the normal way. During favorable conditions, individual measurements typically have noise in the 10–20 mmag range. This is primarily white noise, however, and differential photometry with noise levels of less than 2 mmag is often maintained from dusk to dawn by binning data to simulate 2–4-minute exposures. Absolute magnitude calibration between nights is generally consistent to within 20 mmag.

Images are collected semi-autonomously using MAXMIM DL and CCDCOMMANDER. Additionally, a custom-made software pipeline developed in MATLAB performs the aperture photometry. Although a full discussion of the software pipeline's design is not currently available, Thomas and Guan (2021) provide a somewhat more detailed performance analysis.

Calibration stars are automatically selected based on similarity of magnitude and color. This field of view (FOV) provided approximately 1,000 field stars at the beginning of the campaign and 9,000 at the end, as exposure lengths increased. Typically, the most compatible 50–150 stars are automatically selected and used for differential photometry. This provides relative magnitudes, not the absolute magnitudes required to study this nova over several months. To derive absolute magnitudes, we use Gaia DR2 magnitudes and colors to derive absolute Gaia G-band magnitudes for all usable field stars (Gaia Collab. *et al.* 2016, 2018). To do this, the instrumental magnitude of each star is compared to its Gaia magnitude and used to determine the first order shift to true magnitudes. Then a second order color correction is applied after fitting the magnitude residuals to Gaia B–V colors. Although our photometry is unfiltered and Gaia is G-band, a reliable transformation is possible. Our calibrated magnitudes are compared to Gaia values for a typical night in Figure 1. Although noise is photon-dominated for faint stars, the standard deviation of the difference between our values and Gaia among the brightest third of stars is 15 mmag. For comparison, the quoted errors of Gaia magnitudes in this brightness range are typically 2–6 mmag. It may seem odd to convert our clear unfiltered (CV) data into the Gaia G-band and then to eventually compare those results to AAVSO CV and visual band (V) photometry. We do this because of our overall desire to automatically select and use many calibration stars from among the entire FOV. Usable field stars extend down to 17th magnitude and the authors know of no catalog besides Gaia containing magnitude and color information to this faintness. Additionally, attempting to remain in our native CV would still present compatibility complications since unfiltered data is not a true band because observed flux will depend on sensor sensitivity at various wavelengths. This calibration method will, however, introduce a source of error during the second-order color correction for this target. We use the B–R value of 0.80776 obtained by Gaia prior to the nova's outburst. But in reality, we expect a reddening of the nova during its evolution (Woudt and Ribeiro 2014). Since this color evolution is not known beforehand, we use this fixed value. We will show later that these color correction errors are not overwhelming, especially for our primary objective of analyzing the short-period signal.

4. Analysis and results

AAVSO data were retrieved from the AAVSO International Database (Kafka 2021) on 23 January 2021 and included 44,608 measurements. AAVSO photometry comes from many sources and through a variety of filters. Most (~80%) observations of this target are in CV and in Johnson V (V). The data from the LO, although initially collected as unfiltered, are calibrated to Gaia green (G) and provide an additional 65,462 measurements.

A simple transformation between photometry taken in different colors can usually be determined for a given star. But since this nova is likely changing color over time, making different bands of photometry compatible must be approached more carefully. The light curves in AAVSO's CV and our G-calibrated are shown in Figure 2. The CV and G bands appear highly compatible. There are very few periods in which there are overlapping observations with which to generate a transform. During those few overlapping periods of data we see a disagreement on the order of 0.03 mag. Furthermore, we see similar levels of disagreement between simultaneous AAVSO CV measurements reported by different observers. We therefore choose to incorporate AAVSO CV photometry as directly equivalent to our G-band data.

The majority of AAVSO data are in the V-band and our data (and the relatively equivalent AAVSO CV data) require a transformation. In Figure 3 we see the divergence with time between G/CV and V bands. The V-band brightness fades faster than the G/CV and a time dependent transformation is required to combine the data.

To make these sources compatible, photometry in each color is binned into equal intervals by creating two-minute time steps over the entire time range. All data of a particular color within each interval are averaged to create a single photometric measurement at that color. The deviations between different colors during each interval can then be directly compared since they now have the same time sampling. The deviation between these magnitudes is shown in Figure 4.

The transformation is given as a function of time by the following equation, where JD' is the number of days past epoch, G is the observed magnitude in G-band, and V is the transformed magnitude to the V-band.

$$V = G - (-1.262 \times 10^{-6} \, JD'^4 + 8.716 \times 10^{-5} \, JD'^3 - 1.604 \times 10^{-3} \, JD'^2 - 1.042 \times 10^{-2} \times JD' - 0.131)$$

The transformed light curve from all sources is shown in Figure 5 for the first ten days after epoch. All sources are now highly compatible and show a smooth continuum of photometry.

It is apparent in Figure 5 that short-term variations well beyond the noise levels of the data are present. To determine the period of this activity, we first determine the long-term profile of the fading nova by finding its moving average with a smoothing

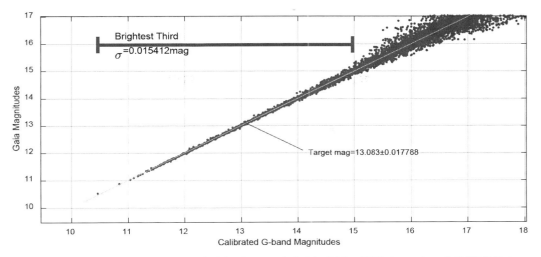

Figure 1. Performance of calibration from unfiltered (CV) to Gaia G-band on a typical night (15 Jan 2021). Approximately 9,000 field stars were transformed to absolute magnitudes. Precision deteriorates for faint stars but appears to be driven by shot noise with no obvious systematic influences. The standard deviation between computed magnitudes and Gaia values is 0.015 mag for the brightest third of stars. This is only about 0.010 mag greater than the errors in the Gaia values themselves and is acceptable for calibrating our data across different nights.

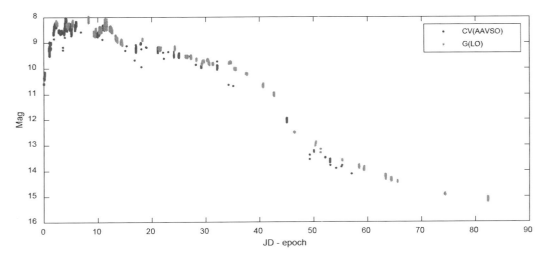

Figure 2. Photometric observations of V1112 Per in CV (from AAVSO) and G (our results). CV and G bands are compatible with each other with no transformation being needed. The epoch is the time of the first observation available from AAVSO (2459179.34333).

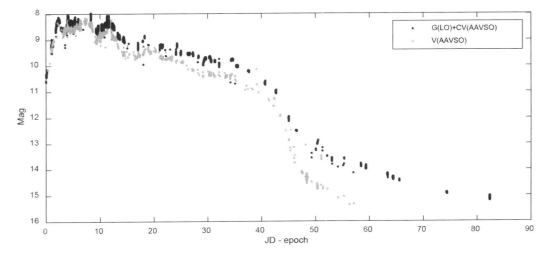

Figure 3. The V-band observations fade more drastically with time than of the G/CV measurements due to the changing color of the nova, requiring a transformation before combining the data.

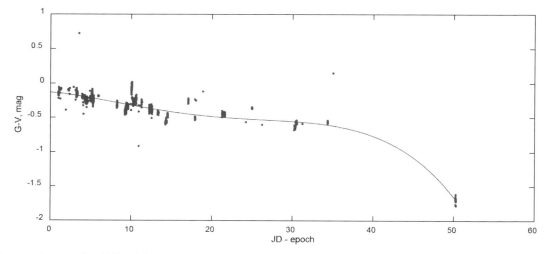

Figure 4. The divergence between G and V-band photometry over time due to the changing color of the nova. A fourth order polynomial fit is also shown and used to transform G-band data into V-band.

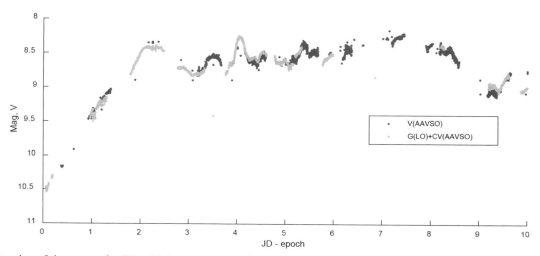

Figure 5. The first ten days of photometry after CV and G data are transformed to V using a time-dependent transform. The three sources are now highly compatible.

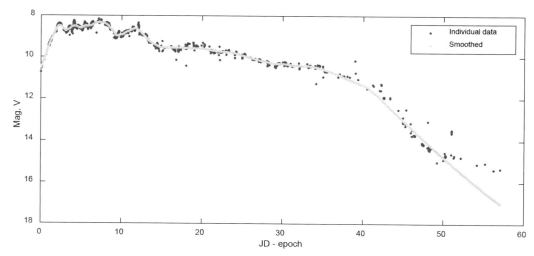

Figure 6. The long-term dimming of the nova (solid) is modelled by smoothing the data with intervals of one day. The short-term variation is then extracted by subtracting this long-term profile.

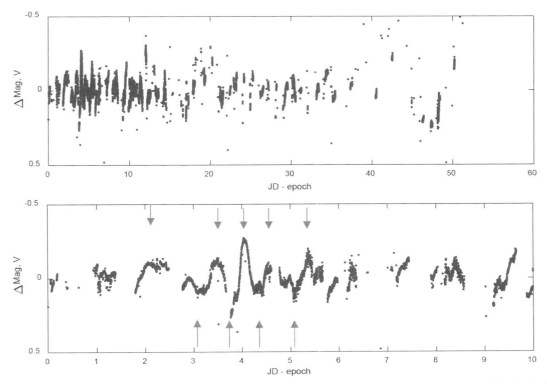

Figure 7. (Top) All photometry after being subtracted from the long-term brightness profile of the nova. (Bottom) Short-term variation in the first ten days. The periodic nature of this variation is clear, although its amplitude is inconsistent. Arrows point to peaks and troughs to highlight the periodic nature of the signal.

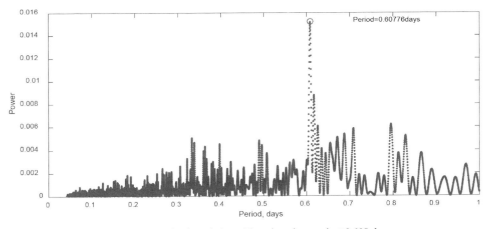

Figure 8. The Lomb-Scargle periodogram of short period magnitude variations. There is a clear peak at 0.608 day.

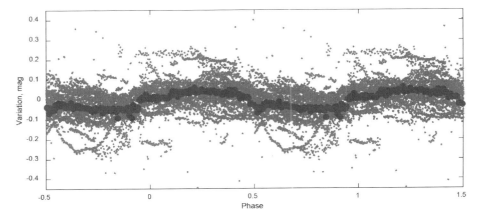

Figure 9. The short period signal phase folded to the detected period. Individual data (blue) are not particularly convincing. When binned by a factor of 200 (red), however, a convincing sinusoidal signal with a mean amplitude of approximately 75 mmag is apparent.

interval of one day. The photometry and the smoothed fit are shown in Figure 6.

The long-term profile is subtracted from the photometry to isolate the variability as shown in Figure 7. The periodic nature of the variability is apparent. Its amplitude ranges approximately a few hundred mmag but varies drastically, even between adjacent cycles.

The Lomb-Scargle periodogram of this signal is given in Figure 8, showing a convincing power peak at a period of 0.608 day (Lomb 1976). The Lomb-Scargle periodogram is a commonly used algorithm to detect periodic signals in unevenly spaced data. It is applied in MATLAB using the PLOMB.m function. It is based on Fourier transform theory and directly returns vectors of matching frequency and power values. A relatively high power at a given frequency indicates a repeating signal. We do not detect the much shorter period (0.09271 day) and smaller amplitude variability previously reported in the I-band (Schmidt 2021). This may not be surprising considering the difference in filters. We have some confidence in our results, however, as one can manually retrieve the 0.608-day period by measuring the peak-to-peak intervals shown in Figure 7. To crudely estimate the error of our derived period we repeat this period search for only the first 20 days and then for days 20–40. The results are 0.6107 and 0.6060 day, respectively. If the period is not actually changing then our error is on the order of the difference between these independent results, or 0.005 day.

This is not a signal that responds well to phase folding due to its irregular amplitudes, as seen in Figure 9. When the phased data are further binned using 200 observations per data point, however, a convincing sinusoidal form is presented with an amplitude of approximately 75 mmag.

5. Conclusion

An 80-day campaign monitored the dimming of Nova Per 2020 and was combined with the AAVSO database to detect a visual variability having a period of 0.608 day and an irregular amplitude that ranges between approximately 50 and 200 mmag. These results may allow for a more detailed understanding of the physical processes at play in this nova.

6. Acknowledgements

We acknowledge with thanks the variable star observations from the AAVSO International Database contributed by observers worldwide and used in this research.
PA#: USAFA-DF-2021-104

References

Bode, M. F., and Evans, A. 2008, *Classical Novae*, Cambridge Astrophys. Ser. 43, Cambridge University Press, Cambridge.

Gaia Collaboration, *et al.* 2016, *Astron. Astrophys.*, **595A**, 1.

Gaia Collaboration, *et al.* 2018, *Astron. Astrophys.*, **616A**, 1.

Kafka, S. 2021, variable star observations from the AAVSO International Database (https://www.aavso.org/aavso-international-database-aid).

Lomb, N. R. 1976, *Astrophys. Space Sci.*, **39**, 447.

Munari, U., Castellani, F., Dallaporta, S., and Andreoli, V. 2020, *Astron. Telegram*, No. 14224, 1.

Schmidt, R. E. 2021, *J. Amer. Assoc. Var. Star Obs.*, **49**, 99.

Thomas, N., and Guan, C., 2021, *Open Eur. J. Var. Stars*, **214**, 1.

Ueda, S. 2020, *Cent. Bur. Astron. Telegrams*, CBAT Transient Object Followup Reports, TCP J04291884+4354232, 2020 11 25.807.

Warner B. 1995, *Cataclysmic Variable Stars*, Cambridge Astrophys. Ser. 28, Cambridge University Press, Cambridge.

Woudt, P. A., and Ribeiro, V. A. R. M. 2014, in *Stella Novae: Past and Future Decades*, eds. P. A. Woudt, V. A. R. M. Ribeiro, Astron. Soc. Pacific Conf. Ser. 490, Astronomical Society of the Pacific, San Francisco.

Spectroscopic and Photometric Study of the Mira Stars SU Camelopardalis and RY Cephei

David Boyd
BAA Variable Star Section, 5 Silver Lane, West Challow OX12 9TX, UK; davidboyd@orion.me.uk

Received June 11, 2021; revised July 14, 2021; accepted July 14, 2021

Abstract Miras are fascinating stars. A kappa-mechanism in their atmosphere drives pulsations which produce changes in their photometric brightness, apparent spectral type, and effective temperature. These pulsations also drive the formation of Balmer emission lines in the spectrum. This behavior can be observed and investigated with small telescopes. We report on a three-year project combining spectroscopy and photometry to analyze the behavior of Mira stars SU Cam and RY Cep, and describe how their brightness, color, spectral type, effective temperature, and Balmer emission vary over four pulsation cycles.

1. Mira stars

Oxygen-rich Miras are pulsating red giant stars with spectral type late K or M and luminosity class III. Mira variables are evolved stars. They begin their thermonuclear lives burning hydrogen in their core on the main sequence of the Hertzsprung-Russell (HR) Diagram. This hydrogen burning produces helium ash which accumulates in the core. Once the hydrogen in the core runs out, the helium-rich core begins to collapse, which heats the hydrogen-rich shell on top of it until the hydrogen ignites in that shell. While hydrogen is burning in this shell, energy is dumped into the outer envelope of the star causing the outer layers to expand and cool, resulting in the star rising to the upper-right on the HR Diagram, the so-called Red Giant Branch. A helium flash at the top of the Red Giant Branch ignites helium burning to start forming carbon and oxygen in the core. Meanwhile hydrogen burning continues in a shell around the core. Eventually helium in the core becomes exhausted. As the star climbs the Asymptotic Giant Branch it contains both helium and hydrogen burning shells surrounding a degenerate core of carbon and oxygen. The star experiences multiple helium and hydrogen shell flashes or thermal pulses separated by many thousands of years.

Meanwhile, a kappa-mechanism of either dust in the atmospheres of these stars (Fleischer *et al.* 1995; Höfner *et al.* 1995) or a hydrogen-ionization zone (Querci 1986) just beneath the visible surface is the likely cause of the approximately annual pulsations which we observe, although this is still the subject of debate (Smith *et al.* 2002). By this stage the star may have expanded to over a hundred times its original size with a very tenuous outer atmosphere. This extended atmosphere makes it difficult to define the radius, which is also changing with time, and the pulsations are continuously driving the loss of gas and dust into the interstellar medium. An informal review of the problems of understanding Mira variables has been given by Wing (1980).

Mira stars are the subset of pulsating giant stars which have visual amplitudes greater than 2.5 magnitudes and pulsation periods of 100–1000 days. They are named after the prototype, omicron Ceti (Mira). The atmospheres of Miras are sufficiently cool that molecules can form, such as TiO in the oxygen-rich Miras. These molecules absorb light in the visual part of the spectrum which would otherwise have escaped from the star. The resulting TiO molecular absorption bands are a prominent feature in the visual spectrum of Miras. During its pulsation cycle the star appears to brighten as more of its light is emitted in the visual and its effective temperature rises, causing some of the molecules to dissociate. Then, as the star fades, it becomes cooler, redder and the molecules reform. The change in the strength of the molecular absorption bands as the star pulsates results in a change in its apparent spectral type. A comprehensive review of our knowledge about Mira stars is given in Willson and Marengo (2012).

2. Balmer emission lines

The transient appearance of hydrogen Balmer emission lines in the spectra of Mira stars has been known for over a hundred years since early objective prism observations at the Harvard College Observatory (Maury and Pickering 1897). They were observed to appear and grow as the star approached maximum light and decline and disappear as it faded. One of the earliest detailed spectroscopic studies of a Mira star was that of omicron Ceti by Joy (1926). These early spectroscopic observations were made with photographic plates which were relatively insensitive to red light and only able to record efficiently the shorter wavelength Balmer lines. The use of digital devices now enables all Balmer lines in the visual spectrum to be recorded and measured.

The formation of emission lines in the spectra of long period variables such as Miras was first explained theoretically by Gorbatskii (1961) as being due to a shock wave produced in the atmosphere of the star. During each pulsation cycle the outward pressure of radiation is countered by the inward pressure of gravity, creating a shock which propagates radially outwards, ionizing hydrogen in the atmosphere and driving mass loss. Many authors, including Willson (1976), Fox *et al.* (1984), and Gillet (1988), have discussed the formation of Balmer emission lines in Mira variables in relation to the production of shock waves. In general, previous observational studies have tended to cover only limited parts of the pulsation cycle or of the visual wavelength range.

3. The project

In a conversation with Arne Henden at the Society for Astronomical Sciences Symposium in 2017, he suggested that observing the pulsation of Miras stars spectroscopically might yield interesting results. This led to a three-year project combining spectroscopy and photometry, the results of which are presented here. Because I wanted to observe Mira stars both spectroscopically and photometrically, I needed to choose stars which at their brightest were not too bright to measure photometrically and at their faintest were not too faint to observe spectroscopically using the equipment available. These constraints together imposed a V magnitude range of approximately 8 to 15 on the stars chosen. As I also wanted to be able to observe them throughout the year from my observatory at 52°N, this imposed a practical lower declination limit of around +75°. A search of the AAVSO International Variable Star Index (VSX; Watson *et al.* 2014) revealed nine stars catalogued as Miras which fulfilled these criteria, although a check of the magnitudes of these stars reported to the AAVSO over the past year indicated that in most cases their current magnitude range was different from that given in VSX.

A further criterion adopted for practical reasons was that the pulsation period should be less than about 300 days so that it would be possible to accumulate data for at least three pulsation periods over a three-year project. I also wanted to be able to observe several consecutive pulsation cycles of both stars on a cadence of 1/20th of their pulsation period, bearing in mind constraints on observing due to weather, without impacting too severely on other ongoing observing projects. In the end I decided to follow two stars which had been under-observed digitally, possibly because they were not on the AAVSO LPV target list, namely SU Cam and RY Cep. It later emerged that they had rather different spectral types which added another interesting dimension to the project.

The project focused on studying how the photometric brightness and color, the apparent spectral type and effective temperature, and the behavior of the Balmer emission lines of these two stars varied over each pulsation cycle and from cycle to cycle. Data analysis was performed using Python software developed by the author which made extensive use of the Astropy package (Astropy Collab. *et al.* 2018).

4. Observations

Spectroscopy was obtained with a 0.28-m Schmidt-Cassegrain Telescope (SCT) operating at f/5 equipped with an auto-guided Shelyak LISA slit spectrograph and a SXVR-H694 CCD camera. The slit width was 23μ, giving a mean spectral resolving power of ~1000. Spectra were processed with the ISIS spectral analysis software (ISIS; Buil 2021). Spectroscopic images were bias, dark, and flat corrected, geometrically corrected, sky background subtracted, spectrum extracted, and wavelength calibrated using the integrated ArNe calibration source. They were then corrected for instrumental and atmospheric losses using spectra of a nearby star with a known spectral profile from the MILES library of stellar spectra (Falcón-Barroso *et al.* 2011) situated as close as possible in airmass to the target star and obtained immediately prior to the target spectra. Typically, 12 five-minute guided integrations were recorded for each spectrum, which gave signal-to-noise ratios ranging from ~100 at maximum brightness to ~10 at minimum. Spectra were calibrated in absolute flux in FLAM units using concurrently measured and transformed V magnitudes as described in Boyd (2020). All spectra were submitted to, and are available from, the BAA spectroscopy database (Br. Astron. Assoc. 2021b).

Photometry was obtained with a 0.35-m SCT operating at f/5 equipped with Astrodon BVRI photometric filters and an SXVR-H9 CCD camera. All photometric observations were made through alternating B and V filters with typically 10 images recorded in each filter. B and V filters were used because B and V magnitudes of comparison stars are available from the AAVSO Photometric All-Sky Survey (APASS; Henden *et al.* 2021). These photometric observations were made concurrently with recording spectra. All photometric images were bias, dark, and flat corrected and instrumental magnitudes obtained by aperture photometry using the software AIP4WIN (Berry and Burnell 2005). An ensemble of five nearby comparison stars was used whose B and V magnitudes were obtained from APASS. Instrumental B and V magnitudes were transformed to the Johnson UBV photometric standard using the measured B–V color index and atmospheric airmass with the algorithm published in Boyd (2011). Times are recorded as Julian Date (JD). All magnitudes were submitted to, and are available from, the BAA photometry database (Br. Astron. Assoc. 2021a).

Figure 1 shows spectra of SU Cam and RY Cep near maximum and minimum, with Balmer emission lines marked and including transmission profiles of the B and V filters used. The spectrum of SU Cam at minimum has been amplified to make it more visible. Figure 1 shows that, at maximum, the spectral type of RY Cep is much earlier than that of SU Cam. At our resolution we are not able to see the detailed structure in the emission lines reported in some higher resolution studies. While neither the Hα or Hβ emission lines contribute significantly in the V band, the Hγ and Hδ emission lines do contribute to the flux recorded in the B filter but calculation shows that this is in most cases much less than 10%. The relative strength of the Balmer lines seen in these spectra will be discussed later.

The distance reported by Gaia EDR3 (Gaia Collab. *et al.* 2016) for SU Cam is 1009 +85 –77 parsecs and for RY Cep is 2578 +150 –138 parsecs. According to Schlafly and Finkbeiner (2011), the galactic extinction towards SU Cam is E(B–V) = 0.097 and towards RY Cep is E(B–V) = 0.158. Both stars lie out of the plane of the galaxy so the real extinction they experience is likely to be close to these values. We use the formulae in Cardelli *et al.* (1989) to compute dereddening profiles used to deredden all our spectra.

5. Photometric brightness and color

Photometric B-band and V-band observations of SU Cam and RY Cep cover the period JD 2457994 to 2459341. Figure 2 shows B and V magnitudes and B–V color index vs JD. Figure 3 shows the V vs B–V color magnitude diagrams in

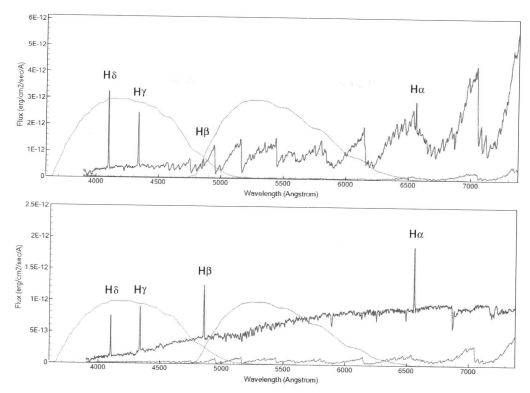

Figure 1. Spectra of SU Cam (upper) and RY Cep (lower) at maximum and minimum brightness, with Balmer emission lines marked and including B (red) and V (green) filter transmission profiles. The spectrum of SU Cam at minimum has been amplified five times to make it more visible. (Note: figures in color are available in the online version of the paper.)

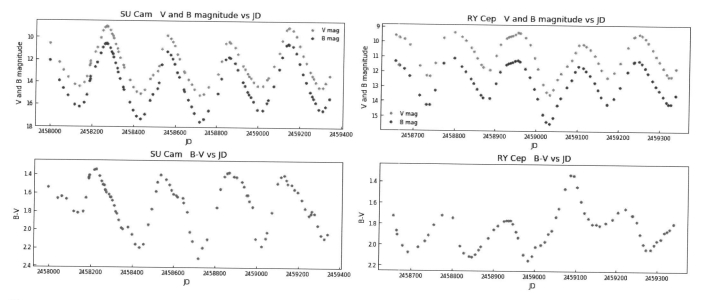

Figure 2. V and B magnitudes and B–V color index vs JD showing four pulsation cycles.

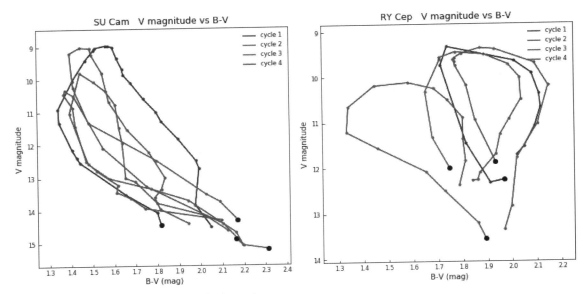

Figure 3. V vs B–V color magnitude diagrams for each pulsation cycle.

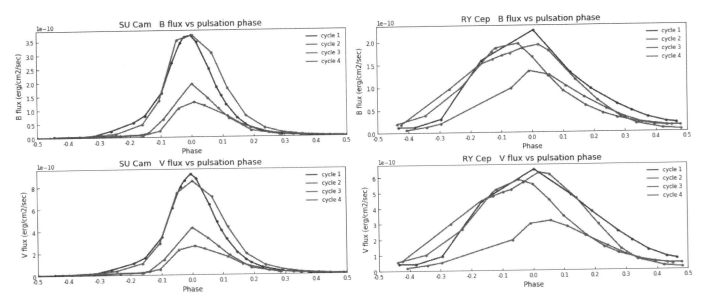

Figure 4. B- and V-band flux vs pulsation phase for each pulsation cycle.

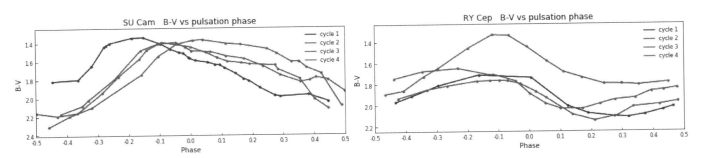

Figure 5. B–V color index vs pulsation phase for each pulsation cycle.

which each color-coded cycle traverses a complex loop. The initial point during each pulsation cycle is marked with a larger black dot to make it easier to follow the trajectories in every cycle. The complex path followed during each cycle in Figure 3 is a consequence of the way the flux profile of the spectrum of each star, as integrated by these filters, changes during its pulsation cycle.

V magnitude measurements around the peak of each of the four recorded pulsation cycles for each star were fitted with a fourth-order polynomial in JD and this was used to find the time of maximum magnitude for each cycle. These times were converted to Heliocentric Julian Date (HJD) and used to derive the following linear ephemerides for times of maximum (ToM) with E ranging from 0 to 3:

SU Cam:
    ToM (HJD) = 2458267.74793(5) + 292.49341(3) * E   (1)

RY Cep:
    ToM (HJD) = 2458798.25577(5) + 152.63684(3) * E   (2)

The mean pulsation period of SU Cam over this time interval is 292.49 days and for RY Cep is 152.64 days. The periods currently listed in VSX are 286.25 days and 149.06 days, respectively. Using these linear ephemerides, all times were converted to phases of the pulsation cycle with phase 0 occurring at or close to the time of maximum brightness.

B and V magnitudes can be converted to absolute flux using photometric zero points derived from CALSPEC spectrophotometric standard stars (Bohlin *et al.* 2014; STScI 2021). The variation of B- and V-band flux with pulsation phase for each pulsation cycle is shown for both stars in Figure 4. While there is considerable variation in the profiles from cycle to cycle, it is noticeable that the B and V flux profiles are more sharply peaked in SU Cam in all cycles compared with the broader peaks in RY Cep.

The reason for the complex behavior of the B–V color magnitude diagrams in Figure 3 becomes clearer when we look at the variation of the B–V color indices over the pulsation phase for each cycle as shown in Figure 5. In most cycles the B–V color index peaks before maximum brightness then becomes redder as the cycle progresses, thereby traversing a clockwise loop in the color-magnitude diagram. This effect is more pronounced in RY Cep compared to SU Cam leading to wider loops in the former.

6. Apparent spectral type and effective temperature

Oxygen-rich giant stars of later spectral type exhibit strong TiO molecular absorption bands in their spectra and the strength of these bands is usually taken as an indication of the spectral type of the star. Because the strength of the molecular bands changes over the pulsation cycle in Mira stars, this relationship is more complex. However, for the purpose of our analysis, we will assume that this relationship can be used to assign a spectral type to Mira stars that changes as they pulsate.

Assigning a spectral type to a spectrum is commonly achieved by comparing it morphologically to a range of standard star spectra in the MK spectral classification system and identifying the closest match (Gray and Corbally 2009). MK standard stars available with the MKCLASS stellar spectral classification system (Gray and Corbally 2014) cover the wavelength range 3800–5600 Å, where atomic absorption lines are concentrated, a legacy of the use of blue-sensitive photographic plates in the early days of the MK standard. In our spectra the flux in this region is relatively low, whereas it is considerably stronger towards the red end of the visual range where the molecular bands are prominent. Given our limited spectral resolution and therefore inability to resolve some of the lines in the blue part of the spectrum used for classification, using the full visual range to classify our spectra offers a more practical way of assigning a spectral type.

As all our SU Cam spectra fell within range of the M spectral type, we decided to use the M giant spectra published in Fluks *et al.* (1994), which are classified on the MK system, to assign an apparent spectral type to each spectrum. The Fluks spectra for spectral types M0 to M10 are defined on the wavelength range 3500–10000 Å at an interval of 1 Å. The spectral type of RY Cep became earlier than M as it approached maximum light in each cycle. This meant finding stars with MK standard spectral type K for which we could also obtain spectra. After considering possible sources, we decided to use stars listed in the Perkins Catalog of stars classified on the revised MK system (Keenan and McNeil 1989) for which there are spectra in the MILES library of stellar spectra (Falcón-Barroso *et al.* 2011). Examination of their spectra showed that they formed a sequence which was both internally consistent and also consistent with the transition to the Fluks M type spectra. The stars used as standard spectral types between K0III and K5III are listed in Table 1.

Prior to using them in our analysis, all spectra being used as standards were interpolated to a wavelength interval of 1 Å and normalized to a mean flux value of unity in the wavelength interval 5610 to 5630 Å, which contains no strong spectral features. All our spectra of SU Cam and RY Cep were similarly interpolated to 1 Å and normalized to unity in the same wavelength interval. After removing emission lines each of our spectra was compared with each of the K and M type standard spectra. The differences in flux at each Angstrom between 4000 and 7000 Å were squared and totalled. This gave a quantitative measure of the difference in profile between each of our spectra and each of the standard spectra. For each spectrum there was one spectral type for which this flux difference was a minimum. By fitting a quadratic polynomial to the flux differences around this minimum, it was possible to assign a spectral sub-type to

Table 1. Stars from the Perkins Catalog (Keenan and McNeil 1989) used as standard spectral types for K0III to K5III.

Spectral Type	HD Number	Common Name
K0III	197989	ε Cyg
K1III	037984	51 Ori
K2III	054719	τ Gem
K3III	171443	α Sct
K4III	069267	β Cnc
K5III	164058	γ Dra

Table 2. Julian Date, cycle number, spectral sub-type, effective temperature (Teff) and Balmer emission line fluxes for each SU Cam spectrum.

Julian Date of Spectrum	Cycle	Spectral Sub-type	Teff (K)	Hα Line Flux (ergs/cm2/sec)	Hβ Line Flux (ergs/cm2/sec)	Hγ Line Flux (ergs/cm2/sec)	Hδ Line Flux (ergs/cm2/sec)
2457994.44120		M7.6	3045	9.99E–13	0.0	5.55E–13	1.86E–12
2458039.37220		M7.8	3011	3.90E–13	0.0	1.60E–13	4.58E–13
2458059.34370		M8.3	2922	2.29E–13	0.0	5.95E–14	2.27E–13
2458082.44680		M8.6	2865	1.35E–13	0.0	5.22E–14	0.0
2458115.43200		M8.9	2806	7.59E–14	0.0	0.0	0.0
2458137.48730	1	M8.8	2826	5.79E–14	0.0	0.0	0.0
2458161.42780	1	M8.5	2885	1.10E–13	0.0	0.0	0.0
2458174.42350	1	M8.5	2885	1.17E–13	0.0	0.0	0.0
2458191.43750	1	M7.6	3045	3.29E–13	0.0	2.92E–13	2.69E–13
2458212.42710	1	M7.5	3061	7.40E–13	0.0	1.23E–12	9.87E–13
2458223.43220	1	M6.6	3199	9.58E–13	9.12E–14	2.35E–12	2.89E–12
2458239.44960	1	M6.4	3227	1.98E–12	2.70E–13	5.13E–12	6.57E–12
2458249.40210	1	M6.0	3281	3.82E–12	7.22E–13	7.28E–12	8.50E–12
2458257.43280	1	M5.7	3320	5.42E–12	1.22E–12	8.88E–12	9.74E–12
2458261.43060	1	M5.6	3332	6.07E–12	1.49E–12	9.18E–12	1.00E–11
2458267.44220	1	M5.4	3357	6.52E–12	1.91E–12	1.00E–11	1.04E–11
2458272.43560	1	M5.7	3320	7.54E–12	1.78E–12	8.20E–12	8.49E–12
2458284.43990	1	M6.0	3281	5.87E–12	9.64E–13	4.58E–12	3.87E–12
2458291.44420	1	M6.3	3241	4.55E–12	6.09E–13	3.72E–12	3.30E–12
2458295.44880	1	M6.6	3199	4.90E–12	4.22E–13	2.53E–12	2.24E–12
2458300.43730	1	M6.5	3213	4.03E–12	3.12E–13	1.75E–12	1.38E–12
2458310.43950	1	M6.7	3184	3.27E–12	2.38E–13	1.26E–12	9.78E–13
2458318.43060	1	M7.3	3093	3.10E–12	2.12E–13	7.37E–13	4.74E–13
2458323.42060	1	M7.8	3011	3.15E–12	1.75E–13	4.57E–13	4.95E–13
2458333.51160	1	M7.7	3028	2.52E–12	1.21E–13	1.57E–13	2.54E–13
2458352.49640	1	M7.8	3011	1.40E–12	0.0	0.0	0.0
2458379.45380	1	M7.8	3011	4.31E–13	0.0	0.0	0.0
2458398.43130	1	M7.9	2994	1.56E–13	0.0	0.0	0.0
2458414.39980	2	M7.8	3011	7.56E–14	0.0	0.0	0.0
2458434.34880	2	M7.7	3028	5.42E–14	0.0	0.0	0.0
2458477.41860	2	M7.9	2994	1.05E–13	0.0	0.0	0.0
2458492.41150	2	M8.3	2922	1.65E–13	0.0	0.0	0.0
2458512.40850	2	M8.2	2941	1.83E–13	0.0	0.0	1.52E–13
2458519.35250	2	M8.0	2977	2.08E–13	0.0	0.0	3.53E–13
2458533.33470	2	M6.7	3184	5.86E–13	0.0	7.22E–13	2.00E–12
2458560.46490	2	M6.0	3281	1.11E–12	0.0	2.66E–12	4.90E–12
2458575.36370	2	M6.2	3254	9.04E–13	0.0	1.63E–12	3.80E–12
2458585.38330	2	M6.3	3241	7.49E–13	0.0	1.02E–12	3.04E–12
2458595.44270	2	M6.7	3184	7.75E–13	0.0	8.43E–13	2.56E–12
2458616.41010	2	M7.7	3028	4.72E–13	0.0	3.68E–13	1.10E–12
2458643.43500	2	M8.2	2941	3.04E–13	0.0	1.29E–13	1.15E–13
2458718.51510	3	M8.5	2885	0.0	0.0	0.0	0.0
2458738.44610	3	M8.4	2904	0.0	0.0	0.0	0.0
2458759.45040	3	M7.6	3045	0.0	0.0	0.0	0.0
2458806.40450	3	M8.1	2959	1.17E–13	0.0	0.0	8.28E–14
2458822.37100	3	M7.7	3028	2.70E–13	0.0	2.06E–13	5.64E–13
2458840.42970	3	M6.5	3213	6.74E–13	0.0	1.64E–12	3.68E–12
2458855.41260	3	M6.5	3213	9.39E–13	0.0	2.42E–12	6.05E–12
2458864.38000	3	M6.6	3199	9.79E–13	0.0	2.41E–12	6.55E–12
2458886.43030	3	M6.6	3199	6.40E–13	0.0	2.04E–12	5.88E–12
2458900.46480	3	M7.6	3045	6.21E–13	0.0	8.74E–13	3.39E–12
2458925.41210	3	M8.0	2977	3.24E–13	0.0	2.34E–13	1.09E–12
2458934.37000	3	M8.3	2922	2.76E–13	0.0	1.90E–13	7.87E–13
2458948.43670	3	M8.1	2959	2.27E–13	0.0	8.00E–14	2.87E–13
2458955.37740	3	M8.2	2941	1.80E–13	0.0	5.76E–14	1.66E–13
2458962.39400	3	M8.1	2959	1.70E–13	0.0	5.00E–14	1.17E–13
2458976.42220	3	M8.2	2941	1.33E–13	0.0	0.0	0.0
2458995.45310	3	M7.8	3011	0.0	0.0	0.0	0.0
2459022.45080	4	M7.7	3028	3.15E–14	0.0	0.0	0.0
2459048.43950	4	M8.2	2941	1.67E–13	0.0	0.0	0.0
2459073.50590	4	M7.8	3011	2.74E–13	0.0	0.0	0.0
2459098.46100	4	M7.6	3045	6.03E–13	0.0	5.94E–13	1.25E–12
2459114.47510	4	M6.5	3213	1.09E–12	0.0	2.80E–12	4.96E–12
2459131.45160	4	M5.5	3345	3.47E–12	9.76E–13	1.05E–11	1.26E–11
2459146.37560	4	M5.5	3345	6.51E–12	2.37E–12	1.52E–11	1.58E–11
2459164.42980	4	M5.7	3320	8.27E–12	3.23E–12	1.39E–11	1.38E–11

Table continued on next page

Table 2. Julian Date, cycle number, spectral sub-type, effective temperature (Teff) and Balmer emission line fluxes for each SU Cam spectrum, cont.

Julian Date of Spectrum	Cycle	Spectral Sub-type	Teff (K)	Hα Line Flux (ergs/cm2/sec)	Hβ Line Flux (ergs/cm2/sec)	Hγ Line Flux (ergs/cm2/sec)	Hδ Line Flux (ergs/cm2/sec)
2459179.32490	4	M6.3	3241	8.10E–12	2.11E–12	7.54E–12	6.99E–12
2459196.39410	4	M6.8	3170	7.25E–12	9.33E–13	2.60E–12	2.16E–12
2459214.35220	4	M7.2	3109	4.59E–12	4.04E–13	7.70E–13	6.63E–13
2459230.33230	4	M7.6	3045	3.09E–12	1.74E–13	2.84E–13	1.93E–13
2459249.47460	4	M7.7	3028	1.60E–12	5.51E–14	0.0	0.0
2459258.40820	4	M7.9	2994	1.21E–12	3.33E–14	0.0	0.0
2459264.41330	4	M7.5	3061	8.19E–13	0.0	0.0	0.0
2459275.34950	4	M8.5	2885	6.54E–13	0.0	0.0	0.0
2459291.42990	4	M8.2	2941	2.63E–13	0.0	0.0	0.0
2459309.36300		M8.6	2865	1.10E–13	0.0	0.0	0.0
2459323.38720		M8.4	2904	1.51E–13	0.0	0.0	0.0
2459341.39820		M8.2	2941	1.51E–13	0.0	0.0	0.0

the nearest tenth to each of our spectra. These assigned K and M spectral sub-types are listed for each SU Cam spectrum in Table 2 and for each RY Cep spectrum in Table 3. The spectral types of SU Cam range from M5 to M8 while those of RY Cep range from K4 to M6.

The relationship between effective temperature and spectral type in Miras is not simple and the literature contains a variety of approaches to this problem. After reviewing the options, we decided to adopt a pragmatic approach and use the data on effective temperature and spectral type for K and M giant stars given in van Belle et al. (1999) and apply these to our Mira spectra. We used a polynomial parameterization of the van Belle data to assign effective temperatures to all our spectra based on their assigned spectral types. These assigned effective temperatures are also listed in Tables 2 and 3.

The variation of effective temperature with pulsation phase over each pulsation cycle is shown in Figure 6. Maximum effective temperature occurs close to the time of maximum brightness in all cycles in both stars. Similar to the B and V flux behavior in Figure 4, the rise in effective temperature around phase 0 is more narrowly peaked in SU Cam than in RY Cep where the effective temperature changes more gradually through the cycle.

Effective temperatures are plotted against concurrently measured V magnitudes for all SU Cam and RY Cep spectra in Figure 7. Different symbols are used to differentiate the rising and falling branches of each cycle. V magnitude and effective temperature are clearly correlated with both rising and falling branches following the same trajectory. Below 13th magnitude the greater scatter is a consequence of the increasing noise in these spectra. Figure 7 also includes the parameters of quadratic fits to the data, including the R-squared value for the correlation. The internal consistency of these plots suggests that our pragmatic approach to assigning effective temperatures was reasonable.

In many stars the B–V color index serves as a useful proxy for effective temperature. This is not the case in Miras, as Figure 8 shows. Different wavelengths probe different depths in their tenuous atmospheres and the varying amounts of molecular material present in the atmosphere during each pulsation cycle change the relative flux in the B and V bands in a complex way which varies during a pulsation cycle and from cycle to cycle. The different path taken by RY Cep during cycle 3 is the result of lower flux in both B and V in this cycle, as shown in Figure 4, which results in lower effective temperature. The flux in V in this cycle is also proportionally lower than the flux in B compared to the other cycles, giving a bluer B–V color index as shown in Figure 5. The combined effect is to move the path taken in this cycle down and to the left in the diagram.

7. Balmer emission lines

Shocks produced in the atmosphere of both stars during each pulsation cycle generate emission lines of the hydrogen Balmer series. The strength of these lines tends to increase as the star brightens and decline after it has passed through maximum brightness. As our spectra have been calibrated in absolute flux, the flux in each emission line above the local continuum can be measured using the software PlotSpectra (Lester 2020). These Balmer line fluxes for each spectrum are listed in Tables 2 and 3, while Figure 9 shows how they vary over each pulsation cycle. Where the flux above the local continuum is too small to be reliably measured, the tables contain the value zero. Many emission lines are asymmetric about phase 0 and have considerable skew in their profiles. There is nevertheless a degree of consistency within cycles, with all lines peaking around the same phase in the same cycle. There also is a noticeable tendency for emission lines to be broader in phase in RY Cep than in SU Cam.

Yao et al. (2017) found that, in oxygen-rich Miras, there is a Balmer increment (Hα < Hβ < Hγ < Hδ) for stars with spectral types M5 to M10 whereas there is a Balmer decrement (Hα > Hβ > Hγ > Hδ) for earlier spectral types. They noted that Hβ is sometimes weak, which is also our experience. Our spectra shown in Figure 1 are consistent with Yao et al.'s conclusions. To investigate this further we used the Hα/Hδ line flux ratio for both stars as a proxy for the Balmer decrement/increment (>1 = decrement, <1 = increment) and effective temperature as a proxy for spectral type. Figure 10 shows that the Hα/Hδ line flux ratio falls as the effective temperature drops and around 3550 K, equivalent to spectral sub-type M3.6, the Balmer decrement changes to an increment.

Previous studies of Mira stars have analyzed the strength and behavior of Balmer emission lines on a per-spectrum basis and

Table 3. Julian Date, cycle number, spectral sub-type, effective temperature (Teff) and Balmer emission line fluxes for each RY Cep spectrum.

Julian Date of Spectrum	Cycle	Spectral Sub-type	Teff (K)	Hα Line Flux (ergs/cm2/sec)	Hβ Line Flux (ergs/cm2/sec)	Hγ Line Flux (ergs/cm2/sec)	Hδ Line Flux (ergs/cm2/sec)
2458655.47320		M0.3	3882	7.12E–12	4.81E–12	3.98E–12	2.83E–12
2458665.45240		M0.8	3834	5.80E–12	3.07E–12	2.53E–12	1.74E–12
2458677.44190		M1.8	3738	4.53E–12	1.65E–12	1.29E–12	8.32E–13
2458690.45110		M3.6	3559	2.88E–12	7.07E–13	5.69E–13	3.46E–13
2458715.43760		M5.7	3320	6.41E–13	3.35E–14	0.0	0.0
2458732.37490	1	M5.9	3294	2.26E–13	0.0	0.0	0.0
2458740.45250	1	M6.2	3254	1.19E–13	0.0	0.0	0.0
2458753.36730	1	M5.1	3393	2.24E–13	0.0	0.0	0.0
2458773.37670	1	M2.5	3670	5.18E–12	2.75E–12	2.70E–12	1.47E–12
2458799.38690	1	K5.6	3949	8.23E–12	6.21E–12	5.20E–12	3.62E–12
2458817.31610	1	M0.1	3901	4.51E–12	1.97E–12	1.66E–12	1.13E–12
2458827.34330	1	M1.2	3796	3.00E–12	9.18E–13	9.55E–13	5.47E–13
2458840.37610	1	M1.8	3738	1.33E–12	3.52E–13	3.24E–13	1.38E–13
2458847.40540	1	M3.6	3559	9.53E–13	1.62E–13	2.58E–13	8.71E–14
2458855.35170	1	M4.5	3462	5.31E–13	0.0	0.0	0.0
2458864.31440	1	M4.9	3417	2.88E–13	0.0	0.0	0.0
2458869.36570	1	M5.6	3332	1.54E–13	0.0	0.0	0.0
2458886.38000	2	M5.2	3381	1.15E–13	0.0	0.0	0.0
2458910.37200	2	M2.8	3640	1.76E–12	7.63E–13	1.17E–12	6.49E–13
2458925.35070	2	M0.1	3901	4.47E–12	2.98E–12	2.74E–12	1.66E–12
2458931.41790	2	M1.1	3805	5.49E–12	3.80E–12	3.90E–12	2.77E–12
2458936.37410	2	K5.7	3939	5.46E–12	4.83E–12	4.50E–12	3.00E–12
2458940.37870	2	M0.6	3853	6.18E–12	4.71E–12	5.10E–12	3.53E–12
2458946.37370	2	M0.1	3901	5.76E–12	4.58E–12	4.74E–12	3.13E–12
2458954.37210	2	K5.8	3929	5.74E–12	3.88E–12	3.26E–12	1.87E–12
2458959.39670	2	M0.4	3872	5.47E–12	2.43E–12	1.85E–12	1.15E–12
2458972.42060	2	M1.4	3777	2.94E–12	7.44E–13	5.36E–13	1.82E–13
2458983.43620	2	M3.4	3580	1.26E–12	2.25E–13	2.62E–13	0.0
2458994.42840	2	M4.9	3417	8.81E–13	0.0	0.0	0.0
2459002.43200	2	M4.8	3428	3.44E–13	0.0	0.0	0.0
2459015.43310	2	M6.0	3281	2.07E–13	0.0	0.0	0.0
2459024.44080	2	M6.5	3213	9.80E–14	0.0	0.0	0.0
2459041.44590	3	M6.5	3213	6.47E–14	0.0	0.0	0.0
2459051.44400	3	M5.9	3294	6.94E–14	0.0	0.0	0.0
2459058.44310	3	M5.6	3332	1.13E–13	0.0	1.93E–13	0.0
2459093.37990	3	M3.0	3620	1.44E–12	7.98E–13	1.87E–12	1.69E–12
2459102.44410	3	M2.5	3670	3.10E–12	2.24E–12	4.20E–12	3.27E–12
2459112.38060	3	M2.6	3660	3.36E–12	2.28E–12	3.46E–12	2.46E–12
2459120.39190	3	M2.8	3640	2.32E–12	1.47E–12	2.26E–12	1.71E–12
2459129.43650	3	M3.5	3569	1.49E–12	6.70E–13	1.24E–12	9.53E–13
2459140.39550	3	M4.5	3462	7.20E–13	2.57E–13	6.25E–13	4.99E–13
2459149.34110	3	M5.2	3381	4.15E–13	0.0	3.53E–13	2.67E–13
2459157.46540	3	M5.7	3320	2.54E–13	0.0	1.62E–13	1.10E–13
2459172.31360	3	M6.3	3241	1.30E–13	0.0	0.0	0.0
2459189.34830	4	M5.6	3332	1.43E–13	0.0	0.0	0.0
2459203.38410	4	M4.7	3440	1.96E–13	0.0	1.94E–13	9.44E–14
2459221.35100	4	M3.7	3549	2.18E–12	1.20E–12	1.77E–12	1.11E–12
2459236.41120	4	K5.7	3939	5.90E–12	5.91E–12	5.20E–12	3.54E–12
2459238.31480	4	K5.9	3920	6.80E–12	6.17E–12	5.38E–12	3.46E–12
2459249.40320	4	K4.2	4084	6.56E–12	6.40E–12	5.45E–12	3.62E–12
2459256.35220	4	M0.1	3901	5.44E–12	4.30E–12	3.70E–12	2.79E–12
2459264.35450	4	M0.7	3843	3.84E–12	2.18E–12	2.19E–12	1.43E–12
2459271.43120	4	M1.5	3767	3.32E–12	1.25E–12	1.01E–12	6.73E–13
2459282.48690	4	M4.4	3473	2.09E–12	3.42E–13	2.72E–13	1.63E–13
2459291.35470	4	M4.3	3484	1.06E–12	1.93E–13	2.18E–13	5.54E–14
2459298.38450	4	M5.2	3381	7.78E–13	0.0	0.0	0.0
2459308.46660	4	M6.0	3281	5.79E–13	0.0	0.0	0.0
2459316.38150	4	M6.2	3254	3.09E–13	0.0	0.0	0.0
2459322.38530	4	M6.4	3227	2.70E–13	0.0	0.0	0.0
2459329.39380	4	M6.5	3213	2.28E–13	0.0	0.0	0.0
2459339.40940		M6.0	3281	3.11E–13	0.0	1.15E–13	0.0

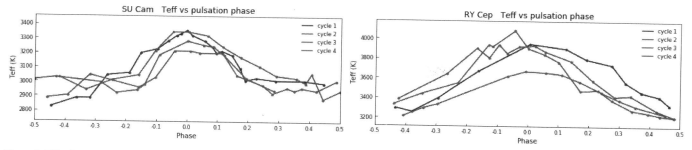

Figure 6. Effective temperature (Teff) vs pulsation phase for each pulsation cycle.

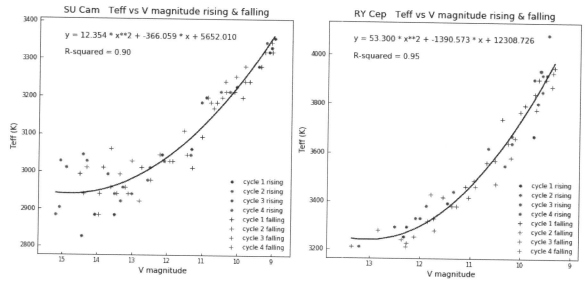

Figure 7. Effective temperature (Teff) vs V magnitude for all spectra plus quadratic fits to the data.

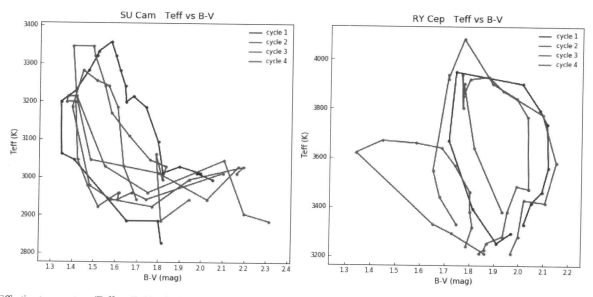

Figure 8. Effective temperature (Teff) vs B–V color index for each pulsation cycle.

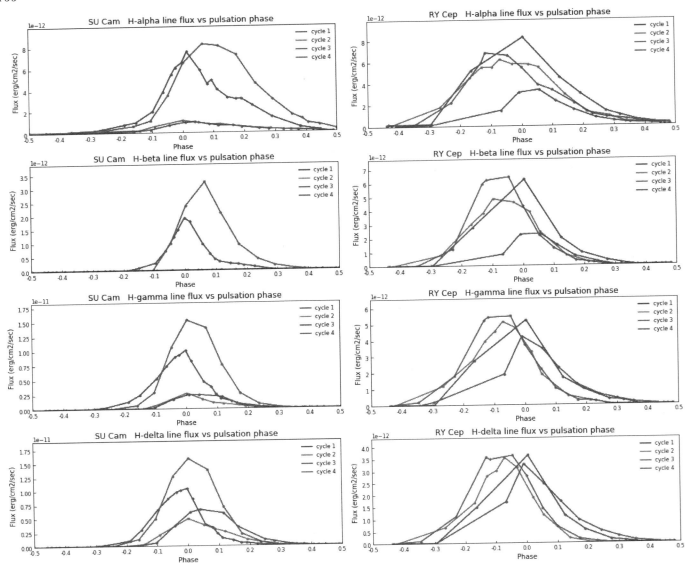

Figure 9. Flux in the Balmer emission lines vs pulsation phase for each pulsation cycle.

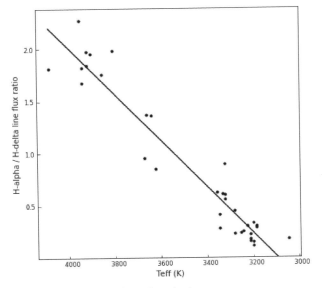

Figure 10. Hα/Hδ line flux ratio vs effective temperature (Teff) plus a linear fit to the data.

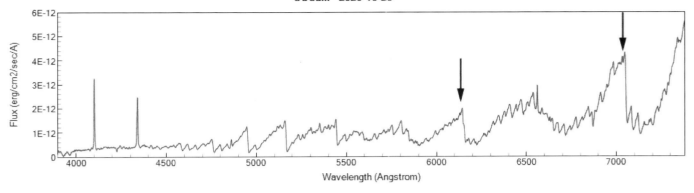

Figure 11. Position of two regions where the mean flux of each spectrum is measured.

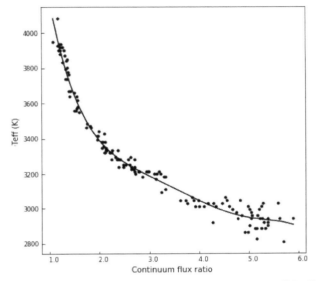

Figure 12. Effective temperature (Teff) vs continuum flux ratio for all SU Cam and RY Cep spectra plus a fitted fifth-order polynomial.

Table 4. Integrated flux emitted in each Balmer line and in the V band during each pulsation cycle in SU Cam.

Cycle	Hα (ergs/cm2)	Hβ (ergs/cm2)	Hγ (ergs/cm2)	Hδ (ergs/cm2)	V band (ergs/cm2)
1	4.71E–05	6.49E–06	4.53E–05	4.80E–05	4.74E–03
2	8.17E–06	0.0	1.08E–05	2.55E–05	2.09E–03
3	7.41E–06	0.0	1.37E–05	3.90E–05	1.59E–03
4	6.72E–05	1.46E–05	7.68E–05	8.29E–05	5.06E–03

Table 5. Integrated flux emitted in each Balmer line and in the V band during each pulsation cycle in RY Cep.

Cycle	Hα (ergs/cm2)	Hβ (ergs/cm2)	Hγ (ergs/cm2)	Hδ (ergs/cm2)	V band (ergs/cm2)
1	3.72E–05	2.09E–05	1.86E–05	1.19E–05	3.75E–03
2	3.10E–05	1.75E–05	1.72E–05	1.06E–05	3.74E–03
3	1.24E–05	7.01E–06	1.37E–05	1.06E–05	1.68E–03
4	2.93E–05	2.01E–05	1.89E–05	1.25E–05	3.08E–03

have covered only part of their pulsation cycle. Because we have comprehensive coverage of several cycles, we can analyze our data on a per-cycle basis. By linearly interpolating and integrating over the emission line profiles we can compute the total flux or energy emitted in each line during each pulsation cycle. Similarly, integrating over the V-band profiles in Figure 4 gives a measure of the total energy emitted in the V band in each cycle. These integrated fluxes are listed in Tables 4 and 5. As noted previously (Fox *et al.* 1984), the shock-induced line flux generally increases with the strength of the pulsation-driven V-band flux. Tables 4 and 5 also show that the integrated Balmer line fluxes in each cycle generally follow an increment in SU Cam and a decrement in RY Cep.

8. Estimating effective temperature from continuum flux ratios

The spectrum of an oxygen-rich M giant star does not represent its true photospheric continuum because of extensive molecular absorption (Fluks *et al.* 1994). Wing (1992) developed a technique for characterising the spectra of red variables by making photometric measurements at three wavelengths and deriving an index of TiO band strength which could be used to estimate spectral type.

As we have a set of Mira spectra for which we have computed effective temperatures, we investigated whether a simple variation on this idea could be used to estimate the effective temperatures of Mira stars. We measured the mean flux in the two wavelength ranges, 6130–6140 Å and 6970–6980 Å, marked in Figure 11. These are adjacent to TiO molecular band heads and are therefore likely to be regions of the spectrum closest to the true photospheric continuum. Because this involves taking a flux ratio, it does require the spectrum to be calibrated in relative flux across this spectral range but not necessarily in absolute flux.

In Figure 12 we plot the effective temperatures of all our SU Cam and RY Cep spectra against the ratio of these mean continuum fluxes. The narrowness of this distribution suggests that it may be possible to estimate effective temperature for Mira stars with spectral types between K4 and M8 by measuring this flux ratio. Fitting a fifth-order polynomial to this distribution gives an R-squared of 0.98 and a rms residual of 45 K.

9. Conclusions

This study shows that, with small telescopes suitably equipped and operated, it is possible, using a combination of spectroscopy and photometry, to monitor the behavior of Mira stars such as SU Cam and RY Cep and to analyze how their brightness, spectra, and Balmer emission vary over multiple pulsation cycles.

We found the following:
• during the time interval of this study the average pulsation periods of SU Cam and RY Cep were 292.49 days and 152.64 days, respectively;
• there is a consistently different pattern of behavior between the earlier spectral type RY Cep and the later SU Cam, with flux in the B and V bands, flux in the Balmer emission lines, and the effective temperature all peaking more sharply around the time of maximum brightness in SU Cam compared to RY Cep;
• maximum effective temperature coincides with maximum brightness in the V band in all pulsation cycles of both stars;
• there is a close correlation between effective temperature and V magnitude over the full brightness range in both rising and falling branches of all cycles of both stars;
• the B–V color index shows large variations over each cycle and from cycle to cycle and is a poor indication of effective temperature in these stars;
• relative Balmer line strengths, as measured by the $H\alpha / H\delta$ flux ratio, change from decrement to increment as the spectral type becomes later, with the transition occurring around spectral sub-type M3.6;
• the ratio of mean fluxes measured at two points near TiO molecular band heads can be used to estimate the effective temperature of a late K or M type Mira star.

10. Acknowledgements

I acknowledge with thanks a constructive and helpful referee's report. I am grateful to Lee Anne Willson for her valuable comments and advice, and to John Percy for his helpful feedback. Richard Gray and Chris Corbally provided advice on MK standard stars and the latter gave useful comments on an early draft. This research made use of the AAVSO Photometric All-Sky Survey (APASS) and the AAVSO Variable Star Index (VSX). The software developed for this project made extensive use of the Astropy package and the efforts of many contributors to this valuable community resource are gratefully acknowledged.

References

Astropy Collaboration, *et al.* 2018, *Astron. J.*, **156**, 123.
Berry, R., and Burnell, J. 2005, *Handbook of Astronomical Image Processing*, Willmann-Bell, Richmond, VA.
Bohlin, R. C., Gordon, Karl D., and Tremblay, P.-E. 2014, *Proc. Astron. Soc. Pacific*, 126, 711.
Boyd, D. 2011, in *Proceedings for the Society for Astronomical Sciences 30th Annual Symposium on Telescope Science*, eds. B. D. Warner, J. Foote, R. Buchheim, Society for Astronomical Sciences, Rancho Cucamonga, CA (http://www.socastrosci.org/Publications.html), 127.
Boyd, D. 2020, "A method of calibrating spectra in absolute flux using V magnitudes" (https://britastro.org/sites/default/files/absfluxcalibration.pdf).
British Astronomical Association. 2021a, BAA Photometry Database (https://britastro.org/photdb).
British Astronomical Association. 2021b, BAA Spectroscopy Database (https://britastro.org/specdb).
Buil, C. 2021, ISIS software (http://www.astrosurf.com/buil/isis-software.html).
Cardelli, J. A., Clayton, G. C., and Mathis, J. S. 1989, *Astrophys. J.*, **345**, 245.
Falcón-Barroso, J., Sánchez-Blázquez, P., Vazdekis, A., Ricciardelli, E., Cardiel, N., Cenarro, A. J., Gorgas, J., and Peletier, R. F. 2011, *Astron. Astrophys.*, **532A**, 95.
Fleischer, A. J., Gauger, A., and Sedlmayr, E. 1995, *Astron. Astrophys.*, **297**, 543.
Fluks, M. A., Plez, B., The, P. S., de Winter, D., Westerlund, B. E., and Steenman, H. C. 1994, *Astron. Astrophys, Suppl. Ser.*, **105**, 311 (http://cdsarc.u-strasbg.fr/cgi-bin/Cat?J/A+AS/105/311).
Fox, M. W., Wood, P. R., and Dopita, M. A. 1984, *Astrophys. J.*, **286**, 337.
Gaia Collaboration, *et al.* 2016 (https://arxiv.org/abs/2012.01533v1).
Gillet, D. 1988, *Astron. Astrophys.*, **192**, 206.
Gorbatskii, V. G. 1961, *Soviet Astron.*, **5**, 192.
Gray, R. O., and Corbally, C. J. 2009, *Stellar Spectral Classification*, Princeton University Press, Princeton, NJ.
Gray, R. O., and Corbally, C. J. 2014, *Astron. J.*, **147**, 80 (http://www.appstate.edu/~grayro/mkclass/mkclassdoc.pdf).
Henden, A. A., *et al.* 2021, AAVSO Photometric All-Sky Survey, data release 10 (https://www.aavso.org/apass).
Hoefner, S., Feuchtinger, M. U., and Dorfi, E. A. 1995, *Astron. Astrophys.*, **297**, 815.
Joy, A. H. 1926, *Astrophys. J.*, **63**, 281.
Keenan, P. C., and McNeil, R. C. 1989, *Astrophys. J., Suppl. Ser.*, **71**, 245.
Lester, T. 2020, PlotSpectra (http://www.spectro-aras.com/forum/viewtopic.php?f=51&t=2448&p=13515&hilit=PlotSpectra#p13515).
Maury, A. C., and Pickering, E. C. 1897, *Ann. Harvard Coll. Obs.*, 28, 1.
Querci, F. R. 1986, in *The M-Type Stars: Monograph Series on Nonthermal Phenomena in Stellar Atmospheres*, eds. H. R. Johnson, F. R. Querci, S. Jordan, R. Thomas, L. Goldb erg, J.-C. Pecker, NASA SP-492, NASA SP, Washington, DC, 1.
Schlafly, E. F., and Finkbeiner, D. P. 2011, *Astrophys. J.*, **737**, 103.
Smith, B. J., Leisawitz, D., Castelaz, M. W., and Luttermoser, D. 2002, *Astron. J.*, **123**, 948.
Space Telescope Science Institute. 2021, CALSPEC (https://www.stsci.edu/hst/instrumentation/reference-data-for-calibration-and-tools/astronomical-catalogs/calspec).

van Belle, G. T., *et al.* 1999, *Astron. J.*, **117**, 521.
Watson, C., Henden, A. A., and Price, C. A. 2014, AAVSO International Variable Star Index VSX (Watson+, 2006–2014, https://www.aavso.org/vsx).
Willson, L. A. 1976, *Astrophys. J.*, **205**, 172.
Willson, L. A., and Marengo, M. 2012, *J. Amer. Assoc. Var. Star Obs.*, **40**, 516.
Wing, R. F. 1980, in *NASA. Goddard Space Flight Center Current Problems in Stellar Pulsation Instabilities* (https://ntrs.nasa.gov/citations/19800016762), 533.
Wing, R. F. 1992, *J. Amer. Assoc. Var. Star Obs.*, **21**, 42.
Yao, Y., Liu, C., Deng, L., de Grijs, R., and Matsunaga, N. 2017, *Astrophys J., Suppl. Ser.*, **232**, 16.

CCD Photometry, Light Curve Modeling, and Period Study of V573 Serpentis, a Totally Eclipsing Overcontact Binary System

Kevin B. Alton
Desert Blooms Observatory, 70 Summit Avenue, Cedar Knolls, NJ 07927; kbalton@optonline.net

Edward O. Wiley
Live Oaks Observatory, 125 Mountain Creek Pass, Georgetown, TX 78633; ewiley@suddenlink.net

Received June 18, 2021; revised July 21, 2021; accepted July 29, 2021

Abstract Precise time-series multi-color light curve data were acquired from V573 Ser at Desert Blooms Observatory (DBO) in 2019 and Live Oaks Observatory (LOO) in 2020. Previously, only monochromatic CCD-derived photometric data were available from automated surveys which employ sparse sampling strategies. New times-of-minimum from data acquired at DBO and LOO, along with other eclipse timings extrapolated from selected surveys, were used to generate a new linear ephemeris. Secular analyses (eclipse timing differences vs. epoch) did not reveal changes in the orbital period of V573 Ser over the past 20 years. Simultaneous modeling of multicolor light curve data during each epoch was accomplished using the Wilson-Devinney code. Since a total eclipse is observed, a unique photometrically derived value for the mass ratio (q_{ptm}) could be determined, which subsequently provided initial estimates for the physical and geometric elements of each variable system.

1. Introduction

Overcontact binaries (OCBs), also known as EW or W UMa-type variables, share a common atmosphere with varying degrees of physical contact. Light curves (LCs) may exhibit eclipse minima with near equal depth that reveal little color change, suggesting they have similar surface temperatures. When the most massive constituent is defined as the primary star the majority of OCBs have mass ratios ($q = m_2/m_1$) that range from unity to as low as 0.065–0.08 (Sriram *et al.* 2016; Mochnacki and Doughty 1972; Paczyński *et al.* 2007; Arbutina 2009). The evolutionary lifetimes of most OCBs are spent in physical contact (Stępień 2006; Gazeas and Stępień 2008; Stępień and Kiraga 2015). Moreover, depending on many factors, including rate of angular momentum loss, mass ratio, total mass, orbital period, and metallicity, OCBs are destined to coalesce into fast rotating stars or to alternatively produce exotic objects such as blue stragglers (Qian *et al.* 2006; Stępień and Kiraga 2015), double degenerate binaries, supernovae, or even double black holes (Almeida *et al.* 2015).

Sparsely sampled monochromatic photometric data for V573 Ser (= NSVS 13459733) were first captured during the ROTSE-I survey between 1999 and 2000 (Akerlof *et al.* 2000; Woźniak *et al.* 2004; Gettel *et al.* 2006). These data can be retrieved from the Northern Sky Variable Survey (NSVS) archives. Other sources which include photometric data from this variable system are the All Sky Automated Survey (ASAS) (Pojmański *et al.* 2005), Catalina Sky Survey (Drake *et al.* 2014), and the All-Sky Automated Survey for SuperNovae (ASAS-SN) (Shappee *et al.* 2014; Kochanek *et al.* 2017; Jayasinghe *et al.* 2018).

No multi-color light curves with Roche modeling have been reported for this OCB so this investigation also provides the first published photometric mass ratio (q_{ptm}) estimates along with preliminary physical and geometric characteristics for V573 Ser.

2. Observations and data reduction

Precise time-series images were acquired at Desert Blooms Observatory (DBO, USA; 31.941 N, 110.257 W) using a QSI 683 wsg-8 CCD camera mounted at the Cassegrain focus of a 0.4-m Schmidt-Cassegrain telescope. A Taurus 400 (Software Bisque) equatorial fork mount facilitated continuous operation without the need to perform a meridian flip. The image (science, darks, and flats) acquisition software (THESKYX Pro Edition 10.50.0; Software Bisque 2019) controlled the main and integrated guide cameras. This focal-reduced (f/7.2) instrument produces an image scale of 0.76 arcsec/pixel (bin=2×2) and a field-of-view (FOV) of 15.9 × 21.1 arcmin. Computer time was updated immediately prior to each session and exposure time for all images adjusted to 75 s.

The equipment at Live Oaks Observatory LOO, USA; 30.98 N, 98.94 W) included an Astrophysics AP900 GEM with a Moravian G2-1600 Mk.1 CCD camera mounted at the Cassegrain focus of a 0.28-m Schmidt-Cassegrain telescope. PDCAPTURE (Miller 2021) controlled the main and integrated guide cameras during image acquisition (science, darks, and flats). This focal-reduced (f/7) instrument produces an image scale of 0.95 arcsec/pixel (bin=1×1) and a field-of-view (FOV) of 16 × 24 arcmin.

Both CCD cameras were equipped with Astrodon B, V, and I_c filters manufactured to match the Johnson-Cousins Bessell specification. Dark subtraction, flat correction, and registration of all images collected at DBO were performed with AIP4WIN v2.4.0 (Berry and Burnell 2005), whereas image calibration at LOO was accomplished with AstroImageJ (Collins *et al.* 2017). Instrumental readings from V573 Ser were reduced to catalog-based magnitudes using APASS DR9 values (Henden *et al.* 2009, 2010, 2011; Smith *et al.* 2011) built into MPO CANOPUS v10.7.1.3 (Minor Planet Observer 2010).

Table 1. Astrometric coordinates (J2000), V mags, and color indices (B–V) for V573 Ser (Figure 1), and the corresponding comparison stars used in this photometric study.

Star Identification	R.A. (J2000)[a] h m s	Dec. (J2000)[a] ° ' "	V mag[b]	(B–V)[b]
(T) V573 Ser	15 59 29.7958	+02 52 21.157	12.846	0.844
(1) GSC 00357-0083	15 59 07.4064	+02 57 27.67	11.635	0.865
(2) GSC 00357-0889	15 59 32.7151	+03 02 03.797	12.368	0.836
(3) GSC 00357-0781	15 59 38.6095	+03 02 26.849	11.886	0.790
(4) GSC 00357-0081	15 59 26.1398	+02 57 19.442	13.276	0.726
(5) GSC 00357-0117	15 59 20.6231	+02 55 25.872	12.79	0.703

a. R.A. and Dec. from Gaia DR2 (Gaia Collab. et al. *2016, 2018*).
b. V-mag and (B–V) for comparison stars derived from APASS DR9 database described by Henden et al. *(2009, 2010, 2011)* and Smith et al. *(2011)*.

Table 2. Summary of image acquisition dates, number of data points and estimated uncertainty (± mag) in each bandpass (BVI$_c$) used for the determination of ToM values and/or Roche modeling.

Target ID	B	B (± mag)	V	V (± mag)	I_c	I_c (± mag)	Location	Dates
V573 Ser	264	0.008	267	0.004	266	0.005	DBO	June 19, 2019–June 25, 2019
V573 Ser	566	0.023	452	0.011	619	0.015	LOO	April 22, 2020–July 20, 2020

3. Results and discussion

Light curves were generated using an ensemble of five comparison stars, the mean of which remained constant (<0.01 mag) throughout each imaging session. The identity, J2000 coordinates, and color indices (B–V) for these stars are provided in Table 1. A CCD image annotated with the location of the target (T) and comparison stars (1–5) is shown in Figure 1. Only data acquired above 30° altitude (airmass <2.0) were included; differential atmospheric extinction was ignored, considering the close proximity of all program stars.

All photometric data can be retrieved from the AAVSO International Database via the International Variable Star Index (Kafka 2021).

3.1. Photometry and ephemerides

Times of minimum (ToM) and associated errors were calculated using the method of Kwee and van Woerden (1956) as implemented in Peranso v2.5 (Paunzen and Vanmunster 2016). Curve fitting all eclipse timing differences (ETD) was accomplished using scaled Levenberg-Marquardt algorithms (QtiPlot 0.9.9-rc9; IONDEV SRL 2021). Photometric uncertainty was calculated according to the so-called "CCD Equation" (Mortara and Fowler 1981; Howell 2006). The acquisition dates, number of data points, and uncertainty for each bandpass used for the determination of ToM values and/or Roche modeling are summarized in Table 2.

Thirteen new ToM measurements were extracted from photometric data acquired at DBO and LOO. These, along with seven other eclipse timings (Table 3), were used to calculate a new linear ephemeris (Figure 2) based on data produced between 1999 and 2020:

$$\text{Min.I}(\text{HJD}) = 2459046.5366\,(5) + 0.3751703\,(1)\,\text{E.} \quad (1)$$

Given the paucity of data, no other underlying variations in the orbital period stand out such as those that might be caused by angular momentum loss/gain, mass transfer, magnetic

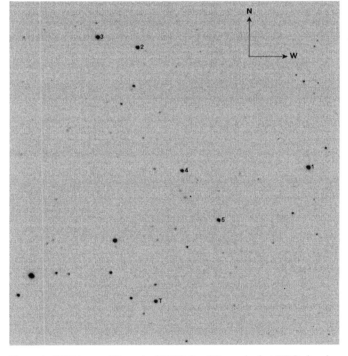

Figure 1. CCD image (V mag) of V573 Ser (T) acquired at DBO showing the location of comparison stars (1–5) used to generate APASS DR9-derived magnitude estimates.

cycles (Applegate 1992), or the presence of an additional gravitationally bound stellar-size body. At a minimum, another decade of precise times of minimum will still be needed to establish whether the orbital period of this system is changing in a predictable fashion.

3.2. Effective temperature estimation

The effective temperature (T_{eff1}) of the more massive, and therefore most luminous component (defined as the primary star herein) was derived from a composite of astrometric (UCAC4; Zacharias et al. 2013) and photometric (2MASS

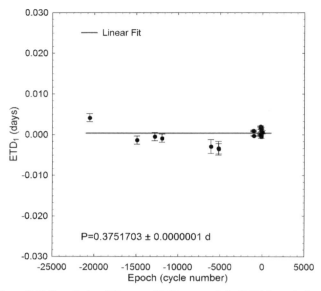

Figure 2. Eclipse timing differences (ETD) vs. epoch for V573 Ser calculated using the updated linear (Equation 1). When available, measurement uncertainty is denoted by the hatched error bars. The solid red line indicates the linear fit.

Table 3. V573 Ser times of minimum (February 23, 1999–July 16, 2020), cycle number and residuals (ETD) between observed and predicted times derived from the updated linear ephemeris (Equation 1).

HJD 2400000+	HJD Error	Cycle No.	ETD[a]	Reference
51321.7837	0.0010	–20590	0.004147	1
53438.8644	0.0010	–14947	–0.001309	1
54239.8539	0.0010	–12812	–0.000457	2
54564.9386	0.0010	–11945.5	–0.000845	2
56751.9920	0.0017	–6116	–0.002867	3
57084.0173	0.0017	–5231	–0.003306	3
57099.9619	0.0013	–5188.5	–0.003445	4
58655.7976	0.0002	–1041.5	0.000924	4
58657.6735	0.0001	–1036.5	0.000970	4
58659.7357	0.0001	–1031	–0.000253	4
58962.8737	0.0003	–223	0.000055	4
58963.8135	0.0002	–220.5	0.001981	4
58964.7493	0.0003	–218	–0.000183	4
58965.8751	0.0003	–215	0.000180	4
58966.8146	0.0003	–212.5	0.001659	4
58987.8234	0.0003	–156.5	0.001015	4
58989.6999	0.0002	–151.5	0.001610	4
58989.8854	0.0004	–151	–0.000465	4
59046.7248	0.0004	0.5	0.000591	4

a. ETD = Eclipse Time Difference.
References: 1. NSVS (Woźniak et al. 2004); 2. CSS (Univ. Arizona 2020); 3. ASAS-SN (Shappee et al. 2014; Kohanek et al. 2017); 4. This study.

and APASS; Skrutskie et al. 2006; Henden et al. 2016) determinations that were as necessary transformed to (B–V)[1,2]. Interstellar extinction (A_V) and reddening (E(B–V) = A_V/3.1) were estimated for targets within the Milky Way Galaxy according to Amôres and Lépine (2005). These models[3] require the Galactic coordinates (l, b) and the distance in kpc (Bailer-Jones 2015). After subtracting out reddening to arrive at a value for intrinsic color, $(B–V)_0$, T_{eff1} estimates were interpolated for each system using the values reported for main sequence dwarf stars by Pecaut and Mamajek (2013). Additional sources used to establish a median value for each T_{eff1} included low resolution spectra obtained from LAMOST-DR5 (Zhao et al. 2012; Wang et al. 2019), the Gaia DR2 release of stellar parameters (Andrae et al. 2018), and an empirical relationship (Houdashelt et al. 2000) based on intrinsic color where $0.32 \leq B–V)_0 \leq 1.35$. The median result ($T_{eff1} = 5365 \pm 220$ K), summarized in Table 4, was adopted for Roche modeling of LCs from V573 Ser.

Table 4. Estimation of effective temperature (T_{eff1}) of the primary star in V573 Ser.

Parameter	V573 Ser
Median combined $(B–V)_0$[a]	0.802 ± 0.023
Galactic reddening E(B–V)[b]	0.043 ± 0.001
Survey T_{eff1}[c] (K)	5300 ± 86
Gaia T_{eff1}[d] (K)	5313 $^{+540}_{-433}$
Houdashelt T_{eff1}[e] (K)	5356 ± 305
LAMOST DR5 T_{eff1}[f] (K)	5492 ± 31
Median Teff1 (K)	5365 ± 220
Spectral Class	G6V[g]

a. Surveys and DBO intrinsic $(B–V)_0$ determined using reddening values (E(B–V)).
b. Model A (http://www.galextin.org).
c. T_{eff1} interpolated from median combined $(B–V)_0$ using Table 4 in Pecaut and Mamajek (2013).
d. Values from Gaia DR2 (Gaia Collab. 2016, 2018; http://vizier.u-strasbg.fr/viz-bin/VizieR?-source=I/345/gaia2).
e. Values calculated with Houdashelt et al. (2000) empirical relationship.
f. Values from LAMOST DR5 v3 (Natnl. Astron. Obs. Chinese Acad. Sci. 2005–2019; (http://dr5.lamost.org/search).
g. Spectral class from LAMOST DR5.

3.3. Roche modeling approach

Roche modeling of LC data during each epoch (2019 and 2020) was initially performed with PHOEBE 0.31a (Prša and Zwitter 2005) and then refined using WDWINT56A (Nelson 2009). Both programs feature a MS Windows-compatible GUI interface to the Wilson-Devinney WD2003 code (Wilson and Devinney 1971; Wilson 1979; Wilson 1990). WDWINT56A incorporates Kurucz's atmosphere models (Kurucz 2002) that are integrated over BVI_c passbands. The final selected model was Mode 3 for an overcontact binary; other modes (detached and semi-detached) never approached the best fit value (χ^2) achieved with Mode 3 using PHOEBE 0.31a. Modeling parameters were adjusted as follows. The internal energy transfer to the stellar surface is driven by convective (7500 K) rather than radiative processes. As a result, the value for bolometric albedo ($A_{1,2} = 0.5$) was assigned according to Ruciński (1969) while the gravity darkening coefficient ($g_{1,2} = 0.32$) was adopted from Lucy (1967). Logarithmic limb darkening coefficients (x_1, x_2, y_1, y_2) were interpolated (van Hamme 1993) following any change in the effective temperature (T_{eff2}) of the secondary star during model fit optimization using differential corrections (DC). All but the temperature of the more massive star (T_{eff1}), $A_{1,2}$, and $g_{1,2}$ were allowed to vary during DC iterations. In general, the best fits for T_{eff2}, i, q, and Roche potentials ($\Omega_1 = \Omega_2$) were collectively refined (method of multiple subsets) by DC using

1. http://www.aerith.net/astro/color_conversion.html. 2. http://brucegary.net/dummies/method0.html. 3. http://www.galextin.org.

the multicolor LC data until a simultaneous solution was found. Not uncommon for OCB systems, LCs from V573 Ser exhibit varying degrees of asymmetry during quadrature (Max I > Max II), which is often called the O'Connell effect (O'Connell 1951). Surface inhomogeneity often attributed to star spots was simulated by the addition of a hot and cool spot on the primary star to obtain the best fit LC models. V573 Ser did not require third light correction ($l_3 = 0$) to improve Roche model fits.

3.4. Roche modeling results

Without radial velocity (RV) data it is generally not possible to unambiguously determine the mass ratio, subtype (A or W), or total mass of an eclipsing binary system. Nonetheless, since a total eclipse is observed, a unique mass ratio value could be found (Terrell and Wilson 2005). Standard errors reported in Tables 5 and 6 are computed from the DC covariance matrix and only reflect the model fit to the observations which assume exact values for any fixed parameter. These errors are generally regarded as unrealistically small considering the estimated uncertainties associated with the mean adopted T_{eff1} values along with basic assumptions about $A_{1,2}$, $g_{1,2}$, and the influence of spots added to the Roche model. Normally, the value for T_{eff1} is fixed with no error during modeling with the W-D code despite measurement uncertainty which can approach 10% relative standard deviation (R.S.D.) without supporting high resolution spectral data. The effect that such uncertainty in T_{eff1} would have on modeling estimates for q, i, $\Omega_{1,2}$, and T_{eff2} has been investigated with other OCBs including A- (Alton 2019; Alton et al. 2020) and W-subtypes (Alton and Nelson 2018). As might be expected, any change in the fixed value for T_{eff1} results in a corresponding change in the T_{eff2}. These findings are consistent whereby the uncertainty in the model fit for T_{eff2} would be essentially the same as that established for T_{eff1}. Furthermore, varying T_{eff1} by as much as 10% did not appreciably affect the uncertainty estimates (R.S.D. < 2.2%) for i, q, or $\Omega_{1,2}$ (Alton 2019; Alton and Nelson 2018; Alton et al. 2020). Assuming that the actual T_{eff1} value falls within 10% of the adopted values used for Roche modeling (a reasonable expectation based on T_{eff1} data provided in Table 4), then uncertainty estimates for i, q, or $\Omega_{1,2}$, along with spot size, temperature, and location, would likely not exceed 2.2% R.S.D.

The fill-out parameter (f) which corresponds to the outer surface shared by each star was calculated according to Equation 2 (Kallrath and Malone 2009; Bradstreet 2005) where:

$$f = (\Omega_{inner} - \Omega_{1,2}) / (\Omega_{inner} - \Omega_{outer}), \quad (2)$$

Table 5. Light curve parameters evaluated by Roche modeling and the geometric elements derived for V573 Ser (2019) assuming it is a W-type W UMa variable.

Parameter[a]	DBO No Spot	DBO Spotted
T_{eff1} (K)[b]	5365	5365
T_{eff2} (K)	5728 (3)	5672 (14)
q (m_2/m_1)	0.367 (1)	0.373 (3)
A[b]	0.50	0.50
g[b]	0.32	0.32
$\Omega_1 = \Omega_2$	2.573 (2)	2.583 (4)
i°	89.7 (4)	83.9 (4)
$A_P = T_S/T_{star}$[c]	—	1.10 (1)
θ_P (spot co-latitude)[c]	—	90 (4)
φ_P (spot longitude)[c]	—	75 (3)
r_P (angular radius)[c]	—	10.2 (1)
$A_P = T_P/T_{star}$[c]	—	0.86 (1)
θ_P (spot co-latitude)[c]	—	90 (2)
φ_P (spot longitude)[c]	—	180 (2)
φ_P (spot longitude)[c]	—	11.3 (1)
$L_1/(L_1+L_2)B$[d]	0.6197 (4)	0.6307 (4)
$L_1/(L_1+L_2)V$	0.6441 (2)	0.6512 (1)
$L_1/(L_1+L_2)I_c$	0.6634 (2)	0.6676 (2)
r_1 (pole)	0.4472 (3)	0.4459 (6)
r_1 (side)	0.4802 (4)	0.4785 (7)
r_1 (back)	0.5094 (6)	0.5076 (8)
r_2 (pole)	0.2848 (4)	0.2850 (15)
r_2 (side)	0.2980 (4)	0.2981 (18)
r_2 (back)	0.3371 (7)	0.3367 (33)
Fill-out factor (%)	16.8	16.9
RMS (B)[e]	0.01398	0.01189
RMS (V)	0.00880	0.00614
RMS (I_c)	0.01063	0.00843

a. All uncertainty estimates for T_{eff2}, q, $\Omega_{1,2}$, i, $r_{1,2}$, and L_1 from WDwint56a (Nelson 2009).
b. Fixed with no error during DC.
c. Spot parameters in degrees (θ_P, φ_P, and r_P) or A_P in fractional degrees (K).
d. L_1 and L_2 refer to scaled luminosities of the primary and secondary stars, respectively.
e. Monochromatic residual mean square error from observed values.

Table 6. Light curve parameters evaluated by Roche modeling and the geometric elements derived for V573 Ser (2020) assuming it is a W-type W UMa variable.

Parameter[a]	LOO No Spot	LOO Spotted
T_{eff1} (K)[b]	5365	5365
T_{eff2} (K)	5649 (2)	5581 (1)
q (m_2/m_1)	0.369 (1)	0.381 (1)
A[b]	0.50	0.50
g[b]	0.32	0.32
$\Omega_1 = \Omega_2$	2.577 (2)	2.592 (3)
i°	86.12 (4)	83.8 (3)
$A_P = T_S/T_{star}$[c]	—	1.09 (1)
θ_P (spot co-latitude)[c]	—	90 (2)
φ_P (spot longitude)[c]	—	60 (4)
r_P (angular radius)[c]	—	10.6 (2)
$A_P = T_P/T_{star}$[c]	—	0.90 (1)
θ_P (spot co-latitude)[c]	—	90 (3)
φ_P (spot longitude)[c]	—	180 (2)
φ_P (spot longitude)[c]	—	11 (2)
$L_1/(L_1+L_2)B$[d]	0.6383 (3)	0.6360 (3)
$L_1/(L_1+L_2)V$	0.6573 (2)	0.6574 (2)
$L_1/(L_1+L_2)I_c$	0.6724 (2)	0.6743 (2)
r_1 (pole)	0.4469 (4)	0.4456 (5)
r_1 (side)	0.4798 (5)	0.4783 (6)
r_1 (back)	0.5092 (6)	0.5082 (7)
r_2 (pole)	0.2855 (4)	0.2882 (11)
r_2 (side)	0.2988 (5)	0.3018 (14)
r_2 (back)	0.3381 (8)	0.3416 (26)
Fill-out factor (%)	16.8	20.0
RMS (B)[e]	0.01261	0.01101
RMS (V)	0.00766	0.00633
RMS (I_c)	0.00970	0.00983

a. All uncertainty estimates for T_{eff2}, q, $\Omega_{1,2}$, i, $r_{1,2}$, and L_1 from WDwint56a (Nelson 2009).
b. Fixed with no error during DC.
c. Spot parameters in degrees (θ_P, φ_P, and r_P) or A_P in fractional degrees (K).
d. L_1 and L_2 refer to scaled luminosities of the primary and secondary stars, respectively.
e. Monochromatic residual mean square error from observed values.

wherein Ω_{outer} is the outer critical Roche equipotential, Ω_{inner} is the value for the inner critical Roche equipotential, and $\Omega = \Omega_{1,2}$ denotes the common envelope surface potential for the binary system. In all cases the systems are considered overcontact since $0 < f < 1$.

LC parameters, geometric elements, and their corresponding uncertainties are summarized in Tables 5 (2019) and 6 (2020). According to Binnendijk (1970) the deepest minimum (Min I) of a W-type overcontact system occurs when a cooler more massive constituent occludes its hotter but less massive binary partner. The flattened-bottom dip in brightness at Min I (Figures 3 and 4) indicates a total eclipse of the secondary star;

Table 7. Fundamental stellar parameters for V573 Ser using the mean photometric mass ratio ($q_{ptm} = m_2/m_1$) from Roche model fits of LC data (2019–2020) and the estimated masses based on empirically derived M-PRs for overcontact binary systems.

Parameter	Primary	Secondary
Mass ($M_\odot$)	1.230 ± 0.023	0.446 ± 0.009
Radius ($R_\odot$)	1.218 ± 0.006	0.768 ± 0.004
a ($R_\odot$)	2.600 ± 0.013	2.600 ± 0.013
Luminosity ($L_\odot$)	1.108 ± 0.257	0.557 ± 0.006
M_{bol}	4.639 ± 0.011	5.385 ± 0.012
Log (g)	4.356 ± 0.009	4.316 ± 0.009

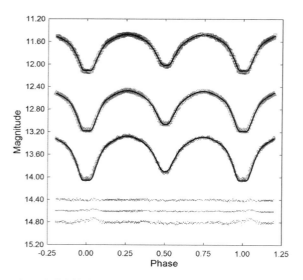

Figure 3. Period folded (0.3751703 ± 0.0000001 d) CCD-derived LCs for V573 Ser produced from photometric data collected at DBO between June 19, 2019 and June 25, 2019 The top (I_c), middle (V) and bottom curve (B) were transformed to magnitudes based on APASS DR9 derived catalog values from comparison stars. In this case, the Roche model assumed a W-subtype overcontact binary with two spots on the most massive star; residuals from the model fits are offset at the bottom of the plot to keep the values on scale.

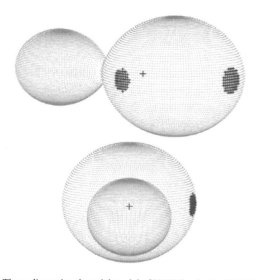

Figure 5. Three-dimensional spatial model of V573 Ser during 2019 illustrating (top) the location of a cool (blue) and hot (red) spot on the primary star and (bottom) the secondary star transit across the primary star face at Min II ($\varphi = 0.5$).

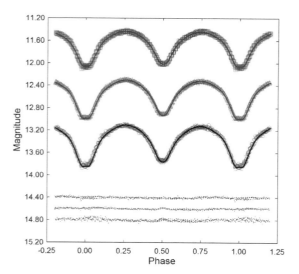

Figure 4. Period folded (0.3751703 ± 0.0000001 d) CCD light curves acquired from V573 Ser at LOO between April 22, 2020 and July 20, 2020. The remaining caption is the same as Figure 3.

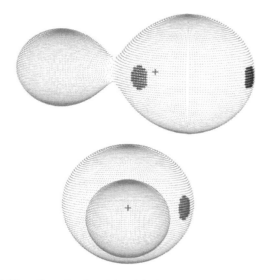

Figure 6. Three-dimensional spatial model of V573 Ser during 2020 illustrating (top) the location of a hot (red) and cool (blue) spot on the primary star and (bottom) the secondary star transit across the primary star face at Min II ($\varphi = 0.5$).

therefore, W-D modeling proceeded under the assumption that V573 Ser is a W-subtype. Since according to the convention used herein whereby the primary star is the most massive ($m_2/m_1 \leq 1$), a phase shift (0.5) was introduced to properly align the LC for subsequent Roche modeling. Even though photometric data were acquired between 2019 and 2020 using different instruments at two sites, the modeled results for V573 Ser compare quite favorably. It would also appear that the surface inhomogeneity modeled with hot and cool spots on the primary star was similar and persisted between June 19, 2019, and July 20, 2020.

Spatial renderings (Figures 5 and 6) were produced with BinaryMaker3 (BM3: Bradstreet and Steelman 2004) using the final WDWint56a modeling results from both epochs (2019 and 2020). A secondary star can be envisioned to completely transit across the primary face during Min II ($\varphi = 0.5$), thereby confirming that the secondary star is totally eclipsed at Min I.

3.5. Preliminary stellar parameters

Mean physical characteristics were estimated for V573 Ser (Table 7) using results from the best fit (spotted) LC simulations from 2019 and 2020. It is important to note that without the benefit of RV data which define the orbital motion, mass ratio, and total mass of the binary pair, these results should be considered "relative" rather than "absolute" parameters and regarded as preliminary.

Calculations are described below for estimating the solar mass and size, semi-major axis, solar luminosity, bolometric V-mag, and surface gravity of each component. Three empirically-derived mass-period relationships (M-PR) for W UMa binaries have been published. The first M-PR was reported by Qian (2003), while two others followed from Gazeas and Stępień (2008) and then Gazeas (2009). According to Qian (2003), when the primary star is less than 1.35 $M_\odot$ or the system is W-type its mass can be determined from Equation 3:

$$\log(M_1) = 0.391\,(59) \cdot \log(P) + 1.96\,(17), \qquad (3)$$

where P is the orbital period in days and leads to $M_1 = 1.126 \pm 0.087\,M_\odot$ for the primary. The M-PR (Equation 4) derived by Gazeas and Stępień (2008):

$$\log(M_1) = 0.755\,(59) \cdot \log(P) + 0.416\,(24), \qquad (4)$$

corresponds to an OCB system where $M_1 = 1.243 \pm 0.099\,M_\odot$. Gazeas (2009) reported another empirical relationship (Equation 5) for the more massive (M_1) star of a contact binary such that:

$$\log(M_1) = 0.725\,(59) \cdot \log(P) - 0.076\,(32) \cdot \log(q) + 0.365\,(32). \qquad (5)$$

from which $M_1 = 1.228 \pm 0.091\,M_\odot$. The mean of three values ($M_1 = 1.230 \pm 0.023\,M_\odot$) estimated from Equations 3–5 was used for subsequent determinations of M_2, semi-major axis a, volume-radii r_L, and bolometric magnitudes (M_{bol}) using the formal errors calculated by WDWint56a (Nelson 2009). The secondary mass = $0.446 \pm 0.009\,M_\odot$ and total mass ($1.676 \pm 0.025\,M_\odot$) were determined using the mean photometric mass ratio ($q_{ptm} = 0.377 \pm 0.006$) derived from the best fit (spotted) models.

The semi-major axis, $a(R_\odot) = 2.600 \pm 0.013$, was calculated from Newton's version (Equation 6) of Kepler's third law where:

$$a^3 = (G \cdot P^2\,(M_1 + M_2))/(4\pi^2). \qquad (6)$$

The effective radius of each Roche lobe (r_L) can be calculated over the entire range of mass ratios ($0 < q < \infty$) according to an expression (Equation 7) derived by Eggleton (1983):

$$r_L = (0.49 q^{2/3})/(0.6 q^{2/3} + \ln(1 + q^{1/3})), \qquad (7)$$

from which values for r_1 (0.4671 ± 0.0002) and r_2 (0.2967 ± 0.0002) were determined for the primary and secondary stars, respectively. Since the semi-major axis and the volume radii are known, the radii in solar units for both binary components can be calculated where $R_1 = a \cdot r_1 = 1.218 \pm 0.006\,R_\odot$ and $R_2 = a \cdot r_2 = 0.768 \pm 0.004\,R_\odot$.

Luminosity in solar units ($L_\odot$) for the primary (L_1) and secondary stars (L_2) was calculated from the well-known relationship derived from the Stefan-Boltzmann law (Equation 8) where:

$$L_{1,2} = (R_{1,2}/R_\odot)^2\,(T_{1,2}/T_\odot)^4. \qquad (8)$$

Assuming that $T_{eff1} = 5365\,K$, $T_{eff2} = 5690\,K$ and $T_\odot = 5772\,K$, then the solar luminosities ($L_\odot$) for the primary and secondary are $L_1 = 1.108 \pm 0.257$ and $L_2 = 0.557 \pm 0.006$, respectively. The Gaia DR2 reported values for radius ($1.026^{+0.22}_{-0.21}$ and luminosity ($1.2^{+0.23}_{-0.23}$) compare very favorably with our estimates for this binary system.

4. Conclusions

New times of minimum for V573 Ser (n = 13) based on multicolor CCD data were determined from LCs acquired at two different locations in 2019 and 2020. These, along with other values extrapolated from multiple sparsely sampled monochromatic surveys, led to a linear ephemeris which suggests that the orbital period for this OCB has not changed significantly over the past 20 years.

The adopted effective temperature ($T_{eff1} = 5365 \pm 220\,K$) was based on a composite of sources that included values from photometric and astrometric surveys, the Gaia DR2 release of stellar characteristics (Andrae et al. 2018), and estimates from LAMOST DR5 spectral data (Zhao et al. 2012; Wang et al. 2019). V573 Ser clearly experiences a total eclipse which is evident as a flattened bottom during Min I, a characteristic of W-subtype variables. It follows that photometric mass ratios determined by Roche modeling should prove to be reliable substitutes for mass ratios derived from RV data. Nonetheless, spectroscopic studies (RV and high resolution classification spectra) will be required to unequivocally determine a total mass and spectral class for each system. Consequently, all parameter values and corresponding uncertainties reported herein should be considered preliminary.

5. Acknowledgements

This research has made use of the SIMBAD database operated at Centre de Données astronomiques de Strasbourg, France. In addition, the Northern Sky Variability Survey hosted by the Los Alamos National Laboratory (https://skydot.lanl.gov/nsvs/nsvs.php), the All Sky Automated Survey Catalogue of Variable Stars (http://www.astrouw.edu.pl/asas/?page=acvs), All-Sky Automated Survey for Supernovae (https://asas-sn.osu.edu/variables), and the International Variable Star Index (AAVSO) were mined for essential information. This work also presents results from the European Space Agency (ESA) space mission Gaia. Gaia data are being processed by the Gaia Data Processing and Analysis Consortium (DPAC). Funding for the DPAC is provided by national institutions, in particular the institutions participating in the Gaia Multilateral Agreement (MLA). The Gaia mission website is https://www.cosmos.esa.int/gaia. The Gaia archive website is https://archives.esac.esa.int/gaia. This paper makes use of data from the first public release of the WASP data as provided by the WASP consortium and services at the NASA Exoplanet Archive, which is operated by the California Institute of Technology, under contract with the National Aeronautics and Space Administration under the Exoplanet Exploration Program. The use of public data from LAMOST is also acknowledged. Guoshoujing Telescope (the Large Sky Area Multi-Object Fiber Spectroscopic Telescope LAMOST) is a National Major Scientific Project built by the Chinese Academy of Sciences. Funding for the project has been provided by the National Development and Reform Commission. LAMOST is operated and managed by the National Astronomical Observatories, Chinese Academy of Sciences. Many thanks to the anonymous referee whose valuable commentary led to significant improvement of this paper.

References

Akerlof, C., *et al.* 2000, *Astron. J.*, **119**, 1901.
Almeida, L. A., *et al.* 2015, *Astrophys. J.*, **812**, 102.
Alton, K. B. 2019, *J. Amer. Assoc. Var. Star Obs.*, **47**, 7.
Alton, K. B., and Nelson, R. H. 2018, *Mon. Not. Roy. Astron. Soc.*, **479**, 3197.
Alton, K. B., Nelson, R. H., and Stępień, K. 2020, *J. Astrophys. Astron.*, **41**, 26.
Amôres, E. B., and Lépine, J. R. D. 2005, *Astron. J.*, **130**, 659.
Andrae, R., *et al.* 2018, *Astron. Astrophys.*, **616A**, 8.
Applegate, J. H. 1992, *Astrophys. J.*, **385**, 621.
Arbutina, B. 2009, *Mon. Not. Roy. Astron. Soc.*, **394**, 501.
Bailer-Jones, C. A. L. 2015, *Publ. Astron. Soc. Pacific*, **127**, 994.
Berry, R., and Burnell, J. 2005, *The Handbook of Astronomical Image Processing*, 2nd ed., Willmann-Bell, Richmond, VA.
Binnendijk, L. 1970, *Vistas Astron.*, **12**, 217.
Bradstreet, D. H. 2005, in *The Society for Astronomical Sciences 24th Annual Symposium on Telescope Science*, Society for Astronomical Sciences, Rancho Cucamonga, CA, 23.
Bradstreet, D. H., and Steelman, D. P. 2004, Binary Maker 3, Contact Software (http://www.binarymaker.com).
Collins, K. A., Kielkopf, J. F., Stassun, K. G., and Hessman, F. V. 2017, *Astron. J.*, **153**, 77 (https://www.astro.louisville.edu/software/astroimagej).
Drake, A. J., *et al.* 2014, *Astrophys. J., Suppl. Ser.*, 213, 9.
Eggleton, P. P. 1983, *Astrophys. J.*, **268**, 368.
Gaia Collaboration, *et al.* 2016, *Astron. Astrophys.*, **595A**, 1.
Gaia Collaboration, *et al.* 2018, *Astron. Astrophys.*, **616A**, 1.
Gazeas, K. D. 2009, *Commun. Asteroseismology*, **159**, 129.
Gazeas, K., and Stępień, K. 2008, *Mon. Not. Roy. Astron. Soc.*, **390**, 1577.
Gettel, S. J., Geske, M. T., and McKay, T. A. 2006, *Astron. J.*, **131**, 621.
Henden, A. A., Levine, S. E., Terrell, D., Smith, T. C., and Welch, D. L. 2011, *Bull. Amer. Astron. Soc.*, **43**, 2011.
Henden, A. A., Terrell, D., Welch, D., and Smith, T. C. 2010, *Bull. Amer. Astron. Soc.*, **42**, 515.
Henden, A. A., Welch, D. L., Terrell, D., and Levine, S. E. 2009, *Bull. Amer. Astron. Soc.*, **41**, 669.
Henden, A. A., *et al.* 2015, AAVSO Photometric All-Sky Survey, data release 9 (https://www.aavso.org/apass).
Houdashelt, M. L., Bell, R. A., and Sweigart, A. V. 2000, *Astron. J.*, **119**, 1448.
Howell, S. B. 2006, *Handbook of CCD Astronomy*, 2nd ed., Cambridge Univ. Press, Cambridge, UK.
IONDEV SRL. 2021, QtiPlot Data Analysis and Scientific Visualisation (https://www.qtiplot.com/index.html).
Jayasinghe, T., *et al.* 2018, *Mon. Not. Roy. Astron. Soc.*, **477**, 3145.
Kafka, S. 2021, Observations from the AAVSO International Database (https://www.aavso.org/data-download).
Kallrath, J., and Milone, E. F. 2009, *Eclipsing Binary Stars: Modeling and Analysis*, Springer-Verlag, New York.
Kochanek, C. S., *et al.* 2017, *Publ. Astron. Soc. Pacific*, **129**, 104502.
Kurucz, R. L. 2002, *Baltic Astron.*, **11**, 101.
Kwee, K. K., and van Woerden, H. 1956, *Bull. Astron. Inst. Netherlands*, **12**, 327.
Lucy, L. B. 1967, *Z. Astrophys.*, **65**, 89.
Miller, C. 2021, PDCapture, version 2.1 (https://groups.io/g/pdcaptureapp).
Minor Planet Observer. 2010, MPO Software Suite (http://www.minorplanetobserver.com), BDW Publishing, Colorado Springs.
Mochnacki, S. W., and Doughty, N. A. 1972, *Mon. Not. Roy. Astron. Soc.*, **156**, 51.
Mortara, L., and Fowler, A. 1981, in *Solid State Imagers for Astronomy*, SPIE Conf. Proc. 290, Society for Photo-Optical Instrumentation Engineers, Bellingham, WA, 28.
National Astronomical Observatories, Chinese Academy of Sciences. 2005–2019, Large Sky Area Multi-Object Fiber Spectroscopic Telescope (LAMOST) Data Release 5 v3 (http://dr5.lamost.org).
Nelson, R. H. 2009, WDwint56a: Astronomy Software by Bob Nelson (https://www.variablestarssouth.org/bob-nelson).
O'Connell, D. J. K. 1951, *Publ. Riverview Coll. Obs.*, **2**, 85.
Paunzen, E., and Vanmunster, T. 2016, *Astron. Nachr.*, **337**, 239.
Pecaut, M., and Mamajek, E. E. 2013, *Astrophys. J., Suppl. Ser.*, **208**, 9.

Paczyński, B., Sienkiewicz, R., and Szczygieł, D. M. 2007, *Mon. Not. Roy. Astron. Soc.*, **378**, 961.

Pojmański, G., Pilecki, B., and Szczygieł, D. 2005, *Acta Astron.*, **55**, 275.

Prša, A., and Zwitter, T. 2005, *Astrophys. J.*, **628**, 426.

Qian, S. 2003, *Mon. Not. Roy. Astron. Soc.*, **342**, 1260.

Qian, S., Yang, Y., Zhu, L., H., He, J., and Yuan, J. 2006, *Astrophys. Space Sci.*, **304**, 25.

Ruciński, S. M. 1969, *Acta Astron.*, **19**, 245.

Shappee, B. J., et al. 2014, *Astrophys. J.*, **788**, 48.

Skrutskie, M. F., et al. 2006, *Astron. J.*, **131**, 1163.

Smith, T. C., Henden, A. A., and Starkey, D. R. 2011, in *The Society for Astronomical Sciences 30th Annual Symposium on Telescope Science*, Society for Astronomical Sciences, Rancho Cucamonga, CA, 121.

Software Bisque. 2019, THESKYX professional edition 10.5.0 (https://www.bisque.com).

Sriram, K., Malu, S., Choi, C.S., and Vivekananda Rao, P. 2016, *Astron. J.*, **151**, 69.

Stępień, K. 2006, *Acta Astron.*, **56**, 199.

Stępień, K., and Kiraga, M. 2015, *Astron. Astrophys.*, **577A**, 117.

Terrell, D., and Wilson, R. E. 2005, *Astrophys. Space Sci.*, **296**, 221.

University of Arizona. 2020, Catalina Sky Survey (http://www.lpl.arizona.edu/css).

van Hamme, W. 1993, *Astron. J.*, **106**, 2096.

Wang, R., et al. 2019, *Pub. Astron. Soc. Pacific*, **131**, 024505

Wilson, R. E. 1979, *Astrophys. J.*, **234**, 1054.

Wilson, R. E. 1990, *Astrophys. J.*, **356**, 613.

Wilson, R. E., and Devinney, E. J., 1971, *Astrophys. J.*, **166**, 605.

Woźniak, P. R., et al. 2004, *Astron. J.*, **127**, 2436.

Zacharias, N., Finch, C. T., Girard, T. M., Henden, A., Bartlett, J. L., Monet, D. G., and Zacharias, M. I. 2013, Astron. J., 145, 44.

Zhao, G., Zhao, Y.-H., Chu, Y.-Q., Jing, Y.-P., and Deng, L.-C. 2012, *Res. Astron. Astrophys.*, **12**, 723.

Distances for the RR Lyrae Stars UU Ceti, UW Gruis, and W Tucanae

Ross Parker
Liam Parker
Hayden Parker
Faraz Uddin
Department of Physical Science and Engineering, Harper College, 1200 W. Algonquin Road, Palatine, IL 60067;
r_parker19@mail.harpercollege.edu

Timothy Banks
Department of Physical Science and Engineering, Harper College, 1200 W. Algonquin Road, Palatine, IL 60067, and Data Science, Nielsen, 200 W. Jackson, Chicago, IL 60606; tim.banks@nielsen.com

Received July 6, 2021; revised November 3, 9, 2021; accepted November 9, 2021

Abstract B, V, i, and z bandpass observations were collected in late 2020 for three RRab type stars: UU Ceti, UW Gruis, and W Tucanae. The period-luminosity (PL) relationships of Catalen, Pritzl, and Smith (2004, *ApJSupp*, 154) and Caceres and Catelan (2008, *ApJSuppl*, 179) were applied to derive distances. These were found to be in reasonable agreement with the *Gaia* Early DR3 distances, lending confidence to use of the PL relationships. Fourier decompositions were applied to data from the *TESS* space telescope to derive, using stepwise linear regression, an empirical relationship between terms of the decomposition and the pulsation period with metallicity [Fe/H]. *TESS* data were available for UU Cet and W Tuc out of the three systems studied. The derived equation gave metallicities in line with the literature for both stars, lending confidence to their usage in the PL-derived distances.

1. Introduction

RR Lyrae stars are low-mass, horizontal branch, short period (< 1 day), pulsating variable stars used as "standard candles" to calculate distances. They have also been used as tracers of the chemical and dynamical properties of old stellar populations within our own and nearby galaxies, and as test objects to validate theories of the evolution of low mass stars and stellar pulsation (Smith 1995).

The European Space Agency's *Gaia* (Gaia Collab. 2018) mission provides the opportunity to compare parallax-derived distances with those based on period-luminosity (PL) relationships. Catelan *et al.* (2004) showed that use of near-infrared band-passes together with PL relationships led to more reliable distance estimates than previous PL relationships, as the PL relationship becomes more linear and more tight. Catelan *et al.* (2004) gave the relation for V as

$$M_V = 2.288 + 0.8824 \log Z + 0.1079 (\log Z)^2, \quad (1)$$

where Z is the metallicity. Caceres and Catelan (2008) provided the first investigations of the RR Lyrae period-luminosity relation in the Sloan Digital Sky Survey (SDSS) system bandpasses. After a review of PL relations in various filter systems, they concluded that the B, V, i, and z filters delivered the most promising results. The paper confirms that redder bandpasses, specifically i and z, identify tight and simple PL relations. The relations for i and z, respectively, are:

$$M_i = 0.908 - 1.035 \log P + 0.220 \log Z \quad (2)$$

$$M_z = 0.839 - 1.295 \log P + 0.211 \log Z, \quad (3)$$

where P is the pulsation period in days. Equation 2 has a standard error of the estimate of 0.045 mag, and Equation 3 0.037 mag. Catelan *et al.* (2004) do not give similar estimates for Equation 1, commenting "...for all equations presented... the statistical errors in the derived coefficients are always very small, of order $10^{-5} - 10^{-3}$."

The aim of this study is firstly to obtain suitable photometric observations for three RR Lyrae stars (UU Cet, W Tuc, and UW Gru), then apply these equations to obtain distance estimates for the stars, and compare these estimates with each other and the published distances such as from *Gaia*. It is part of a wider research effort led by Dr. M. Fitzgerald (Edith Cowan University, Australia) investigating further RR Lyrae stars and the relation between the equations above and parallax-based distances (see, e.g., Jones 2020; Uzpen and Slater 2020; Nicolaides *et al.* 2021).

1.1. UU Cet

UU Cet (RRab, V_{max} = 11.688, V_{min} = 12.237: Gaia Collab. 2018; V_{max} = 11.718, V_{min} = 12.350: Clementini *et al.* 1992) has been documented in many different catalogs. However, only a few papers, by a research group led by Cacciari, narrow down their research to study UU Cet extensively. In Cacciari *et al.* (1992) the authors performed the Baade-Wesselink (BW) method (Baade 1926; Wesselink 1946) on UU Cet using previous observations from published papers. The Infrared Flux (IF) method indicated a distance of 1887 pc, and the Surface Brightness (SB) method a distance of 1825 – 1982 pc with values calculated with both optical colors and (V–K) colors. These numbers are similar to calculations made in the paper Clementini *et al.* (1992), which also performed the BW method on UU Cet using different input variables. Parallax estimates for the star vary across researchers, as shown in Table 4.

Cacciari *et al.* (1992) found metallicity to be −1.0 ± 0.2, however, a concrete value for [Fe/H] does not appear to have been settled on for UU Cet, as it varies in the literature. For example, Chiba and Yoshii (1998) give an [Fe/H] value of −1.32 ± 0.20 while Sandage (1993) calculated it to be −0.79. Cacciari *et al.* (1992) derived 0.606075 d for the period of UU Cet, which is very similar to other findings, such as 0.60608 d found by Lub (1977a) and 0.60606 d from ASAS (Pojmański 1997). A previous observation by Jones (1973) estimated a $(k–b)_2$ value of 0.08 ± 0.019, which is possible evidence of a Blazhko effect, however, no effect was observed in the current paper (although our data are sparse, see Figure 1a).

1.2. UW Gru

UW Gru (RRab) hasn't often been a focal point in many papers as an object of interest. It was first discovered by Hoffmeister (1963), who classified the star as an RR Lyrae with extreme magnitudes between 12 and 13. The next publication on UW Gru was when Alain Bernard collected photoelectric UBV observations over the course of three years (Bernard 1982). The star varied between 12.6 and 13.6 in V (see Figure 1 of Bernard 1982). The Wide-field Infrared Survey Explorer all-sky mission (Gavrilchenko *et al.* 2014) lists UW Gru's period at 0.5650 ± 0.0070 day and a distance of 3282 ± 64 parsecs. The distance and uncertainty were calculated using a mid-IR period-luminosity relation. The WISE period was similar to the period of 0.548210 d found in Bernard and Burnet (1982). Additionally, the [Fe/H] was estimated at −1.6 ± 0.2 (Bernard 1982) and listed at −1.41 dex metallicity on a common [Fe/H] scale (Jurcsik and Kovacs 1996), which reflects the findings of other authors. Blazhko behavior wasn't considered a factor for UW Gru, which we confirm (see Figure 1b). The distance was measured at R = 2900 ± 250 pc from the sun, and 2550 pc below the galactic plane (Bernard and Burnet 1982), although this was based on an assumed absolute magnitude. However, the given parallax from the *Gaia* Data Releases all differ (see Table 4).

1.3. W Tuc

While W Tuc (RRab, V_{max} = 10.96, V_{min} = 12.03: Torrealba *et al.* 2015) is present in many catalogued results for RR Lyrae stars, only a handful of papers have focused on this star as a specific object of interest. These were written primarily by a group led by Cacciari. Cacciari *et al.* (1992) presented JHK light curves for W Tuc, using these together with literature data, such as CORAVEL radial velocities and BVRI photometry (from Cacciari *et al.* 1987; Clementini *et al.* 1990), and the BW method to derive absolute parameters for the star. Using surface brightness methods gave a distance of 1601 to 1667 parsecs (using optical and (V – K) colors, ~0.625 mas), while infrared fluxes indicated a distance of 1555 pc (~0.643 mas). [Fe/H] was estimated as −1.50 (σ = 0.25). No Blazhko effect was evident, which we confirm in this paper (see Figure 1c). Cacciari *et al.* (1992) calculated an ephemeris of 2447490.719 + (0.642235 × N) days, where N is the cycle number. The period is not substantially different from the 0.6422299 day given by both Kukarin *et al.* (1970) and Lub (1977b). The Wide-field Infrared Survey Explorer all-sky mission (Gavrilchenko *et al.* 2014) lists W Tuc's period as 0.5990 ± 0.0040 days and a distance of 1514 ± 20 parsecs (~0.660 mas). Distance estimates for W Tuc in the literature are quite variable, as shown in Table 4 and the distances mentioned above, although there appears to be more of an agreement towards a parallax of ~0.6 rather than ~3–5 mas. Feast *et al.* (2008) provided a later [Fe/H] estimate of −1.57 solar, along with −1.76 from Marsakov *et al.* (2018) and −1.76 from Dambis *et al.* (2013).

2. Method

B, V, i, and z observations were collected for these three systems using the Las Cumbres Observatory (LCO, Brown *et al.* 2013) automated 0.4-m SBIG telescopes over a five-month period (August 2020 to December 2020). This is the first time these stars have been observed using the i and z filters. Up to three or four sets of observations were taken each night, depending on the automated scheduled and observing loads of the network. Exposure times are given in Table 1. Five different observing sites inside the LCO network were used, namely Siding Springs (Australia), Sutherland (South Africa), Cerro Tololo (Chile), Haleakala (Maui), and Teide (Spain). The resulting images were processed through the Our Solar Siblings (OSS) data reduction pipeline (Fitzgerald 2018). OSS performs basic processing such as flat-fielding and cosmic ray removal. These data were then input to the ASTROSOURCE software (ASTROSOURCE Version 1.5.2 is available from https://pypi.org/project/astrosource/) which processed the photometry of the target and comparison stars (see Fitzgerald 2018, and Fitzgerald *et al.* 2021 for further details on this software). ASTROSOURCE has the following procedure:

- It first identifies stars having sufficient signal-to-noise (within the linear range of the imager) which are in all the frames being processed for a given filter.

- Next the variability of these identified stars is calculated in order to identify a subset of the least variable stars, which will be used as the final ensemble set of comparison stars. Sarva *et al.* (2020) explain the selection of comparison stars by ASTROSOURCE. First, the flux of all the potential comparison stars is summed up as if to create a single comparison "star." Then the variability of each comparison star across the observations is compared with the variability of this sum across the same observations. A candidate star with variability greater than three times the standard deviation of the combined variability is removed from consideration. This process loops until the variability of the combined `star' is less than or equal to 0.002 magnitude. The remaining stars are then used as comparison stars for the data reduction, leading to differential photometry of the target star against them. It is possible that the process ends here if no suitable comparison stars are found. The standard errors from this process are reported in Table 2, for each of the stars analyzed in this paper.

- The ensemble set of known stars in the field is calibrated using APASS (Henden *et al.* 2015), SDSS (Alam *et al.* 2015), PanSTARRS (Magnier *et al.* 2016), or Skymapper (Wolf *et al.* 2018), depending on filter selection and declination. Fitzgerald (2018) gives further details on the calibration equations (which include color correction, extinction, and the possibility of time dependent terms if needed), making use of the generalized method for observations across multiple nights as outlined by Harris et al. (1981).

- The software extracts and outputs the photometric estimates, together with diagnostics and charts. Methods based on aperture and point-spread functions (e.g., DAOPhot; Stetson 1987) are available. After testing several methods, for this project the SEK (Source Extractor: KRON; Bertin and Arnouts 1996) method was found to produce calibration estimates with the least variance.
- Finally, ASTROSOURCE calculates periods using the Phase-Dispersion Minimization (PDM) and String-Length algorithms (Altunin et al. 2020).

Information on the number of calibration stars is provided in Table 2, along with the number of science frames, the reference catalogs used, and measures of photometric accuracy. Table 3 gives the calibrated magnitudes for the three target stars, along with the errors as estimated by ASTROSOURCE. Our photometric data for all three stars have been uploaded to and are available in the AAVSO International Database (Kafka 2021).

Table 1. Exposure times (in seconds) of the science frames for each star and filter.

Star	B	V	i	z
UU Cet	185	60	80	360
UW Gru	60	30	60	240
W Tuc	60	30	90	180

Table 2. Calibration information for each star and filter.

Star	Filter	Calibration	Frames	Catalog	SE
UU Cet	B	6	13	APASS	0.0164
	V	7	23	APASS	0.0093
	i	6	23	Skymapper	0.0115
	z	4	21	Skymapper	0.0247
UW Gru	B	6	55	APASS	0.0157
	V	6	63	APASS	0.0105
	i	8	48	Skymapper	0.0109
	z	4	44	Skymapper	0.0083
W Tuc	B	7	108	APASS	0.0139
	V	6	102	APASS	0.0115
	i	8	102	Skymapper	0.0119
	z	7	93	Skymapper	0.0095

Note: Calibration is the number of on-frame calibration stars; Frames are the number of processed images; Catalog is the source of the calibration information (the reference catalog)—Skymapper (Wolf et al. 2018), APASS (Henden et al. 2015); SE—standard error (in magitudes) of the calibration as calculated by ASTROSOURCE.

3. Results

Our data for UW Gru agree well with the BV photometry of Bernard (1982), covering the same ranges bar that our B data covered the dip (just before the star brightens again) which was not covered by Bernard and so we have a fainter magnitude limit for that band. Similarly, our UU Cet data are in good agreement with Clementini et al. (1992). This paper's V range of [10.64, 11.94] for W Tuc is lower than Torealba et al.'s (2015) range of [10.96, 12.03] as well as the intensity mean of 11.43 from the literature compilation of Dambis et al. (2013, compared to our mean magnitude of 11.29). However we note Figure 1b of Clementini et al. (1990), which plots light curves (and colors) of W Tuc, and Table IIIb of the same paper which show a V light curve varying between 10.78 and 11.95, are in closer agreement with our estimates. Figure 1c shows that our lowest magnitude (brightest) is set by a single point, with nearby phases being more dim, bringing our photometry closer to Clementini et al. (1990). Binning the data would have reduced the impact of such apparent outliers. This was not attempted as there were few observations around the peak phase. Perhaps taking the peak magnitude from the Fourier analyses (below) would have been more robust, provided there are sufficient data across the cycle.

In order to calculate distances to the target stars the processed data, together with information from the literature (see Table 5), were used to populate the PL relations from Catelan (2004) and Caceres and Catelan (2008). The period estimates from this study are given in Table 5, and are an arithmetic mean across the four bandpasses and two methods mentioned above. These values are in good agreement with the literature (see the reviews above), bar the WISE estimates, indicating no significant period changes. The calculated distances are given in Table 6 and charted as Figure 2. In general the agreement between the preliminary Gaia EDR3 release and the estimates from this study's light curves is reasonable. As expected, the V band shows the greatest difference to Gaia, indicating closer distances. Distances for the other two band passes are in good agreement, bar the i distance for UW Gru, although its error bar does just overlap that of the z band. Pop and Richney (2021) obtained observations for SX For using the LCO network and processed the data in an manner identical to this study, as did Lester et al. (2021) for YZ Cap. They too found good agreement between the Gaia data and their calculated distances using the PL relations. While it is an extrapolation, these comparisons suggest that the PL relations could be used with some confidence for distances beyond Gaia's capabilities.

Fourier decompositions of light curves have been used to derive relationships between [Fe/H], period, and some

Table 3. Calculated photometric data for the studied stars (in magnitudes).

		UU Cet			UW Gru			W Tuc		
		Min	Max	Mid	Min	Max	Mid	Min	Max	Mid
	B	12.909	12.003	12.456	14.391	12.708	13.549	12.378	10.670	11.524
	V	12.366	11.672	12.019	13.679	12.665	13.272	11.944	10.641	11.293
	i	12.126	11.682	11.903	14.139	12.738	13.438	11.719	10.833	11.276
	z	12.108	11.692	11.900	13.848	12.704	13.276	11.729	10.904	11.317

Note: Max—maximum magnitude numerically (so the brightest for the star); Min—minimum; Mid—arithmetic mean of these two extremes.

of the component sine waves (Simon 1988; Kovacs and Zsoldos 1995). The recent *TESS* mission (Ricker *et al.* 2015) has provided very high accuracy photometry of a number of RR Lyrae stars during its survey (see Figure 3 for an example), which we used to build such a relationship and apply it to W Tuc and UU Cet in a check that the literature reddenings were reasonable. Both Simon (1988) and Kovacs and Zsoldos (1995) used Johnson V for their equations. We did not feel comfortable applying these relationships to the *TESS* data, given their band-pass covers approximately 600 to 1000 nm, and is essentially centered on the Cousins I-band (which has a central wavelength of approximately 787 nm) to the red of Johnson V (central wavelength of approximately 575 nm, with a full-width half maximum of approximately 99 nm). The *TESS* mission was designed as a planet hunter, being optimized to search M dwarfs as possible host stars. The band-pass was chosen to reduce photon counting noise, and to increase the mission's ability to detect small planets transiting late type stars. The long wavelength band-pass end is set by the CCD detectors themselves, being their red limits, and the short wavelength end is set by a coating on the camera lenses. We therefore attempted to build a relationship for *TESS* observations, noting that while the *TESS* data were of high quality, our model would be highly dependent on the quality of the [Fe/H] values used to build it.

No two-minute cadence prepared *TESS* light curves were available for UW Gru, so we were unable to fit this star. Light curves for the other two stars were downloaded from the Mikulski Archive for Space Telescopes (MAST, Jenkins *et al.* 2016). We used the straight Simple Aperture Photometry (SAP) data, applying the PERIOD04 (Lenz and Breger 2005) program for the 12-component decomposition which fitted the following standard equation:

$$f(t) = Z + \sum_{i=1,n} A_i \sin(2\pi(\Omega_i t + \varphi_i)) \quad (4)$$

where t is time. Stepwise linear regression was conducted in R (R Core Team 2017), using period data from the NITRO9 online archive for RR Lyrae Fourier decomposition (https://nitro9.earth.uni.edu/fourier/index.html) and metallicity from the SIMBAD (Wenger *et al.* 2000) system summaries. The set of initial variables were the period, φ_1 to φ_6 inclusive (where φ_i is in the range between 0 and 1 phase inclusive), and $\varphi_{2,1}$ to $\varphi_{6,1}$ inclusive, following the formula $\varphi_{j,1} = \varphi_j - j\varphi_1$ (in the same range). Basic data are given in Table 7. W Tuc and UU Cet were not included in the training data set, which was made up of 21 stars. Regression in both directions (forwards and backwards) settled on the following equation:

$$[Fe/H] = -2.4083 P - 1.2950 \varphi_1 + 1.2888 \varphi_5 - 1.4273 \varphi_6 + 1.341 \varphi_{3,1} \quad (5)$$

where P is the light curve period (in days), with all terms significant at the 1% level or better. The adjusted R^2 value was 0.87, indicating a good model, although by eye it does seem to be over-estimating the metallicity for low [Fe/H] stars. This model was a better fit than one including a constant term. As can be seen by the scatter about the line of perfect agreement in Figure 4, the standard deviation of the residuals is relatively large at 0.49. As a comparison, the relationship derived by Kovacs and Zsoldos (1995) had a prediction accuracy of 0.23 to 0.18 dex. We caution that this should be considered a pilot study, and that these promising results could be built on by a more rigorous follow-up study (although it could be that the wide band-pass itself leads to imprecision).

We then applied this model to the PERIOD04 parameters for UU Cet and W Tuc, finding reasonable agreement with the literature values (see Figure 4), increasing our confidence in the literature values used in the distance estimates for these stars, which in turn are used in the PL relationships given the relationships [M/H] = [Fe/H] + log (0.698 f + 0.362) and log Z = [M/H] – 1.765, where f = $10^{0.3}$. The derived [Fe/H] for UU Cet was –1.46 and that for W Tuc –1.64. Both stars are in the mid-range of the data, and so less affected by questions about the model fit at the extremes. We note that the UU Cet [Fe/H] is not well constrained in the literature (see above), ranging from Chiba *et al.*'s (1998) value of –1.32 ± 0.20 down to Sandage's (1993) value of –0.79. Our calculated value is at the upper end

Table 4. Literature parallaxes (milli-arcseconds) for the studied stars.

Parallax (mas)	UU Cet	UW Gru	W Tuc
ESA	6.48 ± 4.13	—	4.88 ± 1.88
van Leeuwen	5.30 ± 4.06	—	3.33 ± 1.53
Gaia DR1	—	0.0678 ± 0.2287	0.720 ± 0.250
Gaia DR2	0.3823 ± 0.0044	0.2886 ± 0.0204	0.5657 ± 0.0256
Gaia EDR3	0.493 ± 0.0191	0.3299 ± 0.0158	0.5963 ± 0.0133

Note: ESA (Perryman *et al.* 1997) was the original data release for the HIPPARCOS mission, followed by van Leeuwen's (2007) revisions. Clearly, the selected stars are outside the reliable range of HIPPARCOS. The different Gaia Data Releases (DR) are described in Gaia Collab. *et al.* (2016, 2018, 2021). EDR is an early data release, ahead of the later formal one. DR3 is pending.

Table 5. Input parameters for the studied stars.

	UU Cet	UW Gru	W Tuc
Period	0.60608 ± 0.0055	0.548375 ± 0.000625	0.642233 ± 0.00086
E(B–V)	0.022 [1,2,3]	0.021 [3]	0.02 [3]
[Fe/H]	–1.0 ± 0.2 [4]	–1.41 [5]	–1.57 [2]
log Z	–2.551	–2.961	–3.121

Note: Period is average calculation from the four light bands in the current study, using PDM-based estimates. The extinction factor for UU Cet is an average of three values taken from previous observations: [1] Cacciari *et al.* (1992); [2] Feast *et al.* (2008); and [3] Schlafly and Finkbeiner (2011); [4] indicates Cacciari *et al.* (1992) as the source of the information; [5] Jurcsik and Kovacs (1996). log Z were calculated from the [Fe/H] values supplied. The metallicities from the compilation of Dambis *et al.* (2013) are not dramatically different for the three stars, being –1.32, –1.68, and –1.64 for UU Cet, UW Gru, and W Tuc respectively, nor from the metallicities in Table 7 for UU Cet and W Tuc.

Table 6. Calculated distances for the target stars, in parsecs. The mean is across the three filters V, i, and z.

Star	V	i	z	Mean (Viz)
UU Cet	1746 ± 89	1808 ± 63	1807 ± 69	1787 ± 42
UW Gru	3287 ± 132	3745 ± 138	3454 ± 139	3495 ± 79
W Tuc	1345 ± 53	1454 ± 53	1484 ± 46	1428 ± 29

Table 7. Period (in days), [Fe/H] and φ_1 to φ_6 of the Fourier decompositions for TESS data for a selection of RR Lyrae stars. UU Cet and W Tuc were not included in the model building.

Star	[Fe/H]	Period (d)	φ_1	φ_2	φ_3	φ_4	φ_5	φ_6
DX Del	−0.39	0.472611	0.35056	0.40746	0.82712	0.82430	0.30808	0.33452
HH Pup	−0.95	0.390746	0.81737	0.45289	0.31223	0.96700	0.06399	0.78401
AA CMi	−0.15	0.476373	0.99612	0.64010	0.78695	0.36239	0.89092	0.57894
BR Aql	−0.69	0.481878	0.38438	0.84380	0.28869	0.59572	0.40716	0.06225
RR Lyr	−1.39	0.566798	0.29121	0.28716	0.75002	0.34386	0.43074	0.56504
WY Ant	−1.66	0.574337	0.66456	0.31873	0.96257	0.70657	0.31050	0.86966
RS Boo	−0.12	0.377334	0.06832	0.41788	0.39094	0.89740	0.92422	0.58002
VW Scl	−1.46	0.510915	0.27568	0.23861	0.14502	0.23508	0.26144	0.50160
TU UMa	−1.31	0.557650	0.37273	0.58698	0.96831	0.22330	0.50984	0.84274
YY Tuc	−1.82	0.635020	0.92508	0.74387	0.53565	0.35213	0.13144	0.03621
AM Tuc	−1.49	0.405791	0.66800	0.79672	0.11896	0.62165	0.92133	0.28531
MT Tel	−2.58	0.316901	0.69575	0.33340	0.61760	0.81035	0.41737	0.96622
T Sex	−1.76	0.324680	0.99883	0.96056	0.40241	0.67227	0.20861	0.11673
SV Scl	−2.28	0.377359	0.83622	0.84052	0.84700	0.36359	0.17991	0.63991
RV Phe	−2.03	0.596419	0.16356	0.52287	0.68158	0.20848	0.18824	0.55687
RV Oct	−1.34	0.571158	0.01088	0.42023	0.23609	0.33863	0.07027	0.44831
U Lep	−1.78	0.581458	0.27802	0.38164	0.61095	0.80470	0.02459	0.21199
RR Leo	−1.60	0.452403	0.37005	0.59376	0.89671	0.15814	0.49317	0.82571
SV Hya	−1.22	0.478527	0.45564	0.90954	0.34279	0.28423	0.06210	0.70342
XX And	−1.94	0.722747	0.00464	0.16971	0.18725	0.68272	0.03730	0.35984
SW And	−0.07	0.442279	0.11762	0.38491	0.70367	0.41334	0.20898	0.15671
UU Cet	−1.00	0.606081	0.42479	0.78891	0.21132	0.49441	0.33841	0.79084
W Tuc	−1.64	0.642247	0.52457	0.08133	0.55307	0.05375	0.54489	0.84282

of this range. Using our model's predicted value of −1.46, we calculate distances that are ~100 pcs more than our current results, moving closer to the *Gaia* estimates. Additional stars could be included into the model building, perhaps leading to an improved empirical relationship, for instance for the low metallicity stars which do not seem to be so well modelled by the current equation. We also note that our caution about applying the equation of Kovacs and Zsoldos might have been misplaced; using their PL relationship gave metallicities of −1.29 and −1.50 for UU Cet and W Tuc, respectively.

4. Summary

Using ASTROSOURCE to perform photometric analysis, distances were derived for UU Cet, UW Gru, and W Tuc. The average between the V, i, and z (Viz) filters were focused on to compare with the distances found in literature and *Gaia*. The calculated distance for UU Cet was 1787 ± 42 pc, UW Gru's calculated distance was 3495 ± 79 pc, and W Tuc's calculated distance was 1428 ± 29 pc. The distances of all three stars varied compared to the distances in the literature and *Gaia*. Though the calculated distances were not in full agreement with *Gaia*, they were not far off and this is encouraging that further work could demonstrate a closer agreement between the applied PL equations and the parallax-based distance estimates. If shown, then this would lend support for using the PL estimates at distances beyond *Gaia*'s "distance of reliability," noting of course that this would be an extrapolation. There was no apparent pattern found with all three stars: while UU Cet and W Tuc's distances were closer than *Gaia*, UW Gru's distance was farther.

There is considerable variation in literature estimates for metallicity of these stars. The choice of a given metallicity will impact the distance calculated using the PL relations. We therefore have applied Fourier analysis to the high quality light curves from the *TESS* space telescope, and attempted a calibration given the wide bandpass used by this mission. The calibration/model was used to derive metallicities for UU Cet and W Tuc. This led to an upwards revision of some 100 pc for the distances for these stars, bringing them closer to the *Gaia* distances. We believe further work to improve the calibration is worthwhile, particularly given the near full sky coverage of *TESS* means a greater number of RR Lyrae stars will have been observed than have been tested in this paper.

5. Acknowledgements

This work has made use of data from the European Space Agency (ESA) mission *Gaia* (https://www.cosmos.esa.int/gaia), processed by the *Gaia* Data Processing and Analysis Consortium (DPAC; https://www.cosmos.esa.int/web/gaia/dpac/consortium). Funding for the DPAC has been provided by national institutions, in particular the institutions participating in the *Gaia* Multilateral Agreement. We are grateful for observing time on the Las Cumbres Observatory Global Telescope Network, and to Dr. M. Fitzgerald for making available this opportunity. This research has made use of the International Variable Star Index (VSX) database, operated at AAVSO, Cambridge, Massachusetts, USA. It also includes data collected by the *TESS* mission and obtained from the MAST data archive at the Space Telescope Science Institute (STScI). Funding for the *TESS* mission is provided by the NASA's Science Mission Directorate. STScI is operated by the Association of Universities for Research in Astronomy, Inc., under NASA contract NAS 5-26555. We made use of the SIMBAD database, operated at CDS, Strasbourg, France. We thank the anonymous referee for their helpful comments which improved this paper.

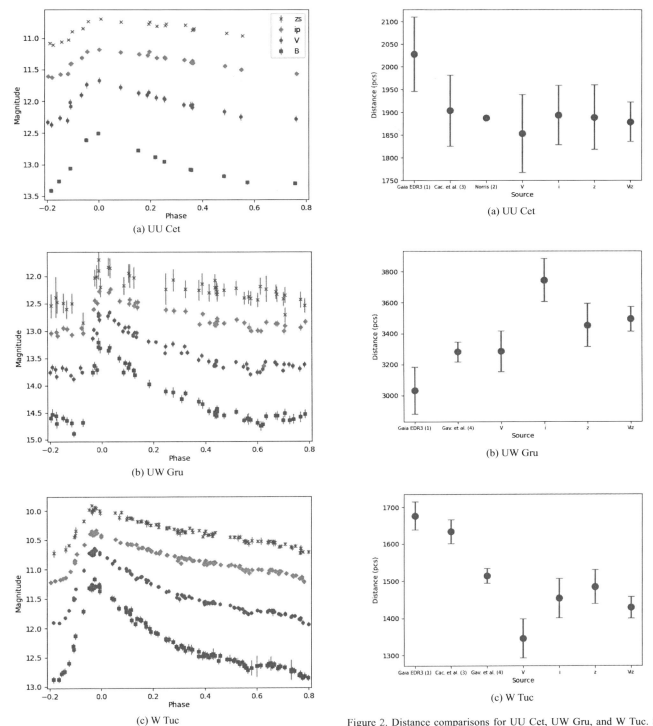

Figure 1. Light curves for the observed RR Lyrae stars. The red points correspond to z band observations, orange to i, green to Johnson V, and blue to Johnson B. z, i, and Johnson B curves were offset by −1, −0.5, and 0.5 magnitudes, respectively.

Figure 2. Distance comparisons for UU Cet, UW Gru, and W Tuc. Some sources are indicated by number for space reasons: (1) Gaia Collab. (2021); (2) Norris (1986), who used Hemenway's (1975) statistical parallax calculation; (3) distances based on the Surface Brightness method by Cacciari et al. 1992; and (4) Gavrilchenko et al. (2014), who used a mid-IR period-luminosity relation. No error value was given for Norris (1986). EDR3 refers to the early data release 3.

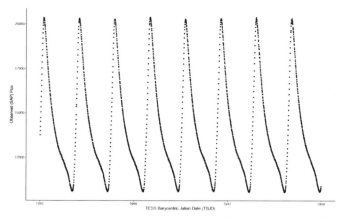

Figure 3. *TESS* observations of RS Boo show the high precision that the mission is capable of. Fluxes were derived using the *TESS* standard "simple aperture photometry" (SAP) data pipeline, and are plotted against *TESS* Barycentric Julian Date (add 2457000 for "normal" Julian Dates).

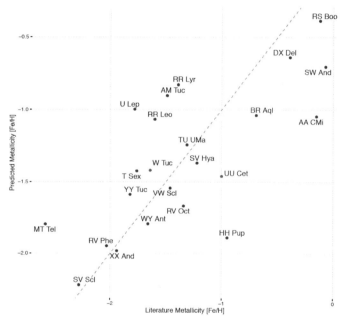

Figure 4. Model vs. Literature Metallicities for an arbitrary selection of RR Lyrae stars observed by the *TESS* mission. The literature metallicity is on the horizontal axis, while the modelled metallicity is on the vertical. The dotted grey line is that of perfect agreement between the literature and the model. The model predictions and literature values for UU Cet and W Tuc are indicated by the red dots. All other systems were used to train the model.

References

Alam, S., *et al.* 2015, *Astrophys. J., Suppl. Ser.*, **219**, 12 (doi:https://doi.org/10.1088/0067-0049/219/1/12).

Altunin, I., Caputo, R., and Tock, K., 2020, *Astron. Theory, Obs., Methods (ATOM)*, 1 (doi:https://doi.org/10.32374/atom.2020.1.1).

Baade, W. 1926, *Astron. Nachr.*, **228**, 359.

Bernard, A. 1982, *Publ. Astron. Soc. Pacific*, **94**, 700.

Bernard, A., and Burnet, M. 1982, *Inf. Bull. Var. Stars*, No. 2072, 1.

Bertin, E., and Arnouts, S. 1996, *Astron. Astrophys., Suppl. Ser.*, **117**, 393.

Brown, T., *et al.* 2013, *Publ. Astron. Soc. Pacific*, **125**, 1031.

Cacciari, C., Clementini, G., and Fernley, J. A. 1992, *Astrophys. J.*, **396**, 219.

Cacciari, C., Clementini, G., Prevot, L., Lindgren, H., Lolli, M., and Oculi, L. 1987, *Astron. Astrophys., Suppl. Ser.*, **69**, 135.

Cáceres, C., and Catelan, M. 2008, *Astrophys. J., Suppl. Ser.*, **179**, 242.

Catelan, M., Pritzl, B., and Smith, H. A. 2004, *Astrophys. J., Suppl. Ser.*, **154**, 633.

Chiba, M., and Yoshii, Y. 1998, *Astron. J.*, **115**, 168.

Clementini, G., Cacciari, C., Fernley, J. A., and Merighi, R. 1992, *Mem. Soc. Astron. Ital.*, **63**, 397.

Clementini, G., Cacciari, C., and Lindgren, H. 1990, *Astron. Astrophys., Suppl. Ser.*, **85**, 865.

Dambis, A. K., Berdnikov, L. N., Kniazev, A. Y., Kravtsov, V. V., Rastorguev, A. S., Sefako, R., and Vozyakova, O. V. 2013, *Mon. Not. Roy. Astron. Soc.*, **435**, 3206.

Feast, M. W., Laney, C. D., Kinman, T. D., van Leeuwn, F., and Whitelock, P. A. 2008, *Mon. Not. Roy. Astron. Soc.*, **386**, 2115.

Fitzgerald, M. T. 2018, *Robotic Telesc., Student Res. Education Proc.*, **1**, 347.

Fitzgerald, M. T., Gomez, E., Salimpour, S., Singleton, J., and Wibowo, R. W. 2021, *J. Open Source Software*, **6**, 2641 (https://doi.org/10.21105/joss.02641).

Gaia Collaboration, Smart, R., Sarro, L. M., Rybizki, J., Reyle, C., Robin, A. C., and Hambly, N. C. 2021, *Astron. Astrophys.*, in press (DOI: 10.1051/0004-6361/202039498).

Gaia Collaboration, *et al.* 2016, Astron. Astrophys., 595A, 2.

Gaia Collaboration, *et al.* 2018, *Astron. Astrophys.*, **616A**, 1.

Gavrilchenko, T., Klein, C. R., Bloom, J. S., and Richards, J. W. 2014, *Mon. Not. Roy. Astron. Soc.*, **441**, 715.

Harris, W. E., Fitzgerald, M. P., and Reed, B. C. 1981, *Publ. Astron. Soc. Pacific*, **93**, 507.

Hemenway, E. K. 1975, *Astron. J.*, **80**, 199.

Henden, A. A., Levine, S., Terrell, D., and Welch, D. L. 2015, Amer. Astron. Soc. Meeting 225, id.336.16.

Hoffmeister, C. 1963, *Astron. Nachr.*, **287**, 169.

Jenkins, J. M., *et al.* 2016, Proc. SPIE, 9913, 3 (doi: 10.1117/12.2233418).

Jones, D. H. P. 1973, *Astrophys. J., Suppl. Ser.*, **25**, 487.

Jones, T. 2020, *Astron. Theory, Obs., Methods (ATOM)*, **1**, 16 (https://rtsre.org/index.php/atom/article/view/42).

Jurcsik, J., and Kovacs, G. 1996, *Astron. Astrophys.*, **312**, 111.

Kafka, S. 2021, Observations from the AAVSO International Database (https://www.aavso.org/data-download).

Kovacs, G., and Zsoldos, E. 1995, *Astron. Astrophys.*, **293**, L57.

Kukarin, B. V., *et al.* 1970, *General Catalogue of Variable Stars*, 3rd ed., Sternberg State Astron. Inst. Moscow State Univ., Moscow.

Lenz, P., and Breger, M. 2005, *Commun. Asteroseismology*, **146**, 53.

Lester, J., Joignant, R., and Mejer, M. 2021, submitted to *J. Amer. Assoc. Var. Star Obs.*

Lub, J. 1977a, *Astron. Astrophys., Suppl. Ser.*, **29**, 345.

Lub, J. 1977b, Unpublished Ph.D. thesis, Univ. of Leiden.

Magnier, E., *et al.* 2016, arXiv preprint, arXiv:1612.05242.

Marsakov, V. A., Gozha, M. L., and Koval, V. V. 2018, *Astron. Rep.*, **62**, 50.

Nicolaides, D, King, D. L., and Cristobal, S. M. 2021, *J. Amer. Assoc. Var. Star Obs.*, **49**, 63.

Norris, J. 1986, *Astrophys. J., Suppl. Ser.*, **61**, 667.

Perryman, M. A. C., European Space Agency Space Science Department, and the Hipparcos Science Team. 1997, *The Hipparcos and Tycho Catalogues*, ESA SP-1200 (VizieR On-line Data Catalog: I/239), ESA Publications Division, Noordwijk, The Netherlands.

Pojmański, G. 1997, *Acta Astron.*, **47**, 467.

Pop, M., and Richey, J. 2021, in preparation.

R Core Team, 2017, R: A language and environment for statistical computing, R Foundation for Statistical Computing, Vienna, Austria, (https://www.R-project.org/).

Ricker, G. R., *et al.* 2015, *J. Astron. Telesc. Instrum. Syst.*, **1**, id. 014003.

Sandage, A. 1993, *Astron. J.*, **106**, 687.

Sarva, J., Freed, R., Fitzgerald, M., and Salimpour, S. 2020, *Astron. Theory, Obs., Methods (ATOM)*, **1**, 34 (https://rtsre.org/index.php/atom/article/view/34).

Schlafly, E. F., and Finkbeiner, D. P. 2011, *Astrophys. J.*, **737**, 103.

Simon, N. R. 1988, *Astrophys. J.*, **328**, 747.

Smith, H. A. 1995, *RR Lyrae Stars*, Cambridge Univ. Press, Cambridge, 1.

Stetson, P. B. 1987, *Publ. Astron. Soc. Pacific*, **99**, 191.

Torrealba, G., *et al.* 2015, *Mon. Not. Roy. Astron. Soc.*, **446**, 2251.

Uzpen, B., and Slater, T. F. 2020, *Astron. Theory, Obs., Methods (ATOM)*, **1**, 54 (https://rtsre.org/index.php/atom/article/view/54).

van Leeuwen, F. 2007, *Astron. Astrophys.*, **474**, 653.

Wenger, M., *et al.* 2000, *Astrophys. J., Suppl. Ser.*, **143**, 9.

Wesselink, A. J. 1946, *Bull. Astron. Inst. Netherlands*, **10**, 91.

Wolf, C., *et al.* 2018, *Publ. Astron. Soc. Australia*, **35**, 10 (doi:https://doi.org/10.1017/pasa.2018.5).

25 New Light Curves and Updated Ephemeris through Analysis of Exoplanet WASP-50 b with EXOTIC

Ramy Mizrachi
Stanford Online High School, 415 Broadway Academy Hall, Floor 2, 8853, Redwood City, CA 94063; ramymizrachi@gmail.com

Dylan Ly
Stanford Online High School, 415 Broadway Academy Hall, Floor 2, 8853, Redwood City, CA 94063; wuyifighter@gmail.com

Leon Bewersdorff
Citizen Scientist, Sandkaulbach 3-5, 52062 Aachen, North Rhine-Westphalia, Germany; leon.bewersdorff@rwth-aachen.de

Kalée Tock
Stanford Online High School, 415 Broadway Academy Hall, Floor 2, 8853, Redwood City, CA 94063; kaleeg@stanford.edu

Received July 21, 2021; revised August 2, 29, September 24, 2021; accepted September 24, 2021

Abstract EXOplanet Transit Interpretation Code (EXOTIC) was used to reduce 75 sets of time-series images of WASP-50 taken by the 6-inch telescope of the Center for Astrophysics | Harvard & Smithsonian MicroObservatory. Of these sets, 25 resulted in clean light curves showing the transit of WASP-50 b, 22 of which had sufficiently low uncertainty to qualify for use in an ephemeris update. We used these results to establish planetary parameters and update WASP-50 b's mid-transit time from 2455558.61237 ± 0.0002 to $2456295.68245 \pm 0.00085$ (BJD_TDB) and its period from 1.9551 ± 5^{-06} to $1.95509584 \pm 0.00000106$ d. The mid-transit time uncertainty of WASP-50 b at the time of projected James Webb Telescope science operations (January 2022) is reduced by a factor of 4.0 using our new ephemeris. We also calculate the planetary size and semi-major axis of WASP-50 b to be approximately $83{,}200 \text{ km} \pm 2{,}230 \text{ km}$ and $0.0294 \text{ AU} \pm 0.0000233 \text{ AU}$, respectively.

1. Introduction

The search for planets outside our solar system has historically been possible with expensive space telescopes. However, even a smaller optical telescope can detect a reduction in the light of a star due to a transiting exoplanet if the host star is bright enough, and if the planet itself is large enough relative to its host. The shorter the period of the orbit, the more often the exoplanet can be observed, thereby preventing the accumulation of uncertainty in transit mid-time. Also, the reduced light curve from each transit can be used to better characterize both the orbit and properties of the planet. Therefore, as citizen scientists make more observations, space telescopes and large ground-based telescopes are able to spend less valuable observation time observing transits whose timing is uncertain.

For a planet with close proximity to the host star, the orbital period is typically on the order of a few days. Since 2001, the Center for Astrophysics | Harvard & Smithsonian MicroObservatory has hosted a campaign to collect images of such large, short-period exoplanets (Sadler *et al.* 2001). Over 75 sets of time-series images of WASP-50 b's transit have accumulated in their archives. The gas giant slightly exceeds the mass of Jupiter at 1.468 Jupiter masses, and it orbits a G-type star. It is a characteristic hot-Jupiter with a short period of around 1.9 days and an orbital radius of 0.0294 AU, or about 3 percent the distance from the Earth to the Sun (Gillon *et al.* 2011).

2. Observations

MicroObservatory hosts a network of automated remote three-foot-tall reflecting telescopes, each with a 6-inch mirror, 560-mm focal length, and KAF1400 CCD with 9-micron pixels. With 2×2 binning, the image size is 650×500 pixels at a pixel scale of 5"/px. MicroObservatory takes images of several exoplanet systems and makes the past month's images publicly available for educational use via their website, at

https://mo-www.cfa.harvard.edu/MicroObservatory/

(Sadler *et al.* 2001).

3. Weather

MicroObservatory uses weather data from NOAA IR satellite images for the region available when the images were taken. The software marks the location where the telescope is, encircles it, and then remaps the pixels within the circle from their 8-bit scale to a 0 to 100 relative scale. The value 0 signifies a complete overcast, whereas a value of 100 would signify that the sky is perfectly clear.

The other metric used by MicroObservatory to determine uncertainty on transits is delta temperature. This metric gives the absolute value difference between the CCD detector and a sensor at the telescope optical tube. Cooled detectors provide a better signal-to-noise ratio because they have less dark current. Dark current is a source of noise from free electrons in the camera sensor arising from thermal energy. To reduce this source of noise, the difference in temperature from the MicroObservatory's ambient temperature should be at least 10° C (Sienkiewicz 2021).

Due to changes in the weather during a transit, some of the images within a transit series might be usable, even if images from other parts of the series are obscured. When this occurs, as it did with the 2013-10-27 transit, it is not possible to fit a reliable light curve that includes all the datapoints. A plot of the weather quality and delta temperature for the 2013-10-27 transit is shown in Figure 1. Unstable seeing indicates poor images, as is shown towards the end of the graph, and an unstable temperature can indicate both a potential change in mechanical focus and efficiency of dark subtraction, as these both depend on a stable temperature.

The final images of the transit are obscured due to weather and have a significantly brighter sky background. Images with a large half-width half maximum were also removed as shown in Figure 2. We chose to remove further outlier images during the transit event that deviated by more than 3 median average distances (MAD) from the surrounding 20 datapoints using HOPS (HOlomon Photometric Software) version 3.0 (Tsiarias 2019).

An image of the original 2013-10-27 light curve is shown in Figure 3. A transit is clearly visible in the initial part of the image sequence, but the poor image quality at the end does not allow for an accurate fit.

The light curve in Figure 4 shows the 2013-10-27 transit after removing the low-quality images. Visually, this is a much better fit. Also, EXOTIC shows the residuals on the bottom sub-plot and reports a "scatter in the residuals" parameter. This parameter is the standard deviation of the residuals in units of percent (relative to baseline flux), which can easily be compared to the transit depth. Once the low-quality images are removed from the 2013-10-27 series, the scatter in the residuals drops from 7.78% to 0.7%, and the mid-transit time shifts from 2456592.927 ± 0.003 d to 2456592.8586 ± 0.0051 d.

Although there is a slight increase in mid-transit uncertainty, it is still under our threshold of 0.007 on mid-transit uncertainty, and is deemed more accurate than the mid-time resulting from the fit shown in Figure 3. This demonstrates the importance of verifying the quality of the observational data themselves, and not solely relying on low fit errors. The process thus produces another usable light curve for WASP-50 b, which would not have been taken into consideration for O–C and ephemeris calculation otherwise, based on the initial review of the image data.

4. Data reduction and photometry

For the photometric evaluation of the data, Exoplanet Watch's EXOTIC (EXOplanet Transit Interpretation Code) software was chosen (Zellem et al. 2020). This software requires the user to select up to ten comparison stars (comp stars). Comparison stars AUD 000-BMD-470 and AUD 000-BMD-471 were selected based on the American Association of Variable Star Observers (AAVSO) Variable Star Plotter (VSP; AAVSO 2021), shown in Figure 5. The AAVSO chart ID X26441FB can be entered on the AAVSO VSP website for retrieval. Apart from giving EXOTIC the option to choose from the two AAVSO-recommended comparison stars, we chose three additional comp star candidates, which are labeled in Figure 6 as C3, C4, and C5, to provide more options for EXOTIC's reduction of our light curves. The manually selected comparison stars were all bright, close to the target, and not too close to other stars in the field.

EXOTIC was created by Exoplanet Watch, a citizen science initiative of NASA Jet Propulsion Laboratory. Its purposes

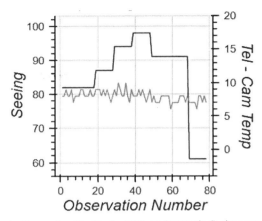

Figure 1. Observation Quality for 2013-10-27 transit. Seeing: avg 86, std 11. Temp diff: avg 8.0, std 0.8. The purple line represents seeing conditions (left axis) and the yellow line represents the ambient temperature in the telescope tube minus the temperature of the CCD detector (right axis).

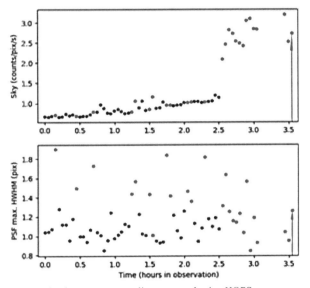

Figure 2. Red points represent outliers removed using HOPS.

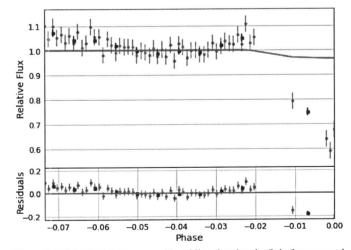

Figure 3. 2013-10-27 light curve with red line showing the fit before removal of outliers.

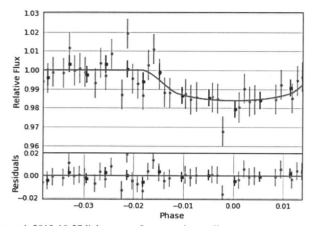

Figure 4. 2013-10-27 light curve after removing outliers.

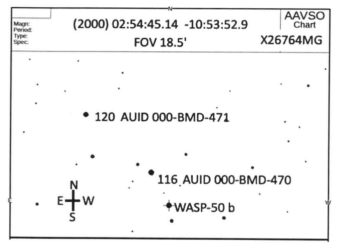

Figure 5. AAVSO VSP view of WASP-50 starfield.

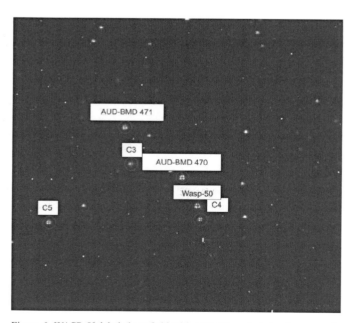

Figure 6. WASP-50 labeled starfield with manually selected extra comps in AstroImageJ (Collins *et al.* 2017).

are both to introduce citizen scientists into astronomy, and to reduce uncertainty of exoplanet transit midpoints in preparation for coming NASA missions, such as the James Webb Space Telescope. EXOTIC can be run locally or by using Google Colaboratory (Colab). We used Colab so that team members could share files and to avoid using local computer resources or space allocation. The script mounts the user's Google Drive account and installs EXOTIC onto a virtual machine in the cloud. It then displays the first image of the series and prompts the user for the target name, the coordinates the target, and up to ten comparison stars. The target name is used to look up parameters in the NASA Exoplanet Archive (NEA) to use as priors in the light curve fit. Then EXOTIC aligns the images and determines the optimal inner and outer photometric apertures. The inner aperture encompasses the star's point spread function (PSF) without including the sky background, which fills the space between the outer and inner apertures. EXOTIC determines the optimal aperture sizes by fitting to a Gaussian PSF model (Fatahi 2021). To account for changes in sky brightness affecting the measured flux, EXOTIC subtracts the background photon count from the star's flux. Finally, the change in flux of the target star is compared to the light emitted by each of the selected comparison stars, and a "quick fit" is performed. The comparison star with the best quick fit is selected for use in the more rigorous fitting routine. For these images, EXOTIC selected one of the AAVSO VSP-recommended comp stars over the manually identified options in 21 of 22 light curve fits, which confirms its agreement with the VSP-recommended stars.

EXOTIC's output included a light curve for each series along with the scatter in the residuals, the midpoint time, transit depth, transit duration, semi-major axis relative to the stellar radius, and planetary versus stellar radius (Winn 2014).

5. Data

EXOTIC's reduction process produced 25 new light curves of WASP-50 b, which are shown in Appendix A. Each plot shows the measured normalized flux with error bars of the host star versus time as the exoplanet transits across its face along with the best possible light curve fit. EXOTIC also outputs planetary radius/stellar radius, transit depth, and semi-major axis over stellar radius. These parameters were all averaged and an uncertainty is reported for these parameters as the standard error of the mean (SEM), or the standard deviation of each parameter divided by the square root of the number of data points. The transit depth is therefore 0.020 ± 0.10, the planetary radius over stellar radius is 0.142 ± 0.00380, and the semi-major axis over stellar radius is 7.49 ± 0.0418. We calculate WASP-50 b to have a radius 14.2% the size of WASP-50, a G-type main sequence star (Gillon *et al.* 2011). From the ratio of the planet to the stellar radius (R_p/R_s) the planetary size can be determined. The literature value of 0.843 solar radii (5.870×10^5 km) for WASP-50 (Chakrabarty and Sengupta 2019) is used for R_s to calculate the radius of the planet in km:

$$r_{km} = R_s * (R_p/R_s) \pm SEM \qquad (1)$$

The orbital distance in Astronomical Units (AU) can also be determined from the ratio between the semi-major axis and the star radius (a/R_s). A planet with a larger semi-major axis thus has a longer transit, which EXOTIC takes into account when fitting this parameter to an individual light curve:

$$d_{AU} = \frac{R_s * (a/R_s)}{1.496 * 10^8 \text{ km}} * 1\text{AU} \pm \frac{\text{SEM}}{1.496 * 10^8 \text{ km}} * 1\text{AU} \quad (2)$$

These two calculations and their respective SEM calculation are performed for each transit reduced with EXOTIC. The results are reported in AU to align with the units used in the literature.

Here the planetary size of WASP-50b is calculated to be approximately 83200 km ± 2230 km or 1.190 ± 0.032 R_J. The same is done for the semi-major axis, which is calculated to be 0.0294 AU ± 0.0000233 AU. The planetary size and semi-major axis are within the uncertainty of those presented in the literature of 1.15 ± 0.05 R_J and 0.0295 ± 0.0009 AU, respectively (Gillon et al. 2011).

The transit mid-times from the MicroObservatory transits calculated using EXOTIC are shown in Table 1.

We produced an O–C plot for WASP-50b using the 22 bolded epochs from Table 1, for which the scatter in the residuals was less than 1.6% and the mid-transit time uncertainty was less than 0.007 day. Using the most recently published values, t_0 = 2455558.61237 BJD (Bonomo et al. 2017) and p = 1.9551 d (Chakrabarty and Sengupta 2019), our data produced the plot shown in Figure 7:

The ephemeris of an exoplanet allows times of transit-minima to be calculated. This calculation includes information about the period of the planet, the transit mid-times, and the uncertainties of measurements. Using image sets of WASP-50b transits, we were able to update the ephemeris using EXOTIC. The orbital ephemeris of WASP-50b is modeled using the following equation:

$$t_{next} = n * P + T_{mid} \quad (3)$$

where t_{next} is a future mid-transit time, P is the period, n is the orbital epoch, and T_{mid} is a reference mid-transit time. The linear ephemeris is optimized using nested sampling to derive posterior distributions for the mid-time and period. (Pearson 2019). The code uses the epochs, mid-transit times, and mid-transit uncertainties for each of the 22 transits and bounds for the mid-transit time and period. The output includes graphs depicting the uncertainties as well as values for the mid-transit time and period. The most recent listing in the NASA Exoplanet Archive cites 1.955100 ± 0.000005 d as the period (Chakrabarty and Sengupta 2019) and 2455558.61237 ± 0.00020 BJD_TDB as the mid-transit time (Bonomo et al. 2017). Based on the ephemeris fitter's analysis of our transits, the updated period and mid-transit time are 1.95509584 ± 0.00000106 d and 2456295.68245 ± 0.00085 d, respectively.

$$t_{next} = n * 1.95509584 + 2456295.68245 \quad (4)$$

Equation 4 represents our proposed new ephemeris. The graphs for the linear ephemeris fit and the residuals versus

Table 1. Transit midtimes for WASP-50b from MOBs data.

Transit Number	Date	Epoch	Mid-transit (BJD_TDB) (2450000+)	Mid-transit Uncertainty (days)	Scatter (%)
1	2013-01-03	377	6295.6795	0.0016	0.85
2	2013-09-14	507	6549.8496	0.0029	1.08
3	2013-10-27	529	6592.8618	0.004	0.62
4	2013-10-31	531	6596.7662	0.0025	0.65
5	2013-11-02	532	6598.7236	0.0027	0.74
6	2013-12-13	553	6639.7807	0.0022	0.82
7	2013-12-15	554	6641.7389	0.0018	1.11
8	2014-11-22	729	6983.8763	0.0033	0.78
9	2014-11-26	731	6987.7864	0.0022	0.76
10	2014-11-30	733	6991.6938	0.0027	0.78
11	2015-12-28	934	7384.6742	0.0029	0.96
12	*2016-12-04*	*1109*	*7726.817*	*0.026*	*1.22*
13	2017-11-15	1286	8072.8631	0.0025	0.79
14	2017-11-17	1287	8074.814	0.0019	0.69
15	2018-01-03	1311	8121.734	0.0037	1.23
16	2018-01-05	1312	8123.6972	0.0022	0.73
17	*2018-09-14*	*1441*	*8375.898*	*0.029*	*1.11*
18	*2018-11-04*	*1467*	*8426.77*	*0.076*	*1.12*
19	2018-11-06	1468	8428.6902	0.0022	0.95
20	2018-12-21	1491	8473.6535	0.0032	1.42
21	2019-10-14	1643	8770.8366	0.0031	1.25
22	2020-09-24	1820	9116.8825	0.0045	1.13
23	2020-09-26	1821	9118.856	0.0032	1.1
24	2020-11-12	1845	9165.767	0.003	0.87
25	2020-12-25	1867	9208.776	0.0022	1.11

Note: Italicized transits are not used in the O–C plot.

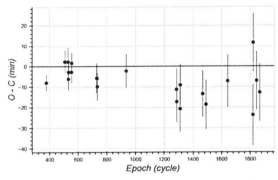

Figure 7. O–C plot for WASP-50b using t_0 = 2455558.61237 and p = 1.9551.

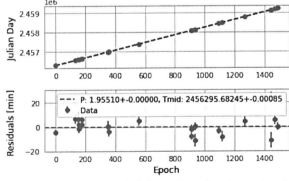

Figure 8. Graph of linear ephemeris fit from ephemeris updater code and graph of residuals against epoch.

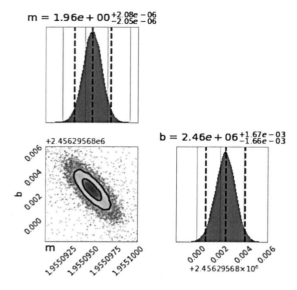

Figure 9. Posterior distributions of the updated mid-transit time and period.

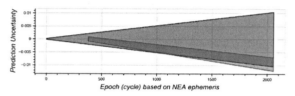

Figure 10. Cone plot comparing our mid-transit prediction (red) to the current NEA prediction (blue) for 1 January 2022. Previous and new ephemerides superimposed with prediction difference offset.

epochs are shown in Figure 8, and the posterior distributions of the mid-transit time and period are shown in Figure 9.

6. Results

The utility of the new ephemeris can be evaluated by playing forward the prediction to 2022-01-01, which is when the James Webb Space Telescope is projected be ready to commence science operations. In Figure 10, the NEA prediction (blue) is compared to our prediction (pink) on that date. As is evident from the figure, our analysis has caused WASP-50 b's mid-transit time uncertainty on 2022-01-01 to decrease by a factor of 4.0 relative to the previous NEA prediction. Our new midpoint prediction for 2022-01-01 is 2459580.24346 ± 0.00262, which is –13.8 minutes different from the NEA prediction of 2459580.25307 ± 0.01049.

7. Conclusion

We present 25 new mid-time values and light curves for WASP-50 b from the MicroObservatory observations and established parameters for WASP-50 b's size and orbit, supporting its classification as a hot Jupiter-type exoplanet. We used the result of the light curve reduction to establish planetary parameters and update the mid-transit time from 2455558.61237 ± 0.0002 to $2456295.68245 \pm 0.00085$ (BJD_TDB) and the period from 1.9551 ± 5^{-6} to $1.95509584 \pm 0.00000106$ d (Gillon *et al.* 2011).

Based on the 22 sets of time-series images taken by the 6-inch MicroObservatory telescope, the uncertainty of the predicted midpoint in January of 2022 has decreased by a factor of 4.0.

8. Future work

With at least five additional light curves and mid-time values, it would be possible to search for TTVs (Transit Timing Variations), which might constitute the signature of another planet in the WASP-50 system.

9. Acknowledgements

Data used here come from the MicroObservatory telescope archives maintained by Frank Sienkiewicz, who also provides information on weather and delta temperature measurements. MicroObservatory is maintained and operated as an educational service by the Center for Astrophysics | Harvard & Smithsonian and is a project of NASA's Universe of Learning, supported by NASA Award NNX16AC65A. Additional MicroObservatory sponsors include the National Science Foundation, NASA, the Arthur Vining Davis Foundations, Harvard University, and the Smithsonian Institution.

Thanks to Martin Fowler for his advice and help with analysis of these data, and to Rob Zellem and his team for fixing and maintaining the EXOTIC software. Also, thanks to Jason Eastman for creating the public time conversion calculator on the "astroutils" website.

This research has made use of the NASA Exoplanet Archive, which is operated by the California Institute of Technology, under contract with the National Aeronautics and Space Administration under the Exoplanet Exploration Program.

This publication makes use of the EXOTIC data reduction package from Exoplanet Watch, a citizen science project managed by NASA's Jet Propulsion Laboratory on behalf of NASA's Universe of Learning. This work is supported by NASA under award number NNX16AC65A to the Space Telescope Science Institute.

References

AAVSO. 2021, Variable Star Plotter
   (VSP; https://www.aavso.org/apps/vsp/).
Bonomo, A. S., *et al.* 2017, *Astron. Astrophys.*, **602A**, 107.
Chakrabarty, A., and Sengupta, S. 2019, *Astron. J.*, **158**, 39.
Collins, K. A., Kielkopf, J. F., Stassun1, K. G., and Hessman, F. V. 2017, *Astron. J.*, **153**, 77
   (https://www.astro.louisville.edu/software/astroimagej)
Fahati, T. 2021, private communication (September 13, 2021).
Gillon, M. *et al.* 2011, *Astron. Astrophys.*, **533A**, 88.
Pearson, K. A. 2019, *Astron. J.*, **158**, 243.
Sadler P., *et al.* 2001, *J. Sci. Education Technol.*, **10**, 39.
Sienkiewicz, F. 2021, private communication (March 25, 2021).
Tsiaras, A. 2019, in *EPSC-DPS Joint Meeting 2019*, id. EPSC-DPS2019-1594.
Winn, J. N. 2014, arXiv:1001.2010v5.
Zellem R., *et al.* 2020, *Publ. Astron. Soc. Pacific*, **132**, 054401.

Appendix A: Light curves of WASP-50 b reduced with EXOTIC

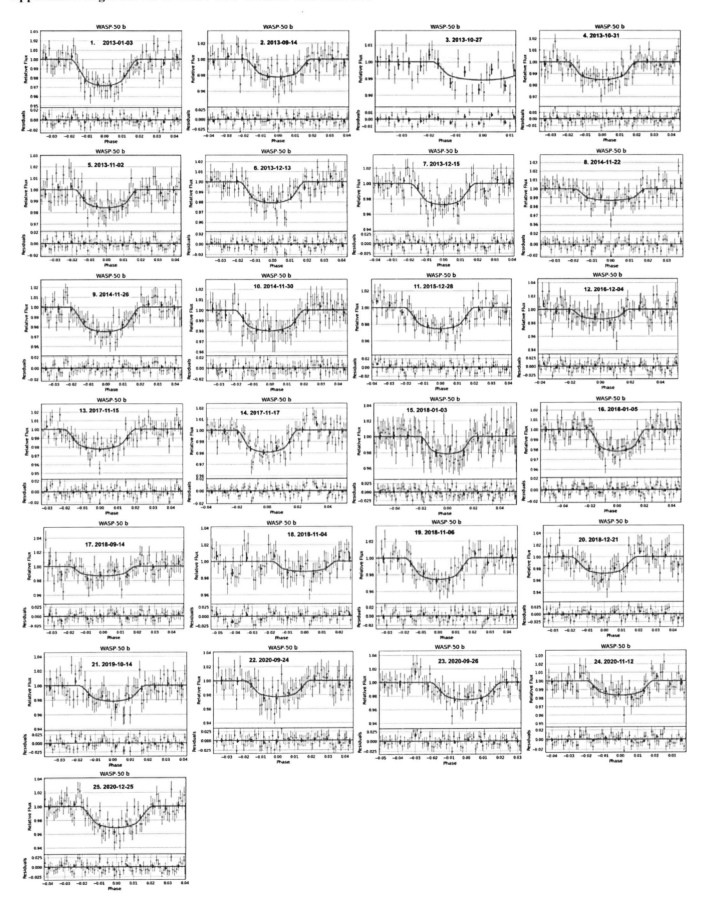

Characterization of NGC 5272, NGC 1904, NGC 3201, and Terzan 3

Paul Hamrick
Avni Bansal
Kalée Tock
Stanford Online High School, 415 Broadway Academy Hall, Floor 2, 8853, Redwood City, CA 94063; paulahamrick@gmail.com; avnibansal2004@gmail.com; kaleeg@stanford.edu

Received July 21, 2021; revised November 9, 2021; accepted November 15, 2021

Abstract Globular clusters are an ideal laboratory for studying and comparing variable stars, since all the variables in a globular cluster formed from the same progenitor gas cloud. Because of this, all of the variables within a given cluster have similar ages, metallicities, reddening, and distances from Earth. Thus we estimated the age, metallicity, reddening, and distance of the clusters NGC 5272, NGC 1904, NGC 3201, and Terzan 3 by doing a visual fit to isochrones. Knowing the characteristics of these clusters as a whole helps us better understand the characteristics of individual variables in these clusters as well. Reddening, metallicity, and age estimates for NGC 1904 and NGC 5272 are consistent with previous literature, but all four of the estimated parameters for Terzan 3 and the metallicity measurement of NGC 3201 differed significantly from literature values. Further research into NGC 3201 and Terzan 3 is recommended.

1. Introduction

Studying the periods of RR Lyraes helps astronomers refine period-metallicity-luminosity relationships (Nemec *et al.* 1994). In addition, knowing the periods of RR Lyraes in a particular globular cluster helps determine how far away the cluster is by the period-luminosity relationship. Specifically, once the period is known, the luminosity can be calculated, and the luminosity can be used to yield the distance by applying the inverse square law and measuring the apparent brightness of the cluster (Catelan *et al.* 2004). In this way, RR Lyraes serve as standard candles.

A challenge in modeling stellar evolution within a cluster is that the temperature and luminosity of stars cannot be measured directly. However, color magnitude diagrams (CMDs) of globular clusters can be used as a proxy for these variables. Within a cluster, the constituent stars can be assumed to have similar age, distance, metallicity, and reddening. This is because all the stars in a given cluster formed together at the same time from the same progenitor gas cloud, so they are roughly the same age and have the same metallicity. Moreover, all stars in a cluster are approximately the same distance from Earth, so their light traverses the same interstellar medium enroute to our telescopes. Therefore, the starlight will be reddened to an extent that depends only on its wavelength.

As a caution, it should be noted that there are some clusters in which there are multiple groups of stars with different ages, such as NGC 6121 (Marino *et al.* 2008). Moreover, clusters with significant reddening tend to show variation in reddening among the stars in them (Bonatto *et al.* 2013). Although there is variation in the reddening and chemical abundances of the stars in some clusters, this is unlikely to affect the model presented here, because the metallicity index that we are using is iron-specific, and the variations in chemical abundances were only found with specific elements which are not iron. In particular, Marino *et al.* found that though other elements or compounds had bimodality in their content, the iron peak-content distribution was found to be homogeneous (Marino *et al.* 2008).

Therefore, assuming that age, metallicity, distance, and reddening are similar for all stars within a cluster, the color magnitude diagram can be used to match the stars to isochrones, which are theoretical models of star populations at a given point in the cluster's evolution. Based on the isochrones, the distance, metallicity, age, and reddening of clusters can be estimated, and these estimates can be refined based on RR Lyrae period-metallicity-luminosity relationships.

Point Spread Function (PSF) photometry was used to identify and measure the stars in NGC 3201, Terzan 3, NGC 5272, and NGC 1904, and create CMDs of these clusters. PSFs are functions that model the brightness profiles of stars in an image. The PSF photometry methods used in this study are Point Spread eXtreme (psx) (Bertin and Arnouts 1996) and DOPhot (Schechter *et al.* 1993). The resulting CMDs were then visually fitted to an appropriate isochrone to estimate the reddening, metallicity, distance, and age of each cluster.

NGC 3201 and Terzan 3 are of particular interest because of their confirmed and hypothesized RR Lyrae populations. NGC 3201 has 160 known RR Lyraes. Astronomers first identified variables in NGC 3201 in 1941 (Wright 1941). Recently, 36 new variables were found (Kaluzny *et al.* 2016). The high incidence of successful variable star searches indicates that NGC 3201 is rich in variables, and further searches may yield yet more variables. Furthermore, even if no new variables are found, it is useful to verify or refine previous period estimates, especially for variables that were discovered around 1941 using older technologies. On the other hand, Terzan 3 does not have any known RR Lyraes, but the preponderance of RR Lyraes in old globular clusters similar to Terzan 3 suggests that a search for RR Lyraes in Terzan 3 may be fruitful.

The four clusters studied here are shown in Figure 1, which was constructed from Las Cumbres Observatory images processed by the Our Solar Siblings pipeline (Fitzgerald *et al.* 2018).

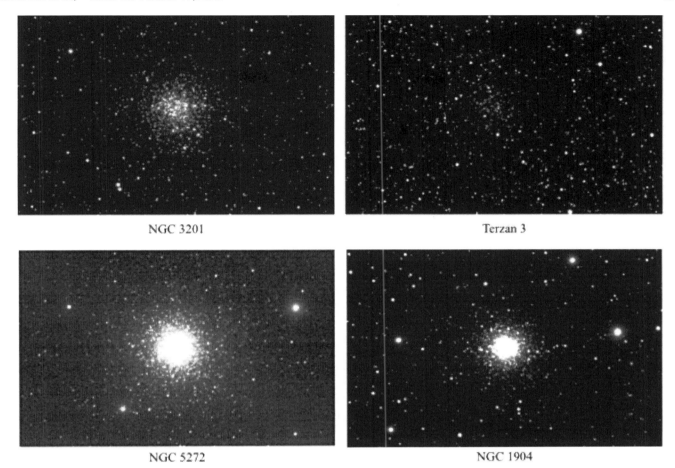

Figure 1. RGB composite images of NGC 3201, Terzan 3, NGC 5272, and NGC 1904.

Table 1. The exposure times of each of the four clusters in each of the eight filters possible with the Las Cumbres Observatory.

Target	SDSS-up	SDSS-gp	SDSS-rp	SDSS-ip	Bessel-B	Bessel-V	PanSTARRS-w	PanSTARRS-z
Terzan 3	300	200	150	120	300	300	120	60
NGC 3201	300	200	150	120	240	300	120	120
NGC 5272	300	200	200	200	300	230	120	150
NGC 1904	300	300	230	200	300	290	120	200

2. Instruments used

The instruments used were the Las Cumbres Observatory (LCO) telescopes in Cerro Tololo, Chile, in Siding Spring, Australia, in Sutherland, South Africa, and in Fort Davis, Texas. Each telescope is 0.4 meter in diameter and is a Meade 16-inch (40cm) RCS tube and three-element optics, mounted in an LCO equatorial C-ring mounting. The optics are a primary, secondary, and Corrector plate (Meade) with an LCO focus mechanism driving corrector plate/secondary. The instruments on the telescopes are the SBIG STL6303 cameras, which have a 19.5 × 29.5' field of view and a pixel scale of 0.591 arcsec/pixel. The images were taken with each of the eight filters provided by the Las Cumbres Observatory: PanSTARRS-w and PanSTARRS-z (PanSTARRS stands for Panoramic Survey Telescope and Rapid Response System), Bessel B and Bessel V, and SDSS-ip, SDSS-rp, SDSS-gp, and SDSS-up (Sloan Digital Sky Survey). The exposure times in each filter are shown in Table 1.

3. Target selection

We used the VizieR Online Data Catalog and the Clements Catalog (Clement 2017) to find currently observable clusters that were bright enough to be seen by the LCO. We initially chose three clusters with known RR Lyraes (NGC 3201, NGC 1904, NGC 5272) and three clusters with no known RR Lyraes (Terzan 3, 2MASS GC01, IC 1257). However, we decided not to investigate 2MASS GC01 because it was not visible at the time of study. Also, IC 1257 was too dim in the PanSTARRS-w filter using the maximum allowable exposure time of 300 seconds.

4. Procedure for modelling clusters

Images of each of the four target clusters in eight different filters were obtained using LCO telescopes. The images were fed into the Our Solar Siblings pipeline (Fitzgerald et al. 2018), which conducted six different types of photometric reduction on each of the images, including three types of aperture photometry and three types of PSF photometry. Since aperture photometry identifies stars based on brightness, without model-fitting, it is often unable to distinguish closely-packed stars in an image. Thus PSF photometry was chosen for this study because of its higher sensitivity and accuracy: specifically the psx and dop methods mentioned above.

5. Calibration

The photometry data were input to PysoChrone, an Our Solar Siblings software tool written by (Fitzgerald *et al.* 2018), which employs an isochrone model based on (Girardi *et al.* 2020). PysoChrone selected reference stars and calibrated their instrumental magnitudes against magnitudes listed in various databases. The calibration stars are selected as follows. First, stars that were listed in the AAVSO Variable Star Index (Watson *et al.* 2014) were rejected from consideration. Stars are rejected that are too bright, which for the CCD cameras of LCO means that they have above 1,000,000 total aperture counts, with signal-to-noise 1,000. Stars are also rejected that are too dim, meaning that they have below 10,000 total aperture counts, with signal-to-noise less than 100. Depending on filter and availability, different calibration catalogues are used, including the AAVSO Photometric All-Sky Survey (APASS; Henden *et al.* 2016), the Sloan Digital Sky Survey (SDSS; Blanton *et al.* 2017), the he Panoramic Survey Telescope and Rapid Response System (PanSTARRS; (Magnier *et al.* 2016),), and SkyMapper (Wolf *et al.* 2018). Stars that are flagged in these catalogues as being imperfectly measured in any way are rejected. Also, stars need a valid magnitude and error in the filter under consideration and also in the complementary color filter (e.g. V for B to make B–V) in order to be considered as possible reference stars.

PysoChrone used the calibrated star magnitudes and photometry from our images as well as catalog data to generate CMDs of each cluster, using ten different color indexes and magnitude measures. We adjusted the age, metallicity, distance, and reddening to until the corresponding Girardi isochrone had the closest visual match to these CMDs.

6. Results

The images below depict the graphs generated in PysoChrone based on the images in the eight filters provided by Las Cumbres Observatories as described above. PysoChrone was run on each of the clusters investigated, which were Terzan 3, NGC 3201, NGC 1904, and NGC 5272. In Figure 2, blue points represent the stars, while the red lines are the Girardi isochrones. The plots without blue stars are graphs for which the corresponding data were unavailable.

Table 2 shows the parameters of the best-fitting isochrones for each cluster.

7. Analysis of results

The best fitting isochrones we found disagree with the values found by previous papers for Terzan 3. Table 3 shows the values found by previous papers for the clusters we investigated.

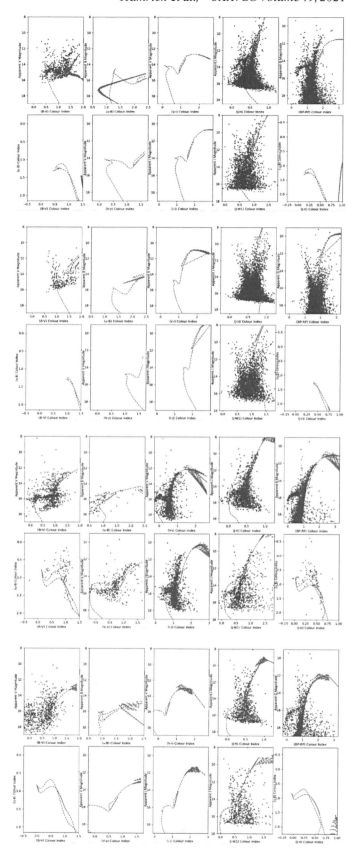

Figure 2. Best fitting isochrones of (from top) NGC 3201, Terzan 3, NGC 5272, and NGC 1904.

Evidently, there is significant disagreement between our values for Terzan 3 (shown in Table 2) and those found using a near infra-red CM diagram by (Valenti *et al.* 2007). While Valenti *et al.* used near infra-red images to derive these results, we have used images in eight different filters ranging from infrared to ultraviolet.

Furthermore, Valenti *et al.* used DAOPhot (Stetson 1987) to analyze images whereas we have used Point Spread eXtractor (Bertin and Arnouts 1996). The difference in photometric reduction technique may account for some of the differences between the values obtained. Photometric reduction of our images with dao found more than twice as many stars as the psx reduction, as measured by the filesize of the resulting photometry text files. However, the dimness of the additional stars identified by dao made their placement on the isochrone uncertain. Therefore, while psx was deemed more reliable in this case, the differing filesizes are indicative of the inherent differences in the two photometric reduction techniques.

In the case of NGC 3201, the magnitude and E(B–V) are reasonably close to the values in the Harris (1996) catalog. However, the metallicity values differ greatly. Furthermore, several other studies using several different methods have all arrived at metallicity values that are closer to the value in the Harris catalog than the value arrived at here, as shown in Table 4. This is likely because the effect of changing metallicity on the isochrone is subtle, and the fit to the isochrone performed here is visual. In other words, insertion of the literature value for the metallicity yields a visual fit that is not significantly different from the one shown in Figure 2.

In the cases of NGC 1904 and NGC 5272, there is strong agreement between our results and those from the Harris catalogue. Minor differences in the metallicity and E(B–V) values can be attributed to the necessarily imprecise nature of visually aligning isochrones to best fit color-magnitude diagrams. Hence, the accumulation of minor inconsistencies in our calibration of metallicity, E(B–V), and age of the cluster may have led to a larger inconsistency in calculating distance in both cases. That this distance measurement is still within 2 kpcs of the Harris catalogue measurement despite error in the case of NGC 1904 should improve confidence in the Harris catalogue values. On the other hand, the distance estimate obtained for NGC 5272 should be taken with a grain of salt.

8. Conclusion

The strong agreement between reddening, metallicity, and age measurements in this study and those previously found for NGC 5272 and NGC 1904 should increase confidence in these parameters. Agreement on the reddening and age values of NGC 3201 should also bolster confidence in those parameters.

Meanwhile, the inexplicable disparity between this study's metallicity values for NGC 3201 and previously found values indicates this cluster's metallicity requires further investigation, ideally using an approach independent from isochrones.

The large disagreement in all four parameters of Terzan 3 calls for further study as well. These disagreements are not surprising since Terzan 3 has been historically understudied. In the absence of a large trove of data it is difficult to pin down the cluster's characteristics. More studies using different techniques to characterize this cluster will eventually result in all the data converging to a narrower, accurate range of characteristics.

Finally, we recommend that the technique of isochrones in estimating distance to clusters should be further analyzed and developed to make this method of finding distances to clusters more reliable and accurate.

Table 2. Characteristics of the four clusters derived by fitting isochrones to CMDs in PysoChrone.

Cluster	Photometry	Distance (kpc)	Magnitude (m–M)	Age (Billions of Years)	Metallicity (Fe/H)	Reddening (E(B–V))
Terzan 3	psx	1.202	10.4	9.5	–0.95	0.60
NGC 3201	psx	4.073	13.05	8.9	–0.4	0.32
NGC 1904	psx	14.8	15.85	8.85	–1.30	0.03
NGC 5272	dop	9.12	14.8	8.7	–1.5	0.23

Table 3. Characteristics of clusters from previous papers.

Cluster	Distance (kpc)	Magnitude (m-M)	Metallicity (Fe/H0)	Reddening E(B-V)	Age (Billions of Years)	Reference
Terzan 3	8.1	14.54	–0.82	0.73	—	Valenti *et al.* (2007)
NGC 3201	4.9	14.2	–1.24	0.24	12	Paust *et al.* (2010); Calamida *et al.* (2008)
NGC 1904	12.9	15.59	–1.37	0.01	11.7	Koleva *et al.* (2008)
NGC 5272	33.9	15.07	–1.34	0.01	11.4	Forbes and Bridges (2010)

Table 4. Characteristics of clusters from previous papers.

Metallicity	Technique	Reference
–1.4	Photoelectric photometry	Zinn (1980)
–1.0	High dispersion spectrometry with echelle spectrograph	Pilachowski *et al.* (1980)
–1.62	121 A mm spectrographs	Zinn and West (1984)

9. Acknowledgements

This research was made possible by AstroImageJ software, which was written by Karen Collins and John Kielkopf, and by PysoChrone, which was written by Michael Fitzgerald.

This work makes use of observations taken by the 0.4-m telescopes of Las Cumbres Observatory Global Telescope Network located in Cerro Tololo, Chile, in Siding Spring, Australia, in Sutherland, South Africa, and in Fort Davis, Texas.

This work utilizes the Harris and Clements catalogs of globular clusters.

This research was made possible through the use of the AAVSO Photometric All-Sky Survey (APASS), funded by the Robert Martin Ayers Sciences Fund, and the International Variable Star Index, maintained by the AAVSO.

References

Bertin, E., and Arnouts, S. 1996, *Astron. Astrophys, Suppl. Ser.*, **117**, 393 (doi: 10.1051/aas:1996164).

Blanton, M. R., *et al.* 2017, *Astron. J.*, **154**, 28 (Sloan Digital Sky Survey IV; https://www.sdss.org/).

Bonatto, C., Campos, F., and Kepler, S. O. 2013, *Mon. Not. Roy. Astron. Soc.*, **435**, 263 (doi: 10.1093/mnras/stt1304).

Calamida, A., *et al.* 2008, in *The Ages of Stars*, Proc. IAU 4, Symp. S258, Cambridge Univ. Press, Cambridge, 189 (doi: 10.1017/S1743921309031846).

Catelan, M., Pritzl, B. J., and Smith, H. A. 2004, *Astrophys. J., Suppl. Ser.*, **154**, 633 (doi: 10.1086/422916).

Clement, C. M. 2017, VizieR On-line Data Catalog: V/150.

Fitzgerald, M. T., McKinnon, D. H., Danaia, L., Cutts, K. R., and Salimpour, M. S. 2018, in *Robotic Telescopes, Student Research and Education Proceedings*, eds. M. Fitzgerald, C. R. Buxner, S. White, RTSRE Proc. 1, RTSRE, San Diego, 217 (https://rtsre.org/index.php/rtsre/issue/view/1).

Forbes, D. A., and Bridges, T. 2010, *Mon. Not. Roy. Astron. Soc.*, **404**, 1203 (doi: 10.1111/j.1365-2966.2010.16373.x).

Girardi, L., Bressan, A., and Bertelli, G., and Chiosi, C. 2000, *Astron. Astrophys, Suppl. Ser.*, **141**, 371 (doi: 10.1051/aas:2000126).

Harris, W. E. 1996, *Astron. J.*, **112**, 1487.

Henden, A. A., Templeton, M., Terrell, D., Smith, T. C., Levine, S., and Welch, D. 2016, VizieR Online Data Catalog: AAVSO Photometric All Sky Survey (APASS) DR9, II/336.

Kaluzny, J., Rozyczka, M., Thompson, I. B., Narloch, W., Mazur, B., Pych, W., and Schwarzenberg-Czerny, A. 2016, *Acta Astron.*, **66**, 31 (https://arxiv.org/abs/1604.01362).

Koleva, M., Prugniel, Ph., Ocvirk, P., Le Borgne, D., and Soubiran, C. 2008, *Mon. Not. Roy. Astron. Soc.*, **385**, 1998 (doi: 10.1111/j.1365-2966.2008.12908.x).

Magnier, E., *et al.* 2016, arXiv preprint, arXiv:1612.05242.

Marino, A. F., Villanova, S., Piotto, G., Milone, A. P., Momany, Y., Bedin, L. R., and Medling, A. M. 2008, *Astron. Astrophys.*, **490**, 625 (doi: 10.1051/0004-6361:200810389).

Nemec, J. M., Nemec, A. F. L., and Lutz, T. E. 1994, *Astron. J.*, **108**, 222 (doi: 10.1086/117062).

Paust, N. E. Q., *et al.* 2010, *Astron. J.*, 139, 476 (doi: 10.1088/0004-6256/139/2/476).

Pilachowski, C. A., Sneden, C., and Canterna, R. 1980, in *Star Clusters*, ed. J. E. Hesser, Proc. IAU Symp. 85, Reidel Publishing Co., Dordrecht, 467.

Schechter, P. L., Mateo, M., and Saha, A. 1993, *Publ. Astron. Soc. Pacific*, **105**, 1342 (doi: 10.1086/133316).

Stetson, P. B. 1987, *Publ. Astron. Soc. Pacific*, **99**, 191 (doi: 10.1086/131977).

Valenti, E., Ferraro, F. R., and Origlia, L. 2007, *Astron. J.*, **133**, 1287 (doi: 10.1086/511271).

Watson, C., Henden, A. A., and Price, C. A. 2014, AAVSO International Variable Star Index VSX (Watson+, 2006–2014; https://www.aavso.org/vsx).

Wright, F. W. 1941, *Bull. Harvard Coll. Obs.*, No. 915, 2.

Wolf, C., *et al.* 2018, *Publ. Astron. Soc. Australia*, **35**, 10 (doi:https://doi.org/10.1017/pasa.2018.5).

Zinn, R. 1980, *Astrophys. J., Suppl. Ser.*, 42, 19 (doi: 10.1086/190643).

Zinn, R., and West, M. J. 1984, *Astrophys. J., Suppl. Ser.*, **55**, 45 (doi: 10.1086/190947).

Retraction of and Re-analysis of the Data from "HD 121620: A Previously Unreported Variable Star with Unusual Properties"

Roy A. Axelsen
P.O. Box 706, Kenmore, Queensland 4069, Australia; reaxelsen@gmail.com

Received August 4, 2021; revised August 17, 2021; accepted August 18, 2021

Abstract The original paper (published in *JAAVSO*, Vol. 48, No. 1, 2020) is retracted, because the photometric data and conclusions in it are erroneous due to varying degrees of saturation of the images of the target star. This paper describes the results of a study of the original data aimed at determining why the erroneous light curves were highly complex.

Editorial note: We concur with this retraction.

1. Introduction

The analysis of time series DSLR photometry on HD 121620 in 2019 led to the conclusion by the author that the star was variable (Axelsen 2020). However, the author's unpublished DSLR observations in 2020 and 2021 showed that the star was constant, with an average magnitude of 7.06 in V. Consultation with Sebastián Otero about this "behavior" led to the suggestion by him that the images taken in 2019 may have been saturated. Investigation by the author confirmed that saturation of HD 121620 was evident in images from both the green and blue channels in line profile plots and in statistics showing peak ADUs. Images of the comparison and check stars were never saturated. Saturated images from 2019 were obtained through an 80-mm refractor, with an exposure time of 180 seconds. Non-saturated images from 2020 were obtained through the same refractor, but with a shorter exposure time of 60 seconds. Further non-saturated images of HD 121620 were captured in 2020 and 2021 through a 200-mm f/2.8 Canon camera lens, with the camera in a fixed position on a tripod.

The original data analysis comprised conversion of DSLR RAW images to FITS, debayering, and calibration with dark frames and flat fields in IRIS (Buil 1999–2021). When aperture photometry was subsequently carried out on the calibrated images in ASTROIMAGEJ (Collins *et al.* 2017), the tabulated output of data showed peculiar negative peak values, disguising the fact of saturation. The author has since found that such spurious values for peak ADUs can be avoided by using IRIS only for conversion to FITS, debayering, and channel separation. Calibration with darks and flats and aperture photometry subsequently carried out in ASTROIMAGEJ will yield accurate data, including the tabulation of recognizable peak ADU values that indicate the presence of saturation.

After it became apparent that images of HD 121620 from 2019 contained saturated pixels, it was recognized that a particular point of interest was the complexity of the light curves as shown in Figures 1 and 2 of the retracted paper. The images from 2019 were therefore re-analyzed.

2. Methods

To investigate the error, all original DSLR RAW images were re-processed with AIP4WIN (Berry 2020) for both calibration and aperture photometry. Plots of peak ADU values for HD 121620 against JD were made for calibrated data from the green and blue channels for all nights of observation. Line profile plots of images of HD 121620 were made in ASTROIMAGEJ for representative non-calibrated images selected after studying the plots of peak ADU values. ASTROIMAGEJ, not AIP4WIN, was used to create the line profiles because the ASTROIMAGEJ output, black lines and text on a white background, is more suitable for figures in publications. Plots of total ADU counts versus JD were drawn for HD 121620, the comparison star, and the check star for selected nights to investigate the apparent "flaring" of the target star.

3. Results and discussion

Figure 1 shows the light curves of HD 121620 and the check star from selected panels of Figure 2 in Axelsen (2020). Each panel represents time series photometry taken through one night. The panels were chosen to reflect the range of "behaviors" of HD 121620. The top panel, for the night of 27–28 May 2019, shows an ascending light curve for HD 121620. Near the beginning there is a temporary, apparent brightening interpreted by the author as a flare. In this panel and the others, the light curve of the check star is horizontal. The middle panel, for the night of 11–12 June, shows slight apparent brightening of HD D121620 for the first one-third of the observing run. There is a near-horizontal curve for the remainder of the night, with an average magnitude of 7.07 in V. The lower panel, for the night of 13–14 June, shows a complex, descending light curve.

To investigate these light curves, images from selected nights were analyzed by examining peak and total ADU values and line profile plots of HD 121620.

The analyses revealed that at least some degree of saturation of the images of HD 121620 occurred every observing night, and that the severity of the saturation (a reflection of the number of pixels saturated) varied between nights, across time within a night, and between the green and blue channels. It is standard practice to defocus images for DSLR photometry, to spread the light from each star across many pixels of the Bayer matrix. The factor that determined differences in the severity of saturation from night to night was the degree of defocussing of the images, with the most severe saturation occurring on the nights when the images were taken closer to focus. On the night of the 13–14 June, when no saturation occurred in either the green or the blue channel during the latter part of the night, the degree of defocus was most pronounced.

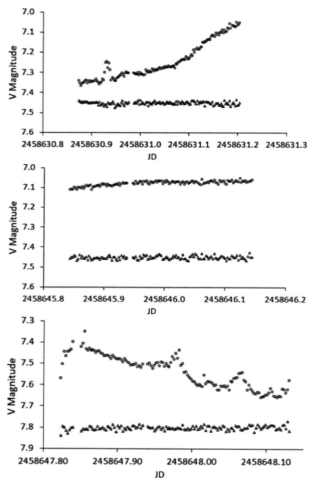

Figure 1. Light curves of HD 121620 and the check star for selected nights in 2019 from Figure 2 of Axelsen (2020), namely, 27–28 May (top panel), 11–12 June (middle panel), and 13–14 June (bottom panel)..

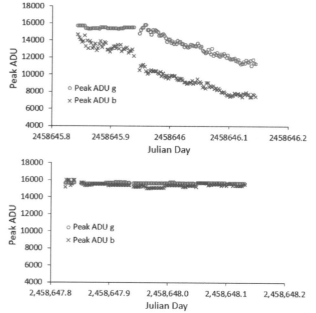

Figure 2. Peak ADU values of HD 121620 for the green (g) and blue (b) channels in calibrated images from the same nights shown in the middle and bottom panels in Figure 1 (the top panel of Figure 1 is discussed toward the end of the paper).

Figure 2 exemplifies these artefacts, showing peak ADU values for HD 121620 from the green and blue channels for the nights of 11–12 June and 13–14 June. Where the plot is a near-horizontal line, at least some pixels are saturated. The author's Canon EOS 500D DSLR camera is routinely set to ISO 400 when taking images for photometry, since the gain is then close to unity. At this setting saturation of pixels occurs at 15,761 ADUs. The saturated regions of the plots in Figure 2 are near-horizontal lines close to this value. The lines are not uniformly at 15,761 because the plots are of data from calibrated images. In Figure 2, saturation of green pixels is evident in the early part of the night of 11–12 June, and for the entire night of 13–14 June. There appear to be no saturated blue pixels on 11–12 June, but blue pixel saturation appears to be present for the entire night of 13–14 June.

Figures 3 and 4 show line profiles of HD 121620 from selected images from the nights of 11–12 June and 13–14 June, respectively. Each panel displays a line profile from one image. In each figure the top panels are from the green channel, and the bottom panels from the blue channel. The panels on the left side of each figure are from an image take early in the night, specifically, the fifth image from the start. The panels on the right side are from an image taken late in the observing run (after midnight), specifically, the fifth image from the end.

The line profiles in Figure 3 show no saturated pixels, although the graphs of peak ADUs for this night in Figure 2 would suggest that there was saturation of at least one or a few pixels in the green channel early in the night. The light curve for this night (Figure 1, middle panel) shows slight apparent brightening early in the night. Later, the light curve of HD 121620 is almost horizontal with an average magnitude in V of 7.07. Since neither the green channel nor the blue channel was saturated late in the observing run, the V light curve here almost certainly shows the true (constant) magnitude of the star. The line profiles of HD 121620 in Figure 4 show extensive green channel saturation both early and late, but only few saturated blue pixels, consistent with the lower panel of Figure 2, which indicates that at least some pixels were saturated in both green and blue channels throughout the entire night of 13–14 June. The presence of a complex light curve for this night presumably reflects variation in the numbers of pixels saturated across the night and non-linearity effects just below saturation. Although not illustrated here, the light curves of HD 121620 drawn from non-transformed green channel magnitudes are essentially the same as the light curves of transformed V magnitudes.

Figure 5 displays total ADUs for HD 121620 for the night of 27–28 May, during which a "flare" was believed by the author to have occurred early in the night (top panel, Figure 1). However, Figure 5 reveals that the fluctuation also affected the comparison and check stars, and was thus due to atmospheric disturbance. The fact that the fluctuation persisted in the final light curve of HD 121620 but not in the light curve of the check star could perhaps be attributed to non-linearity of data just below saturation.

A final point is that data from the All-Sky Automated Survey for Supernovae (ASAS-SN, http://www.astronomy.ohio-state.edu/~assassin/index.shtml) (Shappee et al. 2014; Kochanek et al. 2017) were interpreted by the author in the original paper

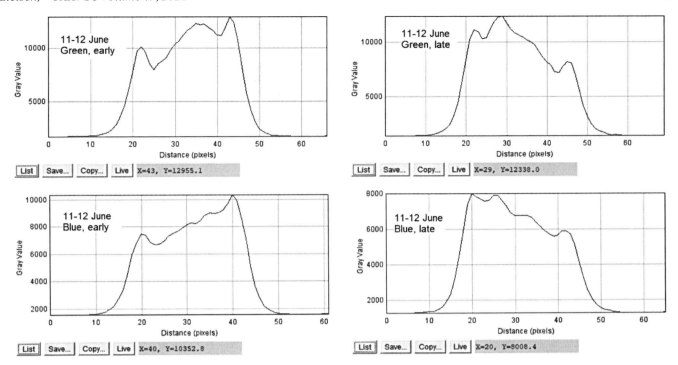

Figure 3. Line profiles of HD 121620 from AstroImageJ in selected uncalibrated images from 11–12 June 2019. Peak ADU values are shown in the grey box at the bottom of each panel. These values should be compared with the graph of peak ADUs in Figure 2 from the same date.

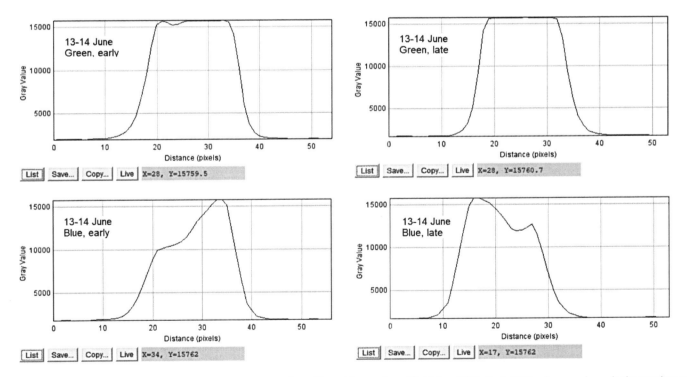

Figure 4. Line profiles of HD 121620 from *AstroimageJ* in selected uncalibrated images from 13–14 June 2019. Peak ADU values are shown in the grey box at the bottom of each panel. These values should be compared with the graph of peak ADUs in Figure 2 from the same date.

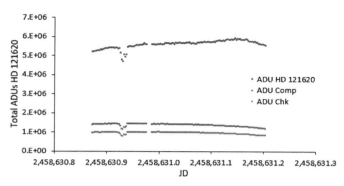

Figure 5. Total green channel ADUs of HD 121620, the comparison star, and the check star for the night of 27–28 May 2019. The light curve of HD 121620 for this night in Axelsen (2020) (reproduced in the upper panel of Figure 1 of this paper) showed a fluctuation in the early part of the light curve for HD 121620, interpreted by the author as a flare, as it was not present in the light curve of the check star. The plot of total ADUs from HD 121620 shown here demonstrates that the fluctuation was due to atmospheric disturbance, as it also affected the comparison and check stars.

to support the proposal that HD 121620 was variable. However, the ASAS-SN images typically saturate at a V magnitude of 10, and algorithms to extend this limit are unlikely to render accurate measurement of 7th magnitude stars.

4. Conclusion

The error reported herein emphasizes the well-known necessity to avoid saturation of images taken for photometry. It also reveals that, when the severity of saturation varies between images in time series studies, between different nights, and between the blue and green channels, the resulting light curves can be complex, rather than simply having a near-horizontal trajectory. Non-linear effects occurring in pixels near the point of saturation could also contribute to artefactual fluctuation in measured signal. These problems can be avoided by using line profiles of the images of stars to check for saturated pixels, and by checking the output from photometry software that tabulates peak ADU values. It is also important for observers to know the upper ADU limit of the linearity of their sensor, which can be determined from a series of images made with different exposure times.

5. Acknowledgements

The author wishes to thank Sebastián Otero for the insight that led to investigation of the erroneous data described above.

References

Axelsen, R. A. 2020, *J. Amer. Assoc. Var. Star Obs.*, **48**, 35.
Berry, R. 2020, Astronomical Image Processing for Windows, AIP4Win@groups.io | Files, Version 2.4.10 Installation File, unregistered version (https://groups.io/g/AIP4Win).
Buil, C. 1999–2021, IRIS An astronomical images processing software (http://www.astrosurf.com/buil/iris-software.html).
Collins, K. A., Kielkopf, J. F., Stassun, K. G., and Hessman, F.V. 2017, *Astron. J.*, **153**, 77.
Kochanek, C. S. *et al.* 2017, *Publ. Astron. Soc. Pacific*, **129**, 104502.
Shappee, B. J. *et al.* 2014, *Astrophys. J.*, **788**, 48.

CCD Photometry, Light Curve Modeling and Period Study of V1073 Herculis, a Totally Eclipsing Overcontact Binary System

Kevin B. Alton
Desert Blooms Observatory, 70 Summit Avenue, Cedar Knolls, NJ 07927; kbalton@optonline.net

John C. Downing
La Ventana Observatory, 28881 Sunset Road, Valley Center, CA 92082; johndowning2014@outlook.com

Received August 9, 2021; revised August 13, September 10, 2021; accepted September 13, 2021

Abstract Precise time-series multi-color (BVI_c) light curve data were acquired from V1073 Her at Desert Blooms Observatory (DBO) in 2020 and La Ventana Observatory (LVO) in 2021. New times of minimum from data acquired at DBO and LVO along with other eclipse timings extracted from selected surveys and the literature were used to generate an updated linear ephemeris. Secular analyses (eclipse timing differences vs. epoch) revealed changes in the orbital period of V1073 Her over the past 22 years. Along with an apparent increase in the orbital period, the residuals after a parabolic fit of the data indicated that there was an underlying sinusoidal-like variability. This behavior suggests the putative existence of a third gravitationally bound object or cycles in magnetic activity, both of which are addressed herein. Simultaneous modeling of multicolor light curve data during each epoch was accomplished using the Wilson-Devinney code. Since a total eclipse is observed, a unique photometrically derived value for the mass ratio (q_{ptm}) could be determined which subsequently provided initial estimates for the physical and geometric elements of V1073 Her.

1. Introduction

Sparsely sampled monochromatic photometric data for V1073 Her (= NSVS 8092487) were first captured during the ROTSE-I survey between 1999 and 2000 (Akerlof *et al.* 2000; Wozniak *et al.* 2004). These data were retrieved from the Northern Sky Variable Survey (NSVS) archives. Blättler and Diethelm (2000) produced a complete unfiltered light curve (LC) for this eclipsing binary star (GSC 2625-1563) along with the first linear ephemeris. Gettel *et al.* (2006) included V1073 Her in their catalog of bright contact binary stars from the ROTSE-I survey. Other sources for photometric data from this variable system include the Catalina Sky Survey (Drake *et al.* 2014), the All-Sky Automated Survey for SuperNovae (ASAS-SN, Shappee *et al.* 2014; Jayasinghe *et al.* 2018), and the SuperWASP Survey (Butters *et al.* 2010). Samec *et al.* (2014) reported the first multi-color (BVR_c and I_c) LCs from V1073 Her which were modeled using the Wilson-Devinney code (Wilson and Devinney 1971; Wilson 1979, 1990). At that time secular analyses suggested the presence of a third gravitationally bound object which the authors proposed to be a brown dwarf. Our investigation of this overcontact binary (OCB) also includes Roche modeling of CCD-derived LCs as well as an in-depth secular analysis of the predicted vs. observed eclipse timing differences (ETD) over the past 22 years.

2. Observations and data reduction

Precise time-series images were acquired at Desert Blooms Observatory (DBO; 31.941 N, 110.257 W) using a QSI 683 wsg-8 CCD camera mounted at the Cassegrain focus of a 0.4-m Schmidt-Cassegrain telescope. A Taurus 400 (Software Bisque) equatorial fork mount facilitated continuous operation without the need to perform a meridian flip. The image (science, darks, and flats) acquisition software (TheSkyX Pro Edition 10.5.0; Software Bisque 2019) controlled the main and integrated guide cameras.

This focal-reduced (f/7.2) instrument produces an image scale of 0.76 arcsec/pixel (bin = 2×2) and a field-of-view (FOV) of 15.9×21.1 arcmin. Computer time was updated immediately prior to each session and exposure time for all images set to 75 s.

The equipment at La Ventana Observatory (LVO; 33.2418 N, 116.9781 W) included an iOptron CEM60 mount with an SBIG Aluma CCD694 camera installed at the Cassegrain focus of a 0.235-m Schmidt-Cassegrain telescope. TheSkyX Pro Edition 10.5.0 controlled the main (30-s exposures) and integrated guide cameras during image acquisition (science, darks, and flats). This focal-reduced (f/7) instrument produces an image scale of 1.14 arcsec/pixel (bin = 2×2) and a field-of-view (FOV) of 26.1×20.9 arcmin.

Both CCD cameras were equipped with Astrodon B, V, R_c, and I_c filters manufactured to match the Johnson-Cousins Bessell specification. Dark subtraction, flat correction, and registration of all images collected at DBO and LVO were performed with AIP4Win v2.4.0 (Berry and Burnell 2005).

Instrumental readings from V1073 Her were reduced to catalog-based magnitudes using APASS DR9 values (Henden *et al.* 2009, 2010, 2011; Smith *et al.* 2011) built into MPO Canopus v 10.7.1.3 (Minor Planet Observer 2010). LC data acquired in 2021 at LVO were only used to supplement ToM values for secular analyses.

3. Results and discussion

Light curves were generated using an ensemble of four comparison stars, each of which remained constant ($< \pm 0.01$ mag) throughout every imaging session. The identity, J2000 coordinates, and color indices (B–V) for these stars are provided in Table 1. A CCD image annotated with the location of the target (T) and comparison stars (1–4) is shown in Figure 1.

Table 1. Astrometric coordinates (J2000), V-mags and color indices (B–V) for V1073 Her (Figure 1), and the corresponding comparison stars used in this photometric study.

Star Identification	R.A. (J2000)	Dec. (J2000)	V-mag[b]	(B–V)[b]
(T) V1073 Her	18 08 35.7571	+33 42 04.755	11.449	0.950
(1) GSC 2629-1797	18 08 55.4755	+33 45 45.118	11.341	0.863
(2) GSC 2629-1443	18 08 39.3063	+33 48 13.847	10.807	1.044
(3) GSC 2625-1672	18 08 22.8761	+33 38 25.825	11.343	0.813
(4) GSC 2625-1752	18 08 41.6918	+33 41 52.148	13.227	0.962

a. R.A. and Dec. from Gaia DR2 (Gaia Collab. et al. 2016, 2018).
b. V-mag and (B–V) for comparison stars derived from APASS DR9 database described by Henden et al. (2009, 2010, 2011) and Smith et al. (2011).

Table 2. Summary of image acquisition dates, number of data points and estimated uncertainty (± mag) in each bandpass (BVI_c) used for the determination of ToM values and/or Roche modeling.

n	Filter	(± mag)	Location	Dates
263	B	0.008	DBO	June 23–June 30, 2020
261	V	0.004	DBO	June 23–June 30, 2020
259	I_c	0.005	DBO	June 23–June 30, 2020
360	V	0.002	LVO	July 14–July 20, 2021

Table 3. Sample table of V1073 Her times-of-minimum (March 21, 1999–July 20, 2021), cycle number and residuals (ETD) between observed and predicted times derived from the updated linear ephemeris (Equation 1).

HJD–2400000	HJD Error	Cycle No.	ETD[a]	Reference
51258.8894	0.0008	–27718	0.0058	1
51277.8726	0.0004	–27653.5	0.0079	1
51746.3660	0.0120	–26061.5	0.0046	2
51746.5125	0.0008	–26061	0.0039	1
51768.4372	0.0003	–25986.5	0.0046	2

a. ETD = Eclipse Time Difference.
1. Blättler and Diethelm (2000); 2. Blättler et al. (2000). Full table available at: ftp://ftp.aavso.org/public/datasets/492-Alton-V1073Her.txt .
All references relevant to the full table that appears on the AAVSO ftp site are included in the References section of this article.

Table 4. Orbital period modulation (P_3) and putative third-body solution to the light-time effect (LiTE) observed from changes in V1073 Her eclipse timings.

Parameter	Units	LiTE[a]
HJD_0–2400000		2451258.8899 ± 0.0007
P_3	(y)	13.678 ± 0.259
A (semi-ampl.)	(d)	0.00309 ± 0.00042
ω	(°)	0
e_3		0 ± 0.1
a'_{12} sin i'	(a.u.)	0.5349 ± 0.0733
$f(M_3)$ (mass func.)	($M_\odot$)	0.00082 ± 0.00004
M_3 (i = 90°)	($M_\odot$)	0.126 ± 0.002
M_3 (i = 60°)	($M_\odot$)	0.146 ± 0.003
M_3 (i = 30°)	($M_\odot$)	0.266 ± 0.005
Q (quad. coeff.)	(10^{-10})	0.3511 ± 0.0001
Sum of squared residuals		0.002542

a. Zasche et al. (2009)—simplex optimization with third body circular orbit.

Table 5. Estimation of effective temperature (T_{eff1}) of the primary star in V1073 Her.

Parameter	V1073 Her
DBO $(B–V)_0$	0.927 ± 0.028
Median combined $(B–V)_0$[a]	0.916 ± 0.023
Galactic reddening E(B–V)[b]	0.0337 ± 0.0001
Survey T_{eff1}[c] (K)	4990 ± 47
Gaia T_{eff1}[d] (K)	5506^{+583}_{-264}
Houdashelt T_{eff1}[e] (K)	5002 ± 360
Median T_{eff1} (K)	5166 ± 201
Spectral Class[f]	K1V-K2V

a. Surveys and DBO intrinsic $(B–V)_0$ determined using reddening values (E(B–V)).
b. Model A (http://www.galextin.org/).
c. T_{eff1} interpolated from median combined $(B–V)_0$ using Table 4 in Pecaut and Mamajek (2013).
d. Values from Gaia DR2 (Gaia Collab. et al. 2016, 2018) (http://vizier.u-strasbg.fr/viz-bin/VizieR?-source=I/345/gaia2).
e. Values calculated with Houdashelt et al. (2000) empirical relationship.
f. Spectral class estimated from Pecaut and Mamajek (2013).

Table 6. Light curve parameters evaluated by Roche modeling and the geometric elements derived for V1073 Her (2019) assuming it is a W-type W UMa variable.

Parameter	DBO No spot	DBO Spotted	Samec et al. (2014) Spotted
T_{eff1} (K)[b]	5166	5166	5150
T_{eff2} (K)	5317 (3)	5296 (3)	5176 (1)
q (m_2/m_1)	0.379 (1)	0.386 (3)	0.404 (4)
A[b]	0.50	0.50	0.50
g[b]	0.32	0.32	0.32
$\Omega_1 = \Omega_2$	2.612 (3)	2.627 (3)	2.640 (4)
i°	84.1 (3)	83.5 (2)	82.3 (1)
$A_P = T_S / T_{star}$[c]	—	0.90 (2)	—
Θ_P (spot co-latitude)[c]	—	104 (5)	—
Φ_P (spot longitude)[c]	—	199 (5)	—
r_P (angular radius)[c]	—	9.0 (2)	—
$A_S = T_S / T_{star}$[c]	—	—	0.861 (5)
Θ_S (spot co-latitude)[c]	—	—	93 (2)
Φ_S (spot longitude)[c]	—	—	244 (1)
r_S (angular radius)[c]	—	—	21.8 (3)
$L_1/(L_1+L_2)$B[d]	0.6649 (2)	0.6674 (2)	0.652 (1)
$L_1/(L_1+L_2)$V	0.6749 (2)	0.6760 (1)	0.658 (1)
$L_1/(L_1+L_2)I_c$	0.6845 (1)	0.6843 (2)	0.663 (1)
$L_1/(L_1+L_2)R_c$	—	—	0.660 (1)
r_1 (pole)	0.4415 (4)	0.4398 (4)	0.440 (1)
r_1 (side)	0.4728 (5)	0.4708 (4)	0.472 (2)
r_1 (back)	0.5009 (6)	0.4989 (5)	0.502 (2)
r_2 (pole)	0.2831 (12)	0.2842 (11)	0.292 (1)
r_2 (side)	0.2957 (15)	0.2969 (13)	0.306 (2)
r_2 (back)	0.3317 (26)	0.3327 (23)	0.345 (4)
Fill-out factor (%)	10	9.5	18
RMS (B)[e]	0.00848	0.00871	—
RMS (V)	0.00789	0.00749	—
RMS (I_c)	0.00658	0.00603	—

a. All DBO uncertainty estimates for T_{eff2}, q, $\Omega_{1,2}$, i, $r_{1,2}$, and L_1 from WDWINT56A (Nelson 2009).
b. Fixed with no error during DC.
c. Spot parameters in degrees ($\Theta_{P,S}$, $\Phi_{P,S}$, and $r_{P,S}$) or $A_{P,S}$ in fractional degrees (K).
d. L_1 and L_2 refer to scaled luminosities of the primary and secondary stars, respectively.
e. Monochromatic residual mean square error from observed values.

Table 7. Fundamental stellar parameters for V1073 Her using the photometric mass ratio ($q_{ptm} = m_2/m_1$) from the spotted Roche model fits of LC data (2020) and the estimated primary star mass based on empirically derived M-PRs for overcontact binary systems.

Parameter	Primary	Secondary
Mass ($M_\odot$)	1.026 ± 0.017	0.396 ± 0.006
Radius ($R_\odot$)	0.970 ± 0.004	0.629 ± 0.003
a ($R_\odot$)	2.094 ± 0.009	2.094 ± 0.009
Luminosity ($L_\odot$)	0.603 ± 0.094	0.280 ± 0.002
Mbol	5.299 ± 0.009	6.132 ± 0.009
Log (g)	4.476 ± 0.008	4.439 ± 0.008

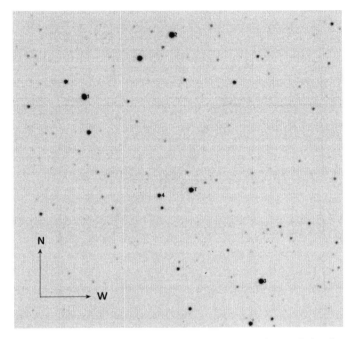

Figure 1. CCD image (V mag) of V1073 Her (T) acquired at DBO showing the location of comparison stars (1–4) used to generate APASS DR9-derived magnitude estimates.

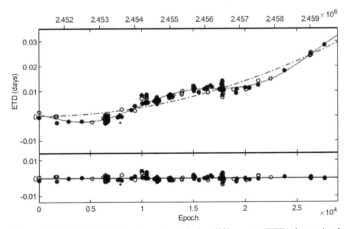

Figure 2. LiTE fit (Table 4) using eclipse timing differences (ETD) determined for V1073 Her between 1999 and 2021. The solid red line in the top panel describes the fit for a circular (e = 0) orbit (P_3 = 13.678 ± 0.259 y) of a putative third body while the dashed blue line defines the quadratic fit from the eclipse timing residuals. Solid circles (●) represent times at Min I whereas open circles (○) indicate times at Min II. The bottom panel illustrates the total residuals remaining after LiTE analysis after subtracting out the quadratic component.

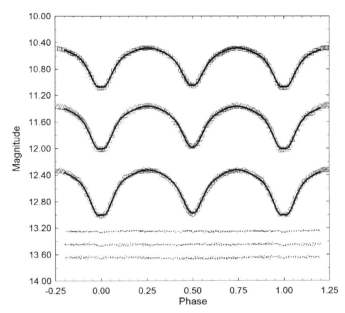

Figure 3. Period folded (0.2942818 ± 0.0000001 d) CCD-derived LCs for V1073 Her produced from photometric data collected at DBO between June 23, 2020 and June 30, 2020 The top (I_c), middle (V), and bottom curve (B) were transformed to magnitudes based on APASS DR9-derived catalog values from comparison stars. In this case, the Roche model assumed a W-subtype overcontact binary with single spot on the most massive star; residuals from the model fits are offset at the bottom of the plot to keep the values on scale.

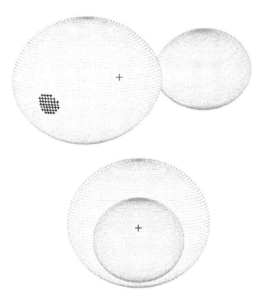

Figure 4. Three-dimensional spatial model of V1073 Her during 2020 illustrating (top) the location of a cool (black) spot on the primary star and (bottom) the secondary star transit across the primary star face at Min II (Φ = 0.5).

Only data acquired above 30° altitude (airmass < 2.0) were included; differential atmospheric extinction was ignored considering the close proximity of all program stars. All photometric data can be retrieved from the AAVSO International Database via the International Variable Star Index (Watson *et al.* 2014).

3.1. Photometry and ephemerides

Times of minimum (ToM) and associated errors were calculated using the method of Kwee and van Woerden (1956) as implemented in Peranso v2.5 (Paunzen and Vanmunster 2016). Curve fitting all eclipse timing differences (ETD) was accomplished using scaled Levenberg-Marquardt algorithms (QtiPlot 0.9.9-rc9; IONDEV SRL 2021). Photometric uncertainty was calculated according to the so-called "CCD Equation" (Mortara and Fowler 1981; Howell 2006). The acquisition dates, number of data points, and uncertainty for each bandpass used for the determination of ToM values and/or Roche modeling are summarized in Table 2.

Six new ToM measurements were extracted from photometric data acquired at DBO and LVO. The SuperWASP survey (Butters *et al.* 2010) provided a wealth of photometric data taken (30-s exposures) at modest cadence that repeats every 9 to 12 min. In some cases (n = 68) these SuperWASP data were amenable to further analysis using the method of Kwee and van Woerden (1956) to estimate ToM values. These, along with 156 other eclipse timings (Table 3) from the literature, were used to calculate a new linear ephemeris based on data produced between 1999 and 2021:

$$\text{Min. I (HJD)} = 2459415.7874(4) + 0.2942818(1) \text{E}. \quad (1)$$

For the purpose of illustration the first five entries in Table 3 are provided herein; all values will be web-archived and made available through the AAVSO ftp site at:

ftp:ftp.aavso.org/public/datasets/492-Alton-V1073Her.txt.

A sinusoidal-like variation was found embedded within the residuals remaining after the initial linear fit (Figure 2). Cyclic changes of eclipse timings can result from the gravitational influence of unseen companion(s), the so-called light-time effect (LiTE). It is not unreasonable to propose that V1073 Her is a ternary system since a significant number (> 50%) of contact binaries observed from the Northern Hemisphere exist as multiple systems (Pribulla and Ruciński 2006). To address this possibility LiTE analyses were performed using the simplex code for MATLAB® reported by Zasche *et al.* (2009).

A quadratic relationship between ETD and epoch takes the general form:

$$\text{ETD}_{\text{fitted}} = c + b \cdot E + Q \cdot E^2 + \tau. \quad (2)$$

When the orbital period change is monotonic, the last term ($\tau = 0$) can be ignored. However, in this case τ from Equation 2 is expanded as follows:

$$\tau = \frac{a_{12} \sin i_3}{c} \left[(1 - e^2) \frac{\sin(v + \omega)}{1 + e \cdot \cos v} + \sin \omega \right] \quad (3)$$

Accordingly, the associated parameters in the LiTE equation (Irwin 1959) were derived, which include parameter values for P_3 (orbital period of star 3 and the 1–2 pair about the barycenter), orbital eccentricity e, argument of periapsis ω, true anomaly v, time of periastron passage T_0, and amplitude $A = a_{12} \sin i_3$. In this case a_{12} is the semi-major axis of the 1–2 pair's orbit about the three-star system center of mass, and i_3 is the orbital inclination of the putative third body in a three-star system.

A single best fit to all the ETD residuals was produced with the Zasche *et al.* (2009) LiTE code using simplex optimization (Table 4). These results are consistent with a putative third body in a circular orbit ($P_3 = 13.678$ y) at a distance no farther than 5.57 ± 0.16 A.U. from the barycenter. The minimum mass of a coplanar ($i_3 = 90°$) orbiting third body was calculated to be $\sim 0.126 \pm 0.002 \, M_\odot$, based on the derived mass function ($f(M_3) = 0.00082 \pm 0.00004$). A brown dwarf is expected to have a mass less than $0.075 \, M_\odot$, therefore our LiTE model results do not support earlier speculation by Samec *et al.* 2014 about an orbiting brown dwarf. The corresponding added luminosity (L_3) of a third main sequence star ($M < 0.126 \, M_\odot$) was estimated to be $\approx 0.22\%$ according to:

$$L_3 \, (\%) \approx \frac{100 \cdot 0.23 \, M_{\min}^{2.3}}{L_1 + L_2 + 0.23 \, M_{\min}^{2.3}}, \quad (4)$$

where $M_{\min}$ is the estimated minimum mass ($i_3 = 90°$) in solar units. This very small percent contribution of light would not be expected nor did it require adjustment by the WD 2003 code third light parameter ($l_3 = 0$) in order to accurately simulate the LC model fits around minimum light (section 3.4).

Modulated changes in the orbital period can also result from magnetic activity cycles attributed to Applegate (1992) or apsidal motion of a binary pair. Since contact binary systems are tidally locked with circular orbits, apsidal motion can be immediately eliminated from consideration. Short-period binaries are magnetically very active due to the formation of photospheric (starspots), chromospheric (plages), and other high energy disturbances (Berdyugina 2005). The corresponding hydromagnetic dynamo can produce changes in the gravitational quadrupole moment of the active star via redistribution of the internal angular momentum with corresponding changes in the magnetic torque within the stellar convective zone. When the gravitational quadrupole moment of the active component increases, its companion experiences a stronger gravitational force which then moves closer to the system barycenter. The orbital period will decrease according to this scenario. By contrast, when the gravitational quadrupole moment of the active star weakens, the orbital period increases. A detailed examination of the energetics ($\Delta E / E_{\text{sec}}$) required to produce the "Applegate effect" was performed according to Völschow *et al.* (2016) and the accompanying "Eclipse Time Variation Calculator" webmodule[1]. $\Delta E / E_{\text{sec}}$ is defined as the energy required to drive the Applegate mechanism divided by the available energy produced in the magnetically active star. This value determines whether the Applegate mechanism is energetically feasible. Solutions are provided from the two-zone model and the constant density model by Völschow *et al.* (2016), along with a solution based on the thin-shell model

---

[1] http://theory-starformation-group.cl/applegate/index.php

by Tian *et al.* (2009). Tian *et al.* (2009) derived a relationship between the energetics necessary to drive the Applegate mechanism and the observed variability in eclipse timings:

$$\frac{\Delta E}{E_{sec}} = 0.233 \cdot \left(\frac{M_{sec}}{M_\odot}\right)^3 \cdot \left(\frac{R_{sec}}{R_\odot}\right)^{-10} \cdot \left(\frac{T_{sec}}{6000\,K}\right)^{-4} \cdot$$
$$\left(\frac{a_{bin}}{R_\odot}\right)^4 \cdot \left(\frac{\Delta P}{s}\right)^2 \cdot \left(\frac{P_{mod}}{yr}\right)^{-1}. \quad (5)$$

The measureables in this case include the secondary mass (M_{sec}), radius (R_{sec}), temperature (T_{sec}), semi-major axis of the binary pair (a_{bin}), the modulation period of the binary pair (P_{mod}), and ΔP where:

$$\frac{\Delta P}{P_{bin}} = 2\pi \left(\frac{O-C}{P_{mod}}\right). \quad (6)$$

Since the $\Delta E/E_{sec}$ value (0.62) is less than one, this would energetically favor orbital period modulations that arise from the Applegate mechanism.

The two-zone model provides two solutions, one requiring more energy and one requiring less energy. Therein the finite shell two-zone model accounts for all essential physics involved with the Applegate effect from main-sequence low mass companions (0.1–0.6 $M_\odot$). Accordingly, the latter energy solution is:

$$\frac{\Delta E^-}{E_{sec}} = k_1 \cdot \frac{M_{sec} R_{sec}^2}{P_{bin}^2 P_{mod} L_{sec}} \cdot$$
$$\left(1 \pm \sqrt{(1-k_2 G) \frac{a_{bin}^2 M_{sec} P_{bin}^2}{R_{sec}^5} \frac{\Delta P}{P_{bin}}}\right)^2, \quad (7)$$

wherein k_1 is assigned a value of 0.133 and k_2 is 3.42. Since the calculated value for $\Delta E/E_{sec}$ is less than unity (0.034), this model also indicates that V1073 Her is a potential candidate for orbital period modulation by magnetic cycles.

The apparent sinusoidal-like behavior is supported by data collected over the past 22 y, which is less than two cycles of orbital period variation. Therefore, some caution should be exercised in that these findings are considered preliminary and not a definitive solution. Furthermore at this time it is not possible to firmly establish whether the gravitational effect of a third body or variations in the quadrapole moment is responsible for cyclic changes in the orbital period of V1073 Her. Unfortunately without other supporting evidence such as might be derived from space-based spectro-interferometry and/or direct imaging, secular analysis still leaves us with two equally plausible but distinctly different phenomenological origins for cyclic modulation of the dominant orbital period.

3.2. Effective temperature estimation

The effective temperature (T_{eff1}) of the more massive, and therefore most luminous component (defined as the primary star herein) was derived from a composite of photometric (USNO-B1, UCAC4, 2MASS, and APASS) determinations that were as necessary transformed to (B–V)[2,3]. Interstellar extinction (A_V) and reddening (E(B–V)=A_V/3.1) were estimated for targets within the Milky Way Galaxy according to Amôres and Lépine (2005). These models[4] require the Galactic coordinates (l, b) and the distance in kpc estimated from Gaia DR2 derived parallax (Bailer-Jones 2015). After subtracting out reddening to arrive at a value for intrinsic color, (B–V)$_0$, T_{eff1} estimates were interpolated for each system using the values reported for main sequence dwarf stars by Pecaut and Mamajek (2013). Additional sources used to establish a median value for each T_{eff1} included the Gaia DR2 release of stellar parameters (Andrae *et al.* 2018), and an empirical relationship (Houdashelt *et al.* 2000) based on intrinsic color where $0.32 \leq (B-V)_0 \leq 1.35$. The median result ($T_{eff1}=5166\pm201$ K), summarized in Table 5, was adopted for Roche modeling of LCs from V1073 Her.

3.3. Roche modeling approach

Roche modeling of LC data from 2020 was initially performed with PHOEBE 0.31a (Prša and Zwitter 2005) and then refined using WDWINT56A (Nelson 2009). Both programs feature a MS Windows-compatible GUI interface to the Wilson-Devinney WD2003 code (Wilson and Devinney 1971; Wilson 1979, 1990). WDWINT56A incorporates Kurucz's atmosphere models (Kurucz 2002) that are integrated over BVI$_c$ passbands. The final selected model was Mode 3 for an overcontact binary; other modes (detached and semi-detached) never approached the best fit value achieved with Mode 3. Modeling parameters were adjusted as follows. The internal energy transfer to the stellar surface is driven by convective (7500 K) rather than radiative processes. As a result, the value for bolometric albedo ($A_{1,2}=0.5$) was assigned according to Ruciński (1969) while the gravity darkening coefficient ($g_{1,2}=0.32$) was adopted from Lucy (1967). Logarithmic limb darkening coefficients (x_1, x_2, y_1, y_2) were interpolated (van Hamme 1993) following any change in the effective temperature (T_{eff2}) of the secondary star during model fit optimization using differential corrections (DC). All but the temperature of the more massive star (T_{eff1}), $A_{1,2}$ and $g_{1,2}$ were allowed to vary during DC iterations. In general, the best fits for T_{eff2}, i, q and Roche potentials ($\Omega_1=\Omega_2$) were collectively refined (method of multiple subsets) by DC using the multicolor LC data until a simultaneous solution was found. Surface inhomogeneity often attributed to star spots was simulated by the addition of a cool spot on the primary star to obtain the best fit LC models around Min I. V1073 Her did not require third light correction ($l_3=0$) to improve Roche model fits.

3.4. Roche modeling results

Without radial velocity (RV) data it is generally not possible to unambiguously determine the mass ratio, subtype (A or W), or total mass of an eclipsing binary system. Nonetheless, since a total eclipse is observed, a unique mass ratio value could be found (Terrell and Wilson 2005). Standard errors reported in Table 6 are computed from the DC covariance matrix and only reflect the model fit to the observations which assume exact values for any fixed parameter. These errors are generally

---

[2] http://www.aerith.net/astro/color_conversion.html; [3] http://brucegary.net/dummies/method0.html; [4] http://www.galextin.org

regarded as unrealistically small considering the estimated uncertainties associated with the mean adopted T_{eff1} values along with basic assumptions about $A_{1,2}$, $g_{1,2}$, and the influence of spots added to the Roche model. Normally, the value for T_{eff1} is fixed with no error during modeling with the WD code despite measurement uncertainty which can approach 10% relative standard deviation (R.S.D.) without supporting high resolution spectral data. The effect that such uncertainty in T_{eff1} would have on modeling estimates for q, i, $\Omega_{1,2}$, and T_{eff2} has been investigated with other OCBs, including A- (Alton 2019; Alton *et al.* 2020) and W-subtypes (Alton and Nelson 2018). As might be expected any change in the fixed value for T_{eff1} results in a corresponding change in the T_{eff2}. These findings are consistent whereby the uncertainty in the model fit for T_{eff2} would be essentially the same as that established for T_{eff1}. Furthermore, varying T_{eff1} by as much as 10% did not appreciably affect the uncertainty estimates (R.S.D. < 2.2%) for i, q, or $\omega_{1,2}$ (Alton 2019; Alton and Nelson 2018; Alton *et al.* 2020). Assuming that the actual T_{eff1} value falls within 10% of the adopted values used for Roche modeling (a reasonable expectation based on T_{eff1} data provided in Table 5), then uncertainty estimates for i, q, or $\Omega_{1,2}$, along with spot size, temperature, and location, would likely not exceed 2.2% R.S.D.

The fill-out parameter (f) which corresponds to the outer surface shared by each star was calculated according to Equation 8 (Kallrath and Milone 2009; Bradstreet 2005) where:

$$f = (\Omega_{inner} - \Omega_{1,2}) / (\Omega_{inner} - \Omega_{outer}), \quad (8)$$

wherein Ω_{outer} is the outer critical Roche equipotential, Ω_{inner} is the value for the inner critical Roche equipotential, and $\Omega = \Omega_{1,2}$ denotes the common envelope surface potential for the binary system. In all cases the systems are considered overcontact since $0 < f < 1$.

LC parameters, geometric elements, and their corresponding uncertainties are summarized in Table 6. According to Binnendijk (1970) the deepest minimum (Min I) of a W-type overcontact system occurs when a cooler more massive constituent occludes its hotter but less massive binary partner. The flattened-bottom dip in brightness at Min I (Figure 3) indicates a total eclipse of the secondary star; therefore, WD modeling proceeded under the assumption that V1073 Her is a W-subtype. Since according to the convention used herein whereby the primary star is the most massive ($m_2/m_1 \leq 1$), a phase shift (0.5) was introduced to properly align the LC for subsequent Roche modeling. Except for spot parameters and the fill-out factors, this investigation and that conducted by Samec *et al.* (2014) provided modeling results for V1073 Her that compare quite favorably (±5%).

Spatial renderings (Figure 4) were produced with Binary Maker 3 (BM3; Bradstreet and Steelman 2004) using the final WDWint56a modeling results from 2020. The smaller secondary can be envisioned to completely transit across the primary face during Min II ($\Phi = 0.5$), thereby confirming that the secondary star is totally eclipsed at Min I.

3.5. Preliminary stellar parameters

Mean physical characteristics were estimated for V1073 Her (Table 7) using results from the best fit (spotted) LC simulations from 2020. It is important to note that without the benefit of RV data which define the orbital motion, mass ratio, and total mass of the binary pair, these results should be considered "relative" rather than "absolute" parameters and regarded as preliminary.

Calculations are described below for estimating the solar mass and size, semi-major axis, solar luminosity, bolometric V-mag, and surface gravity of each component. Three empirically derived mass-period relationships (M-PR) for W UMa binaries were used to estimate the primary star mass. The first M-PR was reported by Qian (2003), while two others followed, from Gazeas and Stępień (2008) and then Gazeas (2009). According to Qian (2003), when the primary star is less than 1.35 $M_\odot$ or the system is W-type its mass can be determined from:

$$\log(M_1) = 0.391\,(59) \cdot \log(P) + 1.96\,(17), \quad (9)$$

where P is the orbital period in days and leads to $M_1 = 0.968 \pm 0.077\,M_\odot$ for the primary. The M-PR derived by Gazeas and Stępień (2008):

$$\log(M_1) = 0.755\,(59) \cdot \log(P) + 0.416\,(24), \quad (10)$$

corresponds to an OCB system where $M_1 = 1.035 \pm 0.094\,M_\odot$. Gazeas (2009) reported another empirical relationship for the more massive (M_1) star of a contact binary such that:

$$\log(M_1) = 0.725\,(59) \cdot \log(P) - 0.076\,(32) \cdot \log(q) + 0.365\,(32). \quad (11)$$

from which $M_1 = 1.026 \pm 0.069\,M_\odot$. The median of three values ($M_1 = 1.026 \pm 0.010\,M_\odot$) estimated from Equations 9–11 is higher than what might be expected (0.85 $M_\odot$) for a K1-K2V star. Notwithstanding, the median value was used for subsequent determinations of M_2, semi-major axis a, volume-radii r_L, and bolometric magnitudes (M_{bol}) using the formal errors calculated by WDWint56a (Nelson 2009). The secondary mass = $0.396 \pm 0.006\,M_\odot$ and total mass ($1.422 \pm 0.018\,M_\odot$) were determined using the mean photometric mass ratio ($q_{ptm} = 0.386 \pm 0.001$) derived from the best fit (spotted) models.

The semi-major axis, $a(R_\odot) = 2.094 \pm 0.009$, was calculated from Newton's version of Kepler's third law where:

$$a^3 = (G \cdot P^2\,(M_1 + M_2)) / (4\pi^2). \quad (12)$$

The effective radius of each Roche lobe (r_L) can be calculated over the entire range of mass ratios ($0 < q < \infty$) according to an expression derived by Eggleton (1983):

$$r_L = (0.49 q^{2/3}) / (0.6 q^{2/3} + \ln(1 + q^{1/3})), \quad (13)$$

from which values for r_1 (0.4631 ± 0.0003) and r_2 (0.3003 ± 0.0002) were determined for the primary and secondary stars, respectively. Since the semi-major axis and the volume radii are known, the radii in solar units for both binary components can be calculated where $R_1 = a \cdot r_1 = 0.970 \pm 0.004\,R_\odot$ and $R_2 = a \cdot r_2 = 0.629 \pm 0.003\,R_\odot$.

Luminosity in solar units ($L_\odot$) for the primary (L_1) and secondary stars (L_2) was calculated from the well-known relationship derived from the Stefan-Boltzmann law (Equation 14) where:

$$L_{1,2} = (R_{1,2}/R_\odot)^2 (T_{1,2}/T_\odot)^4. \quad (14)$$

Assuming that $T_{eff1} = 5166$ K, $T_{eff2} = 5296$ K, and $T_\odot = 5772$ K, then the solar luminosities ($L_\odot$) for the primary and secondary are $L_1 = 0.603 \pm 0.094$ and $L_2 = 0.280 \pm 0.002$, respectively.

4. Conclusions

Six new times of minimum for V1073 Her based on multicolor CCD data were determined from LCs acquired at two different locations in 2020 and 2021. These, along with other values (n=68) extrapolated from the SuperWASP survey (Butters *et al.* 2010), led to an updated linear ephemeris. At this time it is not possible to firmly establish whether the gravitational effect of a third body or variations in the quadrapole moment is responsible for cyclic changes in the eclipse timing residuals from V1073 Her. The adopted effective temperature ($T_{eff1} = 5166 \pm 201$ K) was based on a composite of sources that included values from photometric and astrometric surveys, and the Gaia DR2 release of stellar characteristics (Andrae *et al.* 2018). V1073 Her experiences a total eclipse from our vantage point which is evident as a flattened bottom during Min I, a characteristic of W-subtype variables. The photometric mass ratio ($q_{ptm} = 0.386 \pm 0.001$) determined by Roche modeling is expected to be a reliable substitute for a mass ratio derived from RV data. Nonetheless, spectroscopic studies (RV and high resolution classification spectra) will be required to unequivocally determine a total mass and spectral class for each system. Consequently, all parameter values and corresponding uncertainties reported herein should be considered preliminary.

5. Acknowledgements

This research has made use of the SIMBAD database operated at Centre de Données astronomiques de Strasbourg, France. In addition, the Northern Sky Variability Survey hosted by the Los Alamos National Laboratory (https://skydot.lanl.gov/nsvs/nsvs.php), the All-Sky Automated Survey for Supernovae (https://asas-sn.osu.edu/variables), Catalina Sky Survey (http://nesssi.cacr.caltech.edu/DataRelease/) and the International Variable Star Index (AAVSO) were mined for essential information. This work also presents results from the European Space Agency (ESA) space mission Gaia. Gaia data are being processed by the Gaia Data Processing and Analysis Consortium (DPAC). Funding for the DPAC is provided by national institutions, in particular the institutions participating in the Gaia MultiLateral Agreement (MLA). The Gaia mission website is https://www.cosmos.esa.int/gaia. The Gaia archive website is https://archives.esac.esa.int/gaia. This paper makes use of data from the first public release of the WASP data as provided by the WASP consortium and services at the NASA Exoplanet Archive, which is operated by the California Institute of Technology, under contract with the National Aeronautics and Space Administration under the Exoplanet Exploration Program. Many thanks to the anonymous referee whose valuable commentary led to significant improvement of this paper.

References

Akerlof, C., *et al.* 2000, *Astron. J.*, **119**, 1901.
Alton, K. B. 2019, *J. Amer. Assoc. Var. Star Obs.*, **47**, 7.
Alton, K. B., and Nelson, R. H. 2018, *Mon. Not. Roy. Astron. Soc.*, **479**, 3197.
Alton, K. B., Nelson, R. H., and Stępień, K. 2020, *J. Astrophys. Astron.*, **41**, 26.
Amôres, E. B., and Lépine, J. R. D. 2005, *Astron. J.*, **130**, 659.
Andrae, R., *et al.* 2018, *Astron. Astrophys.*, **616A**, 8.
Applegate, J. H. 1992, *Astrophys. J.*, **385**, 621.
Bailer-Jones, C. A. L. 2015, *Publ. Astron. Soc. Pacific*, **127**, 994.
Berdyugina, S. V. 2005, *Living Rev. Sol. Phys.*, **2**, 8.
Berry, R., and Burnell, J. 2005, *The Handbook of Astronomical Image Processing*, 2nd ed., Willmann-Bell, Richmond, VA.
Binnendijk, L. 1970, *Vistas Astron.*, **12**, 217.
Blättler, E., and Diethelm, R. 2000, *Inf. Bull. Var. Stars*, No. 4975, 1.
Blättler, E., *et al.* 2000, *BBSAG Bull.*, No. 123, 1.
Blättler, E., *et al.* 2001, *BBSAG Bull.*, No. 126, 1.
Blättler, E., *et al.* 2002, *BBSAG Bull.*, No. 128, 1.
Bradstreet, D. H. 2005, in *The Society for Astronomical Sciences 24th Annual Symposium on Telescope Science*, Society for Astronomical Sciences, Rancho Cucamonga, CA, 23.
Bradstreet, D. H., and Steelman, D. P. 2004, BINARY MAKER 3, Contact Software (http://www.binarymaker.com).
Brát, L., Zejda, M., and Svoboda, P. 2007, *Open Eur. J. Var. Stars*, **74**, 1.
Brát, L., *et al.* 2008, *Open Eur. J. Var. Stars*, **94**, 1.
Brát, L., *et al.* 2011, *Open Eur. J. Var. Stars*, **137**, 1.
Butters, O. W., *et al.* 2010, *Astron. Astrophys.*, **520**, L10.
Diethelm, R. 2003, *Inf. Bull. Var. Stars*, No. 5438, 1.
Diethelm, R. 2004, *Inf. Bull. Var. Stars*, No. 5543, 1.
Diethelm, R. 2006, *Inf. Bull. Var. Stars*, No. 5713, 1.
Diethelm, R. 2007, *Inf. Bull. Var. Stars*, No. 5781, 1.
Diethelm, R. 2010, *Inf. Bull. Var. Stars*, No. 5920, 1.
Diethelm, R. 2012, *Inf. Bull. Var. Stars*, No. 6029, 1.
Drake, A. J., *et al.* 2014, *Astrophys. J.*, Suppl. Ser., **213**, 9.
Eggleton, P. P. 1983, *Astrophys. J.*, **268**, 368.
Gaia Collaboration, *et al.* 2016, *Astron. Astrophys.*, **595A**, 1.
Gaia Collaboration, *et al.* 2018, *Astron. Astrophys.*, **616A**, 1.
Gazeas, K. D. 2009, *Commun. Asteroseismology*, **159**, 129.
Gazeas, K., and Stępień, K. 2008, *Mon. Not. Roy. Astron. Soc.*, **390**, 1577.
Gettel, S. J., Geske, M. T., and McKay, T. A. 2006, *Astron. J.*, **131**, 621.
Henden, A. A., Levine, S. E., Terrell, D., Smith, T. C., and Welch, D. L. 2011, *Bull. Amer. Astron. Soc.*, **43**, 2011.
Henden, A. A., Terrell, D., Welch, D., and Smith, T. C. 2010, *Bull. Amer. Astron. Soc.*, **42**, 515.
Henden, A. A., Welch, D. L., Terrell, D., and Levine, S. E. 2009, *Bull. Amer. Astron. Soc.*, **41**, 669.
Hoňková, K., *et al.* 2013, *Open Eur. J. Var. Stars*, **160**, 1.
Hoňková, K., *et al.* 2014, *Open Eur. J. Var. Stars*, **165**, 1.
Houdashelt, M. L., Bell, R. A., and Sweigart, A. V. 2000, *Astron. J.*, **119**, 1448.

Howell, S. B. 2006, *Handbook of CCD Astronomy*, 2nd ed., Cambridge Univ. Press, Cambridge, UK.
Hübscher, J. 2007, *Inf. Bull. Var. Stars*, No. 5802, 1.
Hübscher, J. 2014, *Inf. Bull. Var. Stars*, No. 6118, 1.
Hübscher, J. 2016, *Inf. Bull. Var. Stars*, No. 6157, 1.
Hübscher, J. 2017, *Inf. Bull. Var. Stars*, No. 6196, 1.
Hübscher, J., and Lehmann, P. B. 2012, *Inf. Bull. Var. Stars*, No. 6026, 1.
Hübscher, J., and Lehmann, P. B. 2013, *Inf. Bull. Var. Stars*, No. 6070, 1.
Hübscher, J., and Lehmann, P. B. 2015, *Inf. Bull. Var. Stars*, No. 6149, 1.
Hübscher, J., Lehmann, P. B., and Walter, F. 2012, *Inf. Bull. Var. Stars*, No. 6010, 1.
Hübscher, J., and Monninger, G. 2011, *Inf. Bull. Var. Stars*, No. 5959, 1.
Hübscher, J., Steinbach, H.-M., and Walter, F. 2008, *Inf. Bull. Var. Stars*, No. 5830, 1.
Hübscher, J., Steinbach, H.-M., and Walter, F. 2009, *Inf. Bull. Var. Stars*, No. 5874, 1.
Irwin, J. B. 1959, *Astron. J.*, **64**, 149.
Jayasinghe, T., et al. 2018, *Mon. Not. Roy. Astron. Soc.*, **477**, 3145.
Juryšek, J., et al. 2017, *Open Eur. J. Var. Stars*, **179**, 1.
Kallrath, J., and Milone, E. F. 2009, *Eclipsing Binary Stars: Modeling and Analysis*, Springer-Verlag, New York.
Kurucz, R. L. 2002, *Baltic Astron.*, **11**, 101.
Kwee, K. K., and van Woerden, H. 1956, *Bull. Astron. Inst. Netherlands*, **12**, 327.
Lucy, L. B. 1967, *Z. Astrophys.*, **65**, 89.
Minor Planet Observer. 2010, MPO Software Suite (http://www.minorplanetobserver.com), BDW Publishing, Colorado Springs.
Mortara, L., and Fowler, A. 1981, in *Solid State Imagers for Astronomy*, SPIE Conf. Proc. 290, Society for Photo-Optical Instrumentation Engineers, Bellingham, WA, 28.
Nagai, K. 2010, *Bull. Var. Star Obs. League Japan*, No. 50, 1.
Nagai, K. 2011, *Bull. Var. Star Obs. League Japan*, No. 51, 1.
Nagai, K. 2012, *Bull. Var. Star Obs. League Japan*, No. 53, 1.
Nagai, K. 2016, *Bull. Var. Star Obs. League Japan*, No. 61, 1.
Nagai, K. 2017, *Bull. Var. Star Obs. League Japan*, No. 63, 1.
Nelson, R. H. 2008, *Inf. Bull. Var. Stars*, No. 5820, 1.
Nelson, R. H. 2009, *Inf. Bull. Var. Stars*, No. 5875, 1.
Nelson, R. H. 2009, WDwint56a: Astronomy Software by Bob Nelson (https://www.variablestarssouth.org/bob-nelson).
Nelson, R. H. 2013, *Inf. Bull. Var. Stars*, No. 6050, 1.
Pagel, L. 2018a, *Inf. Bull. Var. Stars*, No. 6244, 1.
Pagel, L. 2018b, *BAV Journal*, No. 31, 1.
Pagel, L. 2021, *BAV Journal*, No. 52, 1.
Paunzen, E., and Vanmunster, T. 2016, *Astron. Nachr.*, **337**, 239.
Pecaut, M., and Mamajek, E. E. 2013, *Astrophys. J., Suppl. Ser.*, **208**, 9.
Pribulla, T., and Ruciński, S. M. 2006, *Astron. J.*, **131**, 2986.
Prša, A., and Zwitter, T. 2005, *Astrophys. J.*, **628**, 426.
Qian, S. 2003, *Mon. Not. Roy. Astron. Soc.*, **342**, 1260.
Ruciński, S. M. 1969, *Acta Astron.*, **19**, 245.
Samec, R. G., Kring, J., Benkendorf, J., Dignan, J., van Hamme, W., and Faulkner, D. R. 2014, *J. Amer. Assoc. Var. Star Obs.*, **42**, 406.
Shappee, B. J., et al. 2014, *Astrophys. J.*, **788**, 48.
Smith, T. C., Henden, A. A., and Starkey, D. R. 2011, in *The Society for Astronomical Sciences 30th Annual Symposium on Telescope Science*, Society for Astronomical Sciences, Rancho Cucamonga, CA, 121.
Software Bisque. 2019, TheSkyX professional edition 10.5.0 (https://www.bisque.com).
Terrell, D., and Wilson, R. E. 2005, *Astrophys. Space Sci.*, **296**, 221.
Tian, Y. P., Xiang, F. Y., and Tao, X. 2009, *Astrophys. Space Sci.*, **319**, 119.
van Hamme, W. 1993, *Astron. J.*, **106**, 2096.
Völschow, M., Schleicher, D. R. G., Perdelwitz, V., and Banerjee, R. 2016, *Astron. Astrophys.*, **587A**, 34.
Watson, C., Henden, A. A., and Price, C. A. 2014, AAVSO International Variable Star Index VSX (Watson+, 2006–2014; https://www.aavso.org/vsx).
Wilson, R. E. 1979, *Astrophys. J.*, **234**, 1054.
Wilson, R. E. 1990, *Astrophys. J.*, **356**, 613.
Wilson, R. E., and Devinney, E. J., 1971, *Astrophys. J.*, **166**, 605.
Woźniak, P. R., et al. 2004, *Astron. J.*, **127**, 2436.
Zasche, P., Liakos, A., Niarchos, P., Wolf, M., Manimanis, V., and Gazeas, K. 2009, *New Astron.*, **14**, 121.

Pulsating Red Giants in a Globular Cluster: 47 Tucanae

John R. Percy
Prateek Gupta
Department of Astronomy and Astrophysics, and Dunlap Institute for Astronomy and Astrophysics, University of Toronto, 50 St. George Street, Toronto, ON M5S 3H4, Canada; john.percy@utoronto.ca

Received August 19, 2021; revised September 13, 2021; accepted September 22, 2021

Abstract We have carried out time-series analysis of a sample of 12 pulsating red giants (PRGs) in the globular cluster 47 Tuc, using observations from the ASAS-SN database, and the AAVSO software package VStar. Most (11/12) of the stars were classified by ASAS-SN as semiregular (SR). We have determined pulsation periods (P) for all 12 of them, and "long secondary periods" (LSPs) for 11 of them. This confirms that LSPs are common in Population II stars. In the context of recent explanations for LSPs, our results imply that many Population II red giants have accreting planetary companions, surrounded by dust. In over half the stars, the period given in the ASAS-SN catalogue is actually the LSP, not the pulsation period. About half the stars show some evidence of a second pulsation period, presumably a second pulsation mode. The amplitudes of the pulsation periods vary by up to a factor of 3.4, on time scales of 10 to 35 pulsation periods (median value 18). The average ratio of LSP to P is 9.0, but the values cluster around 5 and 10. This suggests that some of the stars are pulsating in a lower-order mode, but most are pulsating in a higher-order mode, and half are pulsating in both. The complex variability of the stars in our sample is similar to that of nearby PRGs with a solar composition. The fact that there are about 150 Galactic globular clusters, each with potentially-variable red giants, means that there are many opportunities for studies, like ours, by students and by amateur astronomers with an interest in data analysis, as well as by professional astronomers.

1. Introduction

Globular clusters (GCs), each with hundreds of thousands of stars, are among the oldest objects in our Milky Way (MW) galaxy, ten billion years old, or more. There is a halo of about 150 GCs around the MW. They are not just older than the Sun and most other stars, they also have a much lower abundance of elements heavier than helium—the so-called "metals." They are called "Population II stars."

The brightest stars in GCs are red giants, and red giants are unstable to radial pulsation. Red giant pulsation is complex. Stars can pulsate in one or more of several possible radial modes, and there are also "long secondary periods" (LSPs, Wood 2000). Their cause was unknown until recently (e.g. Takayama and Ita 2020), though the existence of an LSP-luminosity relation parallel to pulsation period-luminosity relations (Wood 2000) was an important clue. Both the pulsation periods and the pulsation amplitudes are variable on time scales of tens of pulsation periods.

In an important development, Soszyński *et al.* (2021) have made a very strong case that LSPs are due to binarity; they are due to the presence of a dusty cloud orbiting the red giant together with a low-mass companion, and obscuring the star once per orbit. The low-mass companion is a former planet which has accreted a significant amount of mass from the envelope of the red giant, and grown into a brown dwarf or low-mass star. The key evidence for this model is the presence of a secondary eclipse in the LSP cycle, seen in the mid-infrared, when the dusty cloud is behind the star.

We were curious to know whether Population II pulsating red giants (PRGs) showed the same complex variability as nearby, more metal-rich stars. This comparison might provide clues as to the cause of the complexities.

GCs have the advantage that the stars in them have very similar compositions, masses, ages, and distances. Variable stars have been studied in these clusters for over a century, but most of the attention has been devoted to RR Lyrae stars. These short-period (0.3 to 1.0 day) variables have well-defined luminosities, so they are a key "standard candle" in the cosmic distance scale. To study these variables, observers made closely-spaced observations for a few days. Clement (2021) maintains a very useful on-line catalogue of variable stars in globular clusters.

PRGs have periods of tens to hundreds of days, and were therefore not well-studied by short runs of closely-spaced observations. However, the All-Sky Automated Survey for Supernovae (ASAS-SN; Shappee *et al.* 2014, Kochanek *et al.* 2017) images the sky every day or two with a network of remote, robotic telescopes around the world. ASAS-SN has observed and catalogued half a million variables, all over the sky, and discovered many thousand new ones. Some of these are in GCs.

The ASAS-SN process for automated analysis, classification, and period determination of variables is not well-suited for the study of PRGs, and often produces incorrect or incomplete results (Percy and Fenaux 2019). In this paper, we do a detailed analysis or re-analysis of ASAS-SN observations of PRGs in one cluster.

We chose 47 Tuc, a bright, populous, well-studied cluster, about 13,000 LY (4 kpc) from the sun. We carefully examined the light curves to look for any of the possible complexities, and then used Fourier and wavelet analysis to study the periods, and changes in the pulsation amplitude. Our project extends the work of Lebzelter and Wood (2005), who obtained several hundred days of observations of several dozen red giants in 47 Tuc; see also Lebzelter *et al.* (2005). They identified many new variables, and determined periods for some known variables. However, their datasets were smaller and shorter than ours.

They determined periods, but not amplitudes (though these could be estimated from the light curves that they published), and did not estimate the values of the LSPs or look for multiperiodic behavior.

Our results are thus complementary to those of Lebzelter and Wood (2005) and of the ASAS-SN team, as well as other results in the literature as compiled by Clement (2021). The ASAS-SN datasets are longer than those of Lebzelter and Wood (2005), but not as long as the AAVSO visual datasets that are often used to study bright nearby PRGs.

Unfortunately ASAS-SN is not able to resolve stars in the dense cores of globular clusters, so it was not possible for us to study most of the stars in the Lebzelter and Wood (2005) sample. Our stars are in the halo of 47 Tuc, but are cluster members on the basis of their distances and proper motions.

This paper is also a "proof of concept" for similar studies of PRGs in GCs using the ASAS-SN database, which could include projects for students and for amateur astronomers with an interest in data analysis. There are many more clusters to be studied!

2. Data and analysis

We analyzed a sample of 12 stars (Table 1) from the ASAS-SN variable star catalogue (Shappee *et al.* 2014, Kochanek *et al.* 2017, Jayasinghe *et al.* 2018, 2019), in the GC 47 Tuc, within 30 arc minutes of the cluster center, and classified by ASAS-SN as Mira stars (visual range greater than 2.5 magnitudes), red semiregular (SR) variables, red irregular (L) variables, or "long secondary periods" (LSP). Almost all were SR. Lebzelter and Obbruger (2009) have shown that, for stars of similar physical properties, there is no essential difference in the pulsation properties of SR and L variables. The ASAS-SN data and light curves are freely available on-line (asas-sn.osu.edu/variables). The error bars on the ASAS-SN observations are 0.02 mag, and this is the noise level in our Fourier analyses.

In addition to very careful analysis of the visual light curve (e.g. Figure 1), we use the Fourier analysis and wavelet routines in the American Association of Variable Star Observers (AAVSO) time-series package VSTAR (Benn 2013). Note that the amplitudes which are given in this paper, including in the tables and figures, are actually semi-amplitudes—the coefficient of the sine curve with the given period—not the full amplitude or range.

Because of the complexity of the variability, and the different time scales involved, visual light curve analysis proved to be especially useful and important. The pulsational variability could be seen clearly, as could the LSP and the variability of the pulsational amplitude (Figures 1, 2, and 3). The intervals between maxima or minima could then be measured and averaged, yielding a period which was accurate to a few percent. Because of the variation of the pulsation amplitude, and the apparent mode switching in a few stars, it was also sometimes useful to inspect the light curve and do Fourier analysis of separate segments or seasons of the dataset.

Fourier analysis was used to confirm and refine the periods but, because of the low amplitudes, and the complexity of the variability, the peaks were often close to the noise level of 0.02 mag (Figure 4). However, the process of estimating the pulsation period(s) from the light curve made it clear which peak in the Fourier spectrum was the correct one. Once the pulsational period had been determined, wavelet analysis was used to study the range and time scale of the pulsational amplitude variability (Figure 5). The wavelet contour diagram was useful for detecting the presence of a second pulsation period and apparent mode-switching (Figure 6).

3. Results

Table 1 lists the ASAS-SN name, identifier (if any) in the Clement (2021) catalogue, type, period, V and K magnitudes, distance in kiloparsecs (from the ASAS-SN website), the primary pulsation period and LSP derived by us, and the ratio of the LSP to the pulsation period. The distance errors are typically 6 to 15 percent. Table 2 gives information about the pulsation amplitude variability—the maximum and minimum amplitudes, their ratio, and the approximate time scale of variability, in units of the pulsation period. This was determined by visual inspection of the amplitude-time graphs (e.g. Figure 5). Table 3 gives information about the stars with probable or possible bimodal behavior—the periods and their ratio. The three stars without an identifier in the Clement (2021) catalogue are presumably new discoveries.

The following are comments on individual stars, including previous periods, from Clement (2021), which appear to be approximations in most cases:

J002516.00-720355.0 This is a large-amplitude Mira star with a period of 193 days, and very little variation in pulsation amplitude. Previous period 192 days.

J002509.10-720215.3 A pulsation period of about 70 days is visible in the light curve (Figure 1), with strongly variable amplitude. There also appears to be an LSP of about 300 days, though the light curve is dominated by the pulsation and its variable amplitude. The Fourier analysis is somewhat uncertain; the LSP is 280 days, but there are pulsation periods of 70 and 78 days with comparable amplitudes. Wavelet analysis suggests that the most likely period is about 72.5 days. Previous period 80 days.

J002258.50-720656.3 A pulsation period of about 50 days and an LSP of about 240 days are visible in the light curve; these are refined to 53.9 and 244 days by Fourier analysis, though the signal is weak. There is also a pulsation period of about 38 days, which shows clearly in some segments of the light curve, and in the wavelet contour diagram. Previous period 40 days.

J002307.35-720029.8 A pulsation period of about 35 days and an LSP of about 380 days are visible in the light curve; these are refined to 35.3 and 377 days by Fourier analysis. There is also a period of 25.1 days which appears in the Fourier spectrum, and in some segments of the light curve.

J002503.68-720931.8 A pulsation period of about 45 days and an LSP of about 750 days are visible in the light curve; the variation in pulsation amplitude is clearly visible. The periods are refined to 45.5 and 769 days by Fourier analysis. Previous period 50 days.

J002355.01-715729.7 A pulsation period of about 40 days and an LSP of about 460 days are visible in the light curve.

Table 1. Period Analysis of ASAS-SN Observations of PRGs.

Name: ASAS-SN-V	ID	Type	PA(d)	V	K	d(kpc)	P(d)	LSP(d)	LSP/P
ASASSN-V J002516.00-720355.0	V3	M	194.4	11.65	6.31	4.834	193.2	—	—
ASASSN-V J002509.10-720215.3	V18	SR	233.3	11.83	6.63	3.626	72.5	280	3.9
ASASSN-V J002258.50-720656.3	V13	SR	255.3	11.54	7.66	4.017	53.9	244	4.5
ASASSN-V J002307.35-720029.8	—	SR	379.9	11.12	7.28	4.590	35.3	377	10.7
ASASSN-V J002503.68-720931.8	V5	SR	45.1	11.64	7.40	4.999	45.5	769	16.9
ASASSN-V J002355.01-715729.7	V17	SR	455.9	11.80	7.31	4.644	39.6	455	11.5
ASASSN-V J002217.84-720612.7	—	SR	37.7	11.75	7.60	4.070	37.7	225	6.0
ASASSN-V J002522.94-721105.1	V16	SR	30.0	11.49	7.23	4.363	30.1	328	10.9
ASASSN-V J002235.85-721110.9	V28	SR	443.2	11.91	6.89	4.066	51.4	444	8.6
ASASSN-V J002452.03-715611.0	LW20	SR	434.5	11.88	7.06	4.142	44.3	435	9.8
ASASSN-V J002422.81-715329.0	V10	SR	412.2	11.71	6.99	4.265	81.0	409	5.0
ASASSN-V J002330.09-722236.3	—	SR	59.3	12.11	6.79	3.886	60.3	701	11.6

Table 2. Amplitude Analysis of ASAS-SN Observations of PRGs.

Name: ASAS-SN-V	Amax	Amin	Amax/Amin	Time Scale/P
ASASSN-V J002516.00-720355.0	1.39	1.31	1.1	8
ASASSN-V J002509.10-720215.3	0.12	0.04	3.4	14
ASASSN-V J002258.50-720656.3	0.07	0.03	2.4	18
ASASSN-V J002307.35-720029.8	0.03	0.01	3.0	29
ASASSN-V J002503.68-720931.8	0.18	0.08	2.2	27
ASASSN-V J002355.01-715729.7	0.05	0.03	1.6	16
ASASSN-V J002217.84-720612.7	0.08	0.03	2.5	18
ASASSN-V J002522.94-721105.1	0.04	0.02	2.0	33
ASASSN-V J002235.85-721110.9	0.14	0.05	2.9	31
ASASSN-V J002452.03-715611.0	0.10	0.03	3.3	21
ASASSN-V J002422.81-715329.0	0.05	0.03	1.6	12
ASASSN-V J002330.09-722236.3	0.20	0.08	2.5	10

Table 3. Analysis of ASAS-SN Observations of Bimodal PRGs.

Name: ASAS-SN-V	Pb(d)	P(d)	Pb/P
ASASSN-V J002258.50-720656.3	37.8	53.9	0.701
ASASSN-V J002307.35-720029.8	25.1	35.3	0.710
ASASSN-V J002355.01-715729.7	39.6	56.7	0.698
ASASSN-V J002522.94-721105.1	30.1	40	0.750
ASASSN-V J002452.03-715611.0	20.8:	44.3	0.470
ASASSN-V J002422.81-715329.0	41.3	81.4	0.510

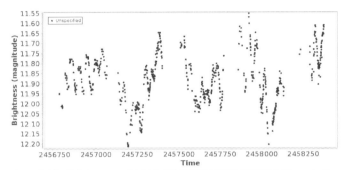

Figure 2. The V light curve of ASAS-SN-V J002452.03-715611.0. The pulsation period and the variability of its amplitude are clearly visible, as is the LSP.

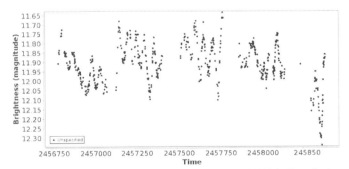

Figure 3. The V light curve of ASAS-SN-V J002235.85-721110.9. The pulsation period and the variability of its amplitude are clearly visible, as is the LSP.

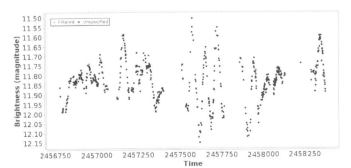

Figure 1. The V light curve of ASAS-SN-V J002509.10-720215.3. The pulsation period and the variability of its amplitude are clearly visible. The LSP is rather weak.

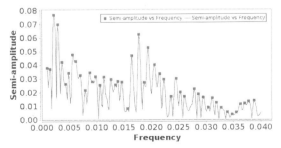

Figure 4. The Fourier spectrum of ASAS-SN-V J002235.85-721110.9, plotting semi-amplitude versus frequency in cycles per day. The pulsation period and LSP are clearly visible, along with their aliases. The light curve is shown in Figure 3.

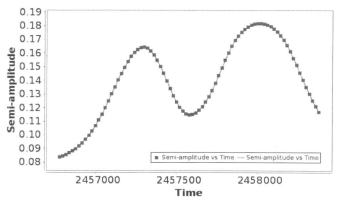

Figure 5. The variable pulsation semi-amplitude of ASAS-SN-V J002503.68-720931.8, as determined by wavelet analysis. The amplitude varies by a factor of two on a time scale of 800–1,000 days.

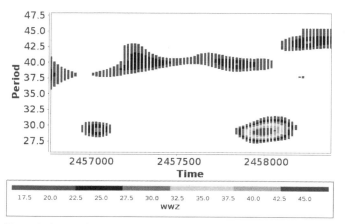

Figure 6. The wavelet contour diagram for ASAS-SN-V J002522.94-721105.1, plotting period in days versus Julian date, with WWZ amplitude in false color. It shows the presence of two pulsation modes with periods of about 30 and 40 days, each variable in amplitude, and apparent mode-switching on a time scale of about 1,000 days.

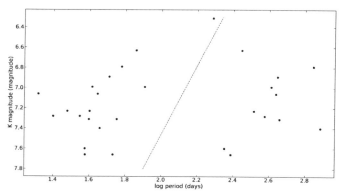

Figure 7. The period-luminosity relation (K magnitude versus log period) for all the periods of the 12 stars in our sample. The dashed line is sequence C (believed to represent the fundamental pulsation mode) for the stars studied by Lebzelter and Wood (2005). The points to the right of the line are LSPs; those to the left are pulsation periods, almost all pulsating in higher-order modes.

These periods are refined to 39.6 and 455 days by Fourier analysis. There is also a possible period of 56.7 days, which is also present in the wavelet contour diagram. Previous period 60 days.

J002217.84-720612.7 A pulsation period of about 35 days and an LSP of about 200–250 days are visible in the light curve, as is the variation in pulsation amplitude. The periods are refined to 37.7 and 225 days by Fourier analysis, though the signals are weak.

J002522.94-721105.1 A pulsation period of about 30 days and an LSP of about 330 days are visible in the light curve; these are refined to 30.1 and 328 days by Fourier analysis, though the amplitude is only 0.02 and the signals are close to the noise level. There is also evidence for a period of about 40 days in the Fourier and wavelet analysis (Figure 6). Previous period 41 days.

J002235.85-721110.9 A pulsation period of about 50 days, and an LSP of about 440 days are visible in the light curve, as is the variation in the pulsation amplitude (Figure 3). The periods are refined to 51.4 and 444 days by Fourier analysis (Figure 4). Previous period 40 days.

J002452.03-715611.0 A pulsation period of about 45 days, with variable amplitude, and an LSP of about 430 days are visible in the light curve. These are refined to 44.4 and 435 days by Fourier analysis. There is also some evidence for a second pulsation period of 20.8 days in the Fourier and wavelet analysis, but it is close to the noise level. There was one discrepant point in the light curve which was not used in the analysis. Previous period 49 days.

J002422.81-715329.0 A pulsation period of about 80 days, with variable amplitude, and an LSP of about 410 days are visible in the light curve. These are refined to 81.0 and 409 days by Fourier analysis. A possible pulsation period of 41.3 days is visible in the Fourier and wavelet analysis, but the amplitude is weak. Known variable, with no period given.

J002330.09-722236.3 A pulsation period of about 60 days and an LSP of about 700 days are clearly visible in the light curve. These periods are refined to 60.3 and 701 days by Fourier analysis. The amplitudes are relatively large.

4. Discussion

The pulsation periods of the 12 stars in Table 1 are a few tens of days, as expected. The stars in Table 3 show probable or possible bimodal behavior. Xiong and Deng (2007) have published pulsation models for red giants. They provide periods and period ratios for low-order radial pulsation modes. The stars in Table 3 with period ratios near 0.7 can be interpreted as pulsating in two adjacent modes. Those with ratios near 0.5 can be interpreted as pulsating in the fundamental and second, or first and third overtone.

LSPs are found in about a third of nearby field PRGs. In our sample, almost all the stars had LSPs. This may be because it is easier to detect them in small-amplitude pulsators, or because they are more common in Population II stars, or because ASAS-SN is more likely to have identified our stars as variable, and hence part of our sample. In any case, our results show that LSPs are very common in PRGs of both Population I and II. This is very interesting in the context of the Soszyński *et al.* (2021) model, since it shows that red giants with former planet companions, and dust clouds around them can exist in Population II stars. Furthermore, some of the binary orbits would be seen face-on, and would therefore not exhibit LSPs, so the incidence of binaries would be greater than the incidence of LSPs.

The ratios of LSP to P average 9.0, but cluster around 5 and 10. The same was found in bright field red giants by Percy and Leung (2017). Wood (2000) and subsequent workers who studied PRGs in stellar systems plotted period-luminosity sequences which they identified as LSP and low-order radial modes. Figure 7 shows such a diagram for our stars; those on the right are LSPs and those on the left are pulsation periods, and the dashed line is Lebzelter and Wood's (2005) sequence for fundamental-mode pulsators. Our LSP/P ratios of 5 and 10 are consistent with the separations of these sequences. Specifically, they would correspond to the pulsation modes being primarily first or second overtone in our sample.

The pulsation amplitudes of the stars in our sample vary by up to a factor of 3.4 on a time scale of 10 to 35 pulsation periods (median value 18). In this respect, these Population II stars behave in the same way as nearby field stars (Percy and Abachi 2013). The cause of these variations is not known. For the bimodal variables, the pulsation amplitude variability can produce apparent "mode switching" which can be seen in the wavelet contour diagram (e.g. Figure 6). Note that the time scale for pulsation amplitude variation is a factor of two longer than the LSPs; there is no evidence that the two phenomena are related.

We note that only half of the pulsation periods that we have determined agree with the ASAS-SN period. In the other cases, the ASAS-SN period is the LSP. The ASAS-SN automated procedure chooses the best (i.e. the dominant) period, which may be either the pulsation period or the LSP. It does not allow for two or more periods.

The periods of these stars may also be variable, as is the case with bright nearby PRGs, whose periods "wander" in a way that can be modelled as random cycle-to-cycle fluctuations. We did not study possible period variability of this kind; our datasets are rather short for this.

This paper is based on a short (100 hours) summer research project by co-author PG, who had just completed the third year of an undergraduate astronomy and physics program. Projects of this kind are an excellent way for students to develop and integrate a wide range of skills in math, physics, and computing, motivated by the knowledge that they are doing real science, with real data. This paper is also an example of the type of project that could be done by skilled amateur astronomers with an interest in variable stars, data mining, and data analysis.

5. Conclusions

We have analyzed in detail the variability of 12 PRGs in the GC 47 Tuc, using ASAS-SN data. We derive pulsation periods for all 12, LSPs for 11, and possible second pulsation periods for 6 of them. The pulsation amplitudes vary by up to a factor of 3.4 on time scales of typically 20 pulsation periods. LSPs are common in these stars. The ratio of LSP to pulsation period averages 9.0, but the values cluster around 5 and 10. The ratio may reflect which mode the star is pulsating in. In all these respects, the PRGs in a metal-poor GC behave in the same way as nearby field variables with a solar composition.

6. Acknowledgements

This paper made use of ASAS-SN photometric data. We thank: the ASAS-SN project team for their remarkable contribution to stellar astronomy, and for making the data freely available on-line; the AAVSO for creating and making available the VSTAR time-series analysis package; and the anonymous referee for several suggestions which have improved this paper. We also acknowledge and thank the University of Toronto Work-Study Program for financial support. The Dunlap Institute is funded through an endowment established by the David Dunlap Family and the University of Toronto.

References

Benn, D. 2013, VSTAR data analysis software (https://www.aavso.org/vstar-overview).

Clement, C. M. 2021, "Catalogue of Variable Stars in Galactic Globular Clusters" (https://www.astro.utoronto.ca/~cclement/read.html).

Jayasinghe, T., et al. 2018, *Mon. Not. Roy. Astron. Soc.*, **477**, 3145.

Jayasinghe, T., et al. 2019, *Mon. Not. Roy. Astron. Soc.*, **486**, 1907.

Kochanek, C. S., et al. 2017, *Publ. Astron. Soc. Pacific*, **129**, 104502.

Lebzelter, T., and Obbruger, M. 2009, *Astron. Nachr.*, **330**, 390.

Lebzelter, T., and Wood, P. R. 2005, *Astron. Astrophys.*, **441**, 1117.

Lebzelter, T., Wood, P. R., Hinkle, K. H., Joyce, R. R., and Fekel, F. C.. 2005, *Astron. Astrophys.*, **432**, 207.

Percy, J. R., and Abachi, R. 2013, *J. Amer. Assoc. Var. Star Obs.*, **41**, 193.

Percy, J. R., and Fenaux, L. 2019, *J. Amer. Assoc. Var. Star Obs.*, **47**, 202.

Percy, J. R., and Leung, H. W.-H. 2017, *J. Amer. Assoc. Var. Star Obs.*, **45**, 30.

Shappee, B. J., et al. 2014, *Astrophys. J.*, **788**, 48.

Soszyński, I., et al. 2021, *Astrophys.J. Lett.*, **911**, L22.

Takayama, M., and Ita, Y. 2020, *Mon. Not. Roy. Astron. Soc.*, 492, 1348.

Wood, P. R. 2000, *Publ. Astron. Soc. Australia*, **17**, 18.

Xiong, D. R., and Deng, L. 2007, *Mon. Not. Roy. Astron. Soc.*, **378**, 1270.

CCD Photometry, Light Curve Modeling, and Period Study of GSC 2624-0941, a Totally Eclipsing Overcontact Binary System

Kevin B. Alton
UnderOak Observatory, 70 Summit Avenue, Cedar Knolls, NJ 07927; kbalton@optonline.net

John C. Downing
La Ventana Observatory, 28881 Sunset Road, Valley Center, CA 92082; johndowning2014@outlook.com

Received Septemver 4, 2021; revised October 26, November 3, 2021; accepted November 3, 2021

Abstract Precise time-series multi-color (BVR$_c$ or I$_c$) light curve (LC) data were acquired from GSC 2624-0941 (= NSVS 8114939 = 2MASS J18275502+3148337) at three different sites between 2018 and 2021. New times of minimum (ToM) from data acquired during this study along with other ToMs extrapolated from the SuperWASP survey were used to generate an updated linear ephemeris. Secular analyses (ToM differences vs. epoch) revealed changes in the orbital period of GSC 2624-0941 over the past 17 years suggesting an apparent increase in the orbital period based on a parabolic fit of the residuals. Simultaneous modeling of multi-color LC data was accomplished using the Wilson-Devinney code. Since a total eclipse is observed, a photometrically derived value for the mass ratio (q_{ptm}) with acceptable uncertainty could be determined which subsequently provided estimates for some physical and geometric elements of GSC 2624-0941.

1. Introduction

Sparsely sampled monochromatic photometric data from GSC 2624-0941 (= NSVS 8114839 = 2MASS J18275502+3148337) were first captured during the ROTSE-I survey between 1999 and 2000 (Akerlof *et al.* 2000; Wozniak *et al.* 2004). Gettel *et al.* (2006) included GSC 2624-0941 in their catalog of bright contact binary stars from the ROTSE-I survey. Other sources of photometric data from this variable system include the All-Sky Automated Survey for SuperNovae (ASAS-SN) (Shappee *et al.* 2014; Jayasinghe *et al.* 2018) and the SuperWASP Survey (Butters *et al.* 2010). Herein, the first multi-color (BVI$_c$) LCs from GSC 2624-0941 with modeling using the Wilson-Devinney code (WD; Wilson and Devinney 1971; Wilson 1979, 1990) are reported. This investigation also includes secular analyses of the predicted vs. observed ToM differences (ETD) over the past 17 years.

2. Observations and data reduction

The imaging system used at UnderOak Observatory (UO, USA; 40.825 N, 74.456 W) during 2018 includes a 0.28-m Schmidt-Cassegrain telescope with an SBIG ST-8XME CCD camera. The focal-reduced (f/6.4) optics for this telescope produce an image scale of 2.06 arcsec/pixel (bin = 2 × 2) and a field-of-view (FOV) of 26.4 × 17.6 arcmin. Additional time-series photometric observations were acquired in 2020 at Desert Blooms Observatory (DBO, USA; 31.941 N, 110.257 W) using a QSI 683 wsg-8 CCD camera mounted at the Cassegrain focus of a 0.4-m Schmidt-Cassegrain telescope. This focal-reduced (f/7.2) instrument produces an image scale of 0.76 arcsec/pixel (bin = 2 × 2) and a field-of-view (FOV) of 15.9 × 21.1 arcmin. The equipment at La Ventana Observatory (LVO, USA; 33.2418 N, 116.9781 W) included an iOptron CEM60 mount with an SBIG Aluma CCD694 camera installed at the Cassegrain focus of a 0.235-m Schmidt-Cassegrain telescope. THESKYX PRO EDITION 10.5.0 controlled the main (30-s exposures) and integrated guide cameras during image acquisition (science, darks, and flats). This focal-reduced (f/7) instrument produces an image scale of 1.14 arcsec/pixel (bin = 2 × 2) and a field-of-view (FOV) of 26.1 × 20.9 arcmin.

All three CCD cameras were equipped with photometric B, V, R$_c$, and/or I$_c$ filters manufactured to match the Johnson-Cousins Bessell specification. Each site used the same image (science, darks, and flats) acquisition software (THESKYX PRO EDITION 10.5.0; Software Bisque 2019) which controlled the main and integrated guide cameras. Computer time was updated immediately prior to each session. Dark subtraction, flat correction, and registration of all images collected at DBO and LVO were performed with AIP4WIN v2.4.0 (Berry and Burnell 2005). Instrumental readings from GSC 2624-0941 were reduced to catalog-based magnitudes using APASS DR9 values (Henden *et al.* 2009, 2010, 2011; Smith *et al.* 2011) built into MPO CANOPUS v 10.7.1.3 (Minor Planet Observer 2010). Since data acquired in 2018 at UO (BVI$_c$) and in 2021 at LVO (BVR$_c$) did not produce total LC coverage, they were only used to supplement ToM values for secular analyses.

LCs were generated using an ensemble of five comparison stars, the average of which remained constant (< ±0.01 mag) throughout every imaging session. The identity, J2000 coordinates, and color indices (B–V) for these stars are provided in Table 1. A CCD image annotated with the location of the target (T) and comparison stars (1–5) is shown in Figure 1. Data acquired below 30° altitude (airmass > 2.0) were excluded; considering the close proximity of all program stars, differential atmospheric extinction was ignored. All photometric data acquired at DBO, LVO, and UO can be retrieved from the AAVSO International Database (Kafka 2021).

3. Results and discussion

Results and detailed discussion about the determination of linear and quadratic ephemerides follow below. Thereafter, discussion about our multi-source approach for estimating

Table 1. Astrometric coordinates (J2000), V-magnitudes and color indices (B–V) for GSC 2624-0941 (Figure 1), and the corresponding comparison stars used in this photometric study.

Star Identification	R.A. (J2000)[a] h m s	Dec. (J2000)[a] ° ′ ″	V-mag.[b]	(B–V)[b]
(1) GSC 2624-2493	18 28 38.6037	+31 48 40.788	11.561	0.286
(2) GSC 2628-0523	18 28 50.8935	+31 53 01.478	11.738	0.551
(3) GSC 2628-0540	18 28 44.3862	+31 54 36.613	11.942	0.344
(4) GSC 2628-2268	18 28 34.4246	+31 54 03.090	13.389	0.434
(5) GSC 2628-2281	18 27 46.3159	+31 54 26.466	11.172	0.657
(T) GSC 2624-0941	18 27 55.0365	+31 48 33.798	11.953	0.350

a. R.A. and Dec. from Gaia DR2 (Gaia Collab. et al. 2016, 2018).
b. V-magnitude and (B–V) for comparison stars derived from APASS DR9 database described by Henden et al. 2009, 2010, 2011, and Smith et al. 2011.

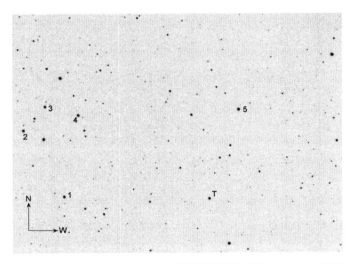

Figure 1. CCD image (V mag; 45 s) of GSC 2624-0941 (T) acquired at DBO (FOV = 15.9 × 21.1 arcmin) showing the location of comparison stars (1–5) used to generate APASS DR9-derived magnitude estimates.

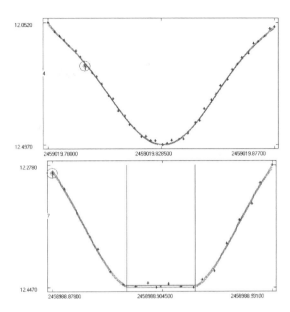

Figure 2. The top panel depicts time of minimum estimates during Min I using polynomial approximation ($\alpha = 6$), while the bottom panel shows the fit achieved with the wall-supported line algorithm during Min II. In both cases, a circled red dot signifies the moment of extremum. The boundary lines which indicate the duration of the Min II total eclipse (0.016398 d) are conveniently calculated by MAVKA.

T_{eff} and Roche-lobe modeling results with the WD code are examined. Finally, preliminary estimates for mass ($M_\odot$) and radius ($R_\odot$) along with corresponding calculations for luminosity ($L_\odot$), surface gravity (log (g)), semi-major axis ($R_\odot$), and bolometric magnitude (M_{bol}) are derived.

3.1. Photometry and ephemerides

Times of minimum (ToM) and associated errors were calculated according to Andrych and Andronov (2019) and Andrych et al. (2020) using the program MAVKA (https://uavso.org.ua/mavka/). Around Min I, simulation of extrema was automatically optimized by finding the most precise degree (α) and best fit algebraic polynomial expression (Figure 2: top panel). During Min II, a "wall-supported line" (WSL) algorithm (Andrych et al. 2017) provided the best fit as the eclipse passes through totality resulting in a flattened bottom (Figure 2, bottom panel). These two, along with seven additional methods featured in MAVKA, are also well suited for other variable star LCs with symmetric or asymmetric extrema. ToM differences (ETD) vs. epoch were fit using scaled Levenberg-Marquardt algorithms (QTIPLOT 0.9.9–rc9; IONDEV SRL 2021). Photometric uncertainty was calculated according to the so-called "CCD Equation" (Mortara and Fowler 1981; Howell 2006). The acquisition dates, number of data points, and uncertainty for each bandpass used for the determination of ToM values and/or WD modeling are summarized in Table 2.

Twelve new ToM measurements were extracted from photometric data acquired at DBO, LVO, and UO. The SuperWASP survey (Butters et al. 2010) provided an abundance of photometric data taken (30-s exposures) at modest cadence that repeats every 9 to 12 min. In some cases (n = 84) these data acquired between 2004 and 2008 were amenable to further

Table 2. Number of data points, estimated uncertainty (±, mag) in each bandpass (BVR_cI_c) and summary of image acquisition dates for GSC 2624-0941.

n (B)	B (± mag.)	n (V)	V (± mag.)	n (R_c)	R_c (± mag.)	n (I_c)	I_c (± mag.)	Location	Dates
351	0.008	355	0.006	—	—	347	0.005	UO	Aug. 6–Sept. 14, 2018
591	0.003	607	0.002	—	—	604	0.003	DBO	July 14–July 20, 2020
219	0.004	216	0.003	218	0.004	—	—	LVO	July 29–Aug. 2, 2021

Table 3. GSC~2624-0941 times-of-minimum (May 12, 2004–August 2, 2021), cycle number and residuals (ETD) between observed and predicted times derived from the updated linear ephemeris (Equation 1).

HJD 2400000+	HJD Error	Cycle No.	ETD[a]	Reference	HJD 2400000+	HJD Error	Cycle No.	ETD[a]	Reference
53137.6162	0.0003	−12714	0.0167	1	54320.4405	0.0002	−10323.5	0.0121	1
53138.6067	0.0003	−12712	0.0177	1	54321.4295	0.0002	−10321.5	0.0115	1
53139.5956	0.0004	−12710	0.0169	1	54322.4187	0.0002	−10319.5	0.0111	1
53141.5738	0.0005	−12706	0.0159	1	54324.3970	0.0005	−10315.5	0.0102	1
53155.6776	0.0004	−12677.5	0.0178	1	54593.5691	0.0004	−9771.5	0.0089	1
53157.6562	0.0003	−12673.5	0.0172	1	54609.6492	0.0004	−9739	0.0078	1
53158.6458	0.0004	−12671.5	0.0172	1	54613.6090	0.0004	−9731	0.0093	1
53162.6046	0.0004	−12663.5	0.0176	1	54618.5566	0.0003	−9721	0.0088	1
53165.5721	0.0003	−12657.5	0.0163	1	54619.5457	0.0002	−9719	0.0083	1
53166.5610	0.0002	−12655.5	0.0155	1	54619.5457	0.0002	−9719	0.0083	1
53167.5512	0.0002	−12653.5	0.0162	1	54620.5354	0.0003	−9717	0.0084	1
53168.5402	0.0002	−12651.5	0.0155	1	54620.5355	0.0003	−9717	0.0085	1
53173.4894	0.0002	−12641.5	0.0167	1	54621.5255	0.0003	−9715	0.0089	1
53177.6945	0.0002	−12633	0.0159	1	54622.5153	0.0004	−9713	0.0090	1
53178.4377	0.0003	−12631.5	0.0169	1	54622.5154	0.0003	−9713	0.0091	1
53179.6743	0.0003	−12629	0.0165	1	54624.4954	0.0003	−9709	0.0099	1
53180.6636	0.0003	−12627	0.0162	1	54625.4840	0.0002	−9707	0.0089	1
53182.6431	0.0003	−12623	0.0165	1	54626.4734	0.0002	−9705	0.0087	1
53183.6324	0.0002	−12621	0.0162	1	54640.5754	0.0004	−9676.5	0.0088	1
53183.6324	0.0002	−12621	0.0162	1	54641.5654	0.0003	−9674.5	0.0092	1
53184.6220	0.0003	−12619	0.0162	1	54650.4715	0.0002	−9656.5	0.0088	1
53185.6122	0.0005	−12617	0.0167	1	54652.4505	0.0002	−9652.5	0.0086	1
53192.5391	0.0004	−12603	0.0164	1	54655.4197	0.0003	−9646.5	0.0090	1
53194.5176	0.0003	−12599	0.0157	1	54660.6147	0.0003	−9636	0.0085	1
53195.5073	0.0003	−12597	0.0158	1	54660.6147	0.0003	−9636	0.0085	1
53196.4963	0.0003	−12595	0.0152	1	54661.6047	0.0005	−9634	0.0089	1
53197.4865	0.0002	−12593	0.0158	1	54663.5846	0.0005	−9630	0.0096	1
53198.4760	0.0004	−12591	0.0156	1	54665.5645	0.0005	−9626	0.0103	1
53199.4658	0.0002	−12589	0.0159	1	54665.5653	0.0007	−9626	0.0111	1
53200.4551	0.0003	−12587	0.0155	1	54666.5540	0.0004	−9624	0.0102	1
53201.4456	0.0003	−12585	0.0164	1	54670.5112	0.0003	−9616	0.0090	1
53223.4641	0.0004	−12540.5	0.0162	1	54671.5011	0.0003	−9614	0.0092	1
53224.4534	0.0003	−12538.5	0.0159	1	54672.4894	0.0002	−9612	0.0079	1
53225.4436	0.0003	−12536.5	0.0164	1	54674.4688	0.0002	−9608	0.0081	1
53227.4218	0.0002	−12532.5	0.0155	1	54675.4585	0.0002	−9606	0.0082	1
53229.4009	0.0003	−12528.5	0.0153	1	54676.4478	0.0002	−9604	0.0080	1
53242.5145	0.0004	−12502	0.0166	1	58354.5658	0.0001	−2170.5	0.0004	2
53243.5044	0.0004	−12500	0.0169	1	58355.5549	0.0001	−2168.5	−0.0001	2
53249.4417	0.0007	−12488	0.0166	1	58366.6879	0.0002	−2146	−0.0002	2
53252.4099	0.0002	−12482	0.0159	1	58988.9042	0.0002	−888.5	0.0001	2
53253.4007	0.0002	−12480	0.0172	1	58990.8831	0.0008	−884.5	−0.0003	2
54296.4409	0.0003	−10372	0.0105	1	59015.8712	0.0003	−834	0.0003	2
54297.4312	0.0003	−10370	0.0112	1	59017.8505	0.0002	−830	0.0003	2
54298.4206	0.0003	−10368	0.0109	1	59019.8289	0.0001	−826	−0.0004	2
54307.5730	0.0005	−10349.5	0.0095	1	59021.8083	0.0001	−822	−0.0003	2
54316.4814	0.0004	−10331.5	0.0114	1	59031.7045	0.0001	−802	−0.0001	2
54318.4604	0.0003	−10327.5	0.0113	1	59424.8262	0.0001	−7.5	−0.0002	2
54318.4604	0.0003	−10327.5	0.0113	1	59428.7853	0.0001	0.5	0.0004	2
54318.4607	0.0003	−10327.5	0.0115	1					

a. ETD = Eclipse Time Difference. b. nr = not reported. References: 1. SuperWASP (Butters et al. 2010); 2. This study.

analysis using MAVKA (Andrych and Andronov 2019; Andrych et al. 2020) to estimate ToM values. All available ToM values are provided in Table 3. A new linear ephemeris based on results obtained between 2018 and 2021 was determined as follows:

$$\text{Min. I (HJD)} = 2459428.5375(2) + 0.4948040(1)\,E. \quad (1)$$

Plotting (Figure 3) the difference (ETD) between observed eclipse times and those predicted by the linear ephemeris against epoch (cycle number) reveals what appears to be a quadratic relationship (Equation 2) where:

$$\text{ETD} = 7.08 \pm 23.97 \cdot 10^{-5} + 2.6021 \pm 0.7704 \cdot 10^{-7}\,E \\ 1.2307 \pm 0.0548 \cdot 10^{-10}\,E^2. \quad (2)$$

Given that the coefficient of the quadratic term (Q) is positive, this result would suggest that the orbital period has been increasing at the rate (dP/dt = 2Q/P) of $0.0157 \pm 0.0007\,\text{s} \cdot \text{y}^{-1}$. The absolute rate is similar to many other overcontact systems reported in the literature (Latković et al. 2021). Secular period change described by a parabolic expression is often attributed to mass transfer or by angular momentum loss (AML) due to magnetic stellar wind (Qian 2001, 2003; Li et al. 2019).

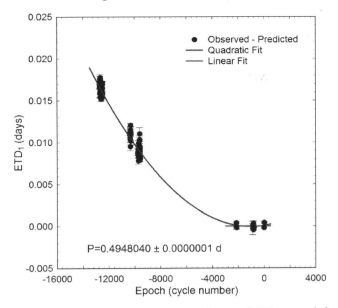

Figure 3. Linear and quadratic fits of ToM differences (ETD) vs. epoch for GSC 2624-0941 calculated using the new linear ephemeris (Equation 1). Measurement uncertainty is denoted by the error bars.

Table 4. Estimation of effective temperature (T_{eff1}) of the primary star in GSC 2624-0941.

Parameter	Value
DBO $(B–V)_0$	0.268 ± 0.021
Median combined $(B–V)_0$[a]	0.269 ± 0.023
Galactic reddening $E(B–V)$[b]	0.0809 ± 0.0032
Survey T_{eff1}[c] (K)	7350 ± 160
Gaia T_{eff1}[d] (K)	6925^{+625}_{-488}
Houdashelt T_{eff1}[e] (K)	7107 ± 170
Median T_{eff1} (K)	7127 ± 190
Spectral Class[f]	A0V-F9V

a. Surveys and DBO intrinsic $(B–V)_0$ determined using reddening values $(E(B–V))$.
b. NASA/IPAC Infrared Science Archive (2021) (https://irsa.ipac.caltech.edu/applications/DUST/).
c. T_{eff1} interpolated from median combined $(B–V)_0$ using Table 4 in Pecaut and Mamajek (2013).
d. Values from Gaia DR2 (Gaia Collab. 2016, 2018) (http://vizier.u-strasbg.fr/viz-bin/VizieR?-source=I/345/gaia2).
e. Values calculated with Houdashelt et al. (2000) empirical relationship
f. Spectral class estimated from Pecaut and Mamajek (2013).

Ideally when AML dominates, the net effect is a decreasing orbital period. If conservative mass transfer from the most massive to a less massive secondary star prevails, then the orbital period can also speed up. Separation increases when conservative mass transfer from the less massive to a more massive component takes place or spherically symmetric mass loss from either body (e.g. a wind but not magnetized) occurs. In mixed situations (e.g. mass transfer from less massive star, together with AML) the orbit evolution depends on which process dominates.

3.2. Effective temperature estimation

The effective temperature (T_{eff1}) of the more massive, and therefore most luminous component (defined as the primary

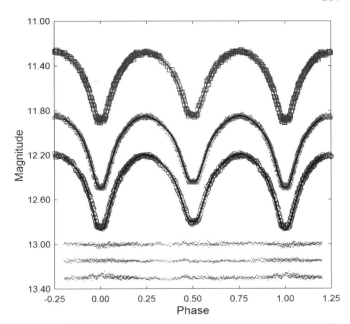

Figure 4. Period-folded (0.4948025 ± 0.0000001 d) CCD-derived LCs for GSC 2624-0941 produced from photometric data collected at DBO between July 14, 2020, and July 20, 2020. The top (I_c), middle (V), and bottom curves (B) were transformed to magnitudes based on APASS DR9-derived catalog values from comparison stars. In this case, the model assumed an A-subtype overcontact binary with single spot on the primary star; residuals from the model fits are offset at the bottom of the plot to keep the values on scale.

Table 5. Light curve parameters evaluated by WD modeling and the geometric elements derived for GSC 2624-0941 (2020) assuming it is an A-type W UMa variable.

Parameter[a]	DBO No Spot	DBO Spotted
T_{eff1} (K)[b]	7127 (190)	7127 (190)
T_{eff2} (K)	7065 (188)	6939 (185)
$q (m_2/m_1)$	0.396 (1)	0.417 (3)
A[b]	0.50	0.50
g[b]	0.32	0.32
$\Omega_1 = \Omega_2$	2.622 (2)	2.663 (2)
Ω_{inner}	2.670 (2)	2.711 (4)
Ω_{outer}	2.428 (1)	2.458 (3)
$i°$	85.43 (3)	83.4 (1)
$A_P = T_S / T_{star}$[c]	—	0.80 (1)
Θ_P (spot co-latitude)[c]	—	80 (2)
φ_P (spot longitude)[c]	—	189 (1)
r_P (angular radius)[c]	—	12 (1)
$L_1/(L_1+L_2)$ B[d]	0.7033 (2)	0.7089 (1)
$L_1/(L_1+L_2)$ V	0.7020 (1)	0.7051 (1)
$L_1/(L_1+L_2)$ I_c	0.7006 (1)	0.7010 (1)
r_1 (pole)	0.4426 (3)	0.4384 (2)
r_1 (side)	0.4748 (4)	0.4696 (2)
r_1 (back)	0.5050 (4)	0.5002 (3)
r_2 (pole)	0.2914 (9)	0.2951 (7)
r_2 (side)	0.3052 (11)	0.3091 (8)
r_2 (back)	0.3452 (19)	0.3487 (15)
Fill-out factor (%)	19.8 (1.1)	19.0 (1.5)
RMS (B)[e]	0.01067	0.00832
RMS (V)	0.00846	0.00595
RMS (I_c)	0.00919	0.00794

a. All DBO uncertainty estimates for T_{eff2}, q, $\Omega_{1,2}$, i, $r_{1,2}$, and L_1 from WDwint56A (Nelson 2009). b. Fixed with no error during DC. c. Spot parameters in degrees (Θ_P, φ_P, and r_P) or A_P in fractional degrees (K). d. L_1 and L_2 refer to scaled luminosities of the primary and secondary stars, respectively. e. Monochromatic residual mean square error from observed values.

star herein) was derived from a composite of photometric (2MASS and APASS) determinations that were as necessary transformed to (B–V) (http://www.aerith.net/astro/color_conversion.html; http://brucegary.net/dummies/method0.html). Interstellar extinction (A_V) and reddening ($E(B-V) = A_V/3.1$) was estimated (image size = 2°) according to a galactic dust map model derived by Schlafly and Finkbeiner (2011). Additional sources used to establish a median value for each T_{eff1} included the Gaia DR2 release of stellar parameters (Andrae *et al.* 2018), and an empirical relationship (Houdashelt *et al.* 2000) based on intrinsic color. The median result ($T_{eff1} = 7127 \pm 190$ K), summarized in Table 4, was adopted for WD modeling of LCs from GSC 2624-0941.

3.3. Modeling approach with the Wilson-Devinney Code

Modeling of LC data from 2020 (Figure 4) was initially performed with PHOEBE 0.31a (Prša and Zwitter 2005) and then refined using WDWINT56A (Nelson 2009). Both programs feature a graphical interface to the Wilson-Devinney WD2003 code (Wilson and Devinney 1971; Wilson 1979, 1990). WDWINT56A incorporates Kurucz's atmosphere models (Kurucz 2002) that are integrated over BVI_c passbands. The final selected model was Mode 3 for an overcontact binary; other modes (detached and semi-detached) never achieved an improved LC simulation as defined by the model residual mean square error. Internal energy transfer to the stellar surface is driven by convective (7200 K) rather than radiative processes (Bradstreet and Steelman 2004). Therefore, bolometric albedo ($A_{1,2} = 0.5$) was assigned according to Ruciński (1969) while the gravity darkening coefficient ($g_{1,2} = 0.32$) was adopted from Lucy (1967). During model fit optimization with differential corrections (DC), logarithmic limb darkening coefficients (x_1, x_2, y_1, y_2) were interpolated (van Hamme 1993) following any change in the effective temperature. All but the temperature of the more massive star (T_{eff1}), $A_{1,2}$ and $g_{1,2}$ were allowed to vary during DC iterations. In general, the best fits for T_{eff2}, i, q and Roche potentials ($\Omega_1 = \Omega_2$) were collectively refined (method of multiple subsets) by DC using the multicolor LC data until a simultaneous solution was found. In this case, surface inhomogeneity often attributed to star spots was simulated by the addition of a cool spot on the primary star to obtain the best fit LC models around Min II. GSC 2624-0941 did not require third light correction ($l_3 = 0$) to improve WD model fits.

3.4. Wilson-Devinney modeling results

Without radial velocity (RV) data it is generally not possible to unambiguously determine the mass ratio or total mass of an eclipsing binary system. A total eclipse is observed at Min II, suggesting that GSC 2624-0941 is an A-subtype overcontact binary system (Binnendijk 1970). Like GSC 2624-0941, other A-type OCBs tend to have relatively hot (spectral class A-F) component stars and orbital periods between 0.4 and 0.8 d. Since the proposed T_{eff1} (7127 K) for the primary approached the generally regarded boundary (7200 K) between convective and radiative energy transfer, we attempted to model the LCs using gravity-brightening ($g_1 = 1$ and $g_{1,2} = 1$) and albedo ($A_1 = 1$ and $A_{1,2} = 1$) values associated with a radiative star. These changes always produced inferior LC fits compared to those obtained when assuming GSC 2624-0941 was a purely convective system ($g_{1,2} = 0.32$ and $A_{1,2} = 0.5$).

Standard errors reported in Table 5 are computed from the DC covariance matrix and only reflect the model fit to the observations which assume exact values for any fixed parameter. These formal errors are generally regarded as unrealistically small, considering the estimated uncertainties associated with the mean adopted T_{eff1} values along with basic assumptions about $A_{1,2}$, $g_{1,2}$ and the influence of spots added to the WD model. Normally, the value for T_{eff1} is fixed with no error during modeling with the WD code. When T_{eff1} is varied by as much as ±10%, investigations with other OCBs, including A- (Alton 2019; Alton *et al.* 2020) and W-subtypes (Alton and Nelson 2018), have shown that uncertainty estimates for i, q, or $\Omega_{1,2}$ were not appreciably (<2.5%) affected. Assuming that the actual T_{eff1} value falls within ±10% of the adopted values used for WD modeling (a reasonable expectation based on T_{eff1} data provided in Table 4), then uncertainty estimates for i, q, or $\Omega_{1,2}$ along with spot size, temperature, and location would likely not exceed this amount.

The fill-out parameter (f) which corresponds to the outer surface shared by each star was calculated according to Equation 2 (Kallrath and Milone 2009; Bradstreet 2005) where:

$$f = (\Omega_{inner} - \Omega_{1,2}) / (\Omega_{inner} - \Omega_{outer}), \qquad (3)$$

wherein Ω_{outer} is the outer critical Roche equipotential, Ω_{inner} is the value for the inner critical Roche equipotential, and $\Omega = \Omega_{1,2}$ denotes the common envelope surface potential for the binary system. In this case GSC 2624-0941 is considered overcontact since $0 < f < 1$.

Spatial renderings (Figure 5) were produced with BINARY MAKER 3 (BM3; Bradstreet and Steelman 2004) using the final WDWINT56A modeling results from 2020. The smaller secondary is shown to completely transit across the primary face during the deepest minimum ($\varphi = 0.0$), thereby confirming that the secondary star is totally eclipsed at Min II.

3.5. Preliminary stellar parameters

Mean physical characteristics were estimated for GSC 2624-0941 (Table 6) using results from the best fit (spotted) LC simulations from 2020. It is important to note that without the benefit of RV data which define the orbital motion, mass ratio, and total mass of the binary pair, these results should be considered "relative" rather than "absolute" parameters and regarded as preliminary. Nonetheless, since the photometric mass ratio (q_{ptm}) is derived from a totally eclipsing OCB, there is a reasonable expectation that DC optimization with the WD2003 code would have arrived at a solution with acceptable uncertainty for q (Terrell and Wilson 2005).

Calculations are described below for estimating the solar mass and size, semi-major axis, solar luminosity, bolometric V-mag, and surface gravity of each component. Three empirically derived mass-period relationships (M-PR) for W UMa-binaries were used to estimate the primary star mass. The first M-PR was reported by Qian (2003), while two others followed, from Gazeas and Stępień (2008) and then Gazeas (2009). According to Qian (2003), when the primary star is

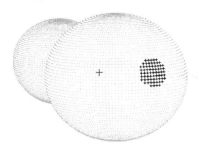

Figure 5. Three-dimensional spatial model of GSC 2624-0941 during 2020 illustrating (top) the location of a cool (black) spot on the primary star and (bottom) the secondary star transit across the primary star face at Min I ($\varphi = 0.0$).

Table 6. Fundamental stellar parameters for GSC 2624-0941 using the photometric mass ratio ($q_{ptm} = m_2/m_1$) from the spotted WD model fits of LC data (2020) and the estimated primary star mass based on empirically derived M-PRs for overcontact binary systems.

Parameter	Primary	Secondary
Mass ($M_\odot$)	1.532 ± 0.045	0.638 ± 0.019
Radius ($R_\odot$)	1.555 ± 0.012	1.044 ± 0.008
a ($R_\odot$)	3.409 ± 0.025	3.409 ± 0.025
Luminosity ($L_\odot$)	5.631 ± 0.606	2.276 ± 0.245
M_{bol}	2.873 ± 0.117	3.857 ± 0.117
Log (g)	4.240 ± 0.014	4.206 ± 0.014

greater than 1.35 $M_\odot$ or the system is A-type, its mass can be determined from:

$$M_1 = 0.761(150) + 1.82(28) \cdot P ; \qquad (4)$$

where P is the orbital period in days and leads to $M_1 = 1.662 \pm 0.204\, M_\odot$ for the primary. The M-PR derived by Gazeas and Stępień (2008):

$$\log(M_1) = 0.755(59) \cdot \log(P) + 0.416(24) ; \qquad (5)$$

corresponds to an OCB system where $M_1 = 1.532 \pm 0.106\, M_\odot$. Gazeas (2009) reported another empirical relationship for the more massive (M_1) star of a contact binary such that:

$$\log(M_1) = 0.725(59) \cdot \log(P) - 0.076(32) \cdot \log(q) + 0.365(32). \qquad (6)$$

from which $M_1 = 1.487 \pm 0.114\, M_\odot$. The median of three values ($M_1 = 1.532 \pm 0.045\, M_\odot$) estimated from Equations 4–6 was used for subsequent determinations of M_2, semi-major axis a, volume-radii r_L, and bolometric magnitudes (M_{bol}) using the formal errors calculated by WDWINT56A (Nelson 2009). The secondary mass = $0.638 \pm 0.019\, M_\odot$ and total mass ($2.170 \pm 0.049\, M_\odot$) were determined using the mean photometric mass ratio ($q_{ptm} = 0.417 \pm 0.003$) derived from the best fit (spotted) models.

The semi-major axis, $a(R_\odot) = 3.409 \pm 0.025$, was calculated from Newton's version of Kepler's third law where:

$$a^3 = (G \cdot P^2 (M_1 + M_2)) / (4\pi^2). \qquad (7)$$

The effective radius of each Roche lobe (r_L) can be calculated over the entire range of mass ratios ($0 < q < \infty$) according to an expression derived by Eggleton (1983):

$$r_L = (0.49 q^{2/3}) / (0.6 q^{2/3} + \ln(1 + q^{1/3})), \qquad (8)$$

from which values for r_1 (0.4562 ± 0.0002) and r_2 (0.3063 ± 0.0002) were determined for the primary and secondary stars, respectively. Since the semi-major axis and the volume radii are known, the radii in solar units for both binary components can be calculated where $R_1 = a \cdot r_1 = 1.555 \pm 0.012\, R_\odot$ and $R_2 = a \cdot r_2 = 1.044 \pm 0.008\, R_\odot$.

Luminosity in solar units ($L_\odot$) for the primary (L_1) and secondary stars (L_2) was calculated from the well-known relationship derived from the Stefan-Boltzmann law (Equation 9) where:

$$L_{1,2} = (R_{1,2}/R_\odot)^2 (T_{1,2}/T_\odot)^4. \qquad (9)$$

Assuming that $T_{eff1} = 7127$ K, $T_{eff2} = 6939$ K, and $T_\odot = 5772$ K, then the solar luminosities ($L_\odot$) for the primary and secondary are $L_1 = 5.631 \pm 0.606$ and $L_2 = 2.276 \pm 0.245$, respectively.

4. Conclusions

This first detailed photometric investigation of GSC 2624-0941 has added valuable information to a ever growing list of OCBs that have been physically and geometrically characterized with a reliable mass ratio. Although we did not uncover anything strikingly remarkable about this system, the proposed effective temperature ($T_{eff1} = 7127 \pm 190$ K) of the primary star proved to be within the top 8% hottest in a catalog of 687 individually studied W UMa stars (Latković et al. 2021). LCs from this variable star exhibit a flattened bottom during Min II, a characteristic of a totally eclipsing A-subtype OCB. Twelve new times of minimum for GSC 2624-0941 based on multicolor CCD data were determined from LCs acquired at three different locations between 2018 and 2021. These, along with other values (n = 84) extrapolated from the SuperWASP survey (2004–2008), led to updated linear and quadratic ephemerides. Secular analyses suggested that the orbital period of GSC 2624-0941 is changing at a rate ($0.0157\,\mathrm{s \cdot y^{-1}}$) consistent with other similarly classified OCBs. The photometric mass ratio ($q_{ptm} = 0.417 \pm 0.003$) determined by WD modeling is expected to correspond closely to a mass ratio derived from RV data. Nonetheless, spectroscopic studies (RV and high resolution classification spectra) will be required to unequivocally determine a total mass and spectral class for each system. Consequently, all parameter values and corresponding uncertainties reported herein should be considered preliminary.

5. Acknowledgements

This research has made use of the SIMBAD database operated at Centre de Données astronomiques de Strasbourg, France. In addition, the Northern Sky Variability Survey hosted by the Los Alamos National Laboratory (https://skydot.lanl.gov/nsvs/nsvs.php), the All-Sky Automated Survey for Supernovae (https://asas-sn.osu.edu/variables), and the AAVSO International Database were mined for essential information. This work also presents results from the European Space Agency (ESA) space mission Gaia. Gaia data are being processed by the Gaia Data Processing and Analysis Consortium (DPAC). Funding for the DPAC is provided by national institutions, in particular the institutions participating in the Gaia MultiLateral Agreement (MLA). The Gaia mission website is https://www.cosmos.esa.int/gaia. The Gaia archive website is https://archives.esac.esa.int/gaia. This paper makes use of data from the first public release of the WASP data as provided by the WASP consortium and services at the NASA Exoplanet Archive, which is operated by the California Institute of Technology, under contract with the National Aeronautics and Space Administration under the Exoplanet Exploration Program. Many thanks to the anonymous referee whose valuable commentary led to significant improvement of this paper.

References

Akerlof, C., *et al.* 2000, *Astron. J.*, **119**, 1901.
Alton, K. B. 2019, *J. Amer. Assoc. Var. Star Obs.*, **47**, 7.
Alton, K. B., and Nelson, R. H. 2018, *Mon. Not. Roy. Astron. Soc.*, **479**, 3197.
Alton, K. B., Nelson, R. H. and Stępień, K. 2020, *J. Astrophys. Astron.*, **41**, 26.
Andrae, R., *et al.* 2018, *Astron. Astrophys.*, **616**, A8.
Andrych, K. D., and Andronov, I. L. 2019, *Open Eur. J. Var. Stars*, **197**, 65.
Andrych, K. D., Andronov, I. L., and Chinarova, L. L. 2017, *Odessa Astron. Publ.*, **30**, 57.
Andrych, K. D., Andronov, I. L., and Chinarova, L. L. 2020, *J. Phys. Stud.*, **24**, 1902.
Berry, R., and Burnell, J. 2005, *The Handbook of Astronomical Image Processing*, 2nd ed., Willmann-Bell, Richmond, VA.
Binnendijk, L. 1970, *Vistas Astron.*, **12**, 217.
Bradstreet, D. H. 2005, in *The Society for Astronomical Sciences 24th Annual Symposium on Telescope Science*, Society for Astronomical Sciences, Rancho Cucamonga, CA, 23.
Bradstreet, D. H., and Steelman, D. P. 2004, Binary Maker 3, Contact Software (http://www.binarymaker.com).
Butters, O. W., *et al.* 2010, *Astron. Astrophys.*, **520**, L10.
Eggleton, P. P. 1983, *Astrophys. J.*, **268**, 368.
Gaia Collaboration, *et al.* 2016, *Astron. Astrophys.*, **595A**, 1.
Gaia Collaboration, *et al.* 2018, *Astron. Astrophys.*, **616A**, 1.
Gazeas, K. D. 2009, *Commun. Asteroseismology*, **159**, 129.
Gazeas, K., and Stępień, K. 2008, *Mon. Not. Roy. Astron. Soc.*, **390**, 1577.
Gettel, S. J., Geske, M. T., and McKay, T. A. 2006, *Astron. J.*, **131**, 621.
Henden, A. A., Levine, S. E., Terrell, D., Smith, T. C., and Welch, D. L., 2011, *Bull. Amer. Astron. Soc.*, **43**, 2011.
Henden, A. A., Terrell, D., Welch, D., and Smith, T. C. 2010, *Bull. Amer. Astron. Soc.*, **42**, 515.
Henden, A. A., Welch, D. L., Terrell, D., and Levine, S. E. 2009, *Bull. Amer. Astron. Soc.*, **41**, 669.
Houdashelt, M. L., Bell, R. A., and Sweigart, A. V. 2000, *Astron. J.*, **119**, 1448.
Howell, S. B. 2006, *Handbook of CCD Astronomy*, 2nd ed., Cambridge University Press, Cambridge.
Jayasinghe, T., *et al.* 2018, *Mon. Not. Roy. Astron. Soc.*, **477**, 3145.
Kafka, S. 2021, Observations from the AAVSO International Database (https://www.aavso.org/data-download).
Kallrath, J., and Milone, E. F. 2009, *Eclipsing Binary Stars: Modeling and Analysis*, Springer, New York.
Kurucz, R. L. 2002, *Baltic Astron.*, **11**, 101.
Latković, O., Čeki, A., and Lazarević, S. 2021, *Astrophys. J., Suppl. Ser.*, **254**, 10.
Li, K., *et al.* 2019, *Res. Astron. Astrophys.*, **19**, 147.
Lucy, L. B. 1967, *Z. Astrophys.*, **65**, 89.
Minor Planet Observer. 2010, MPO Software Suite (http://www.minorplanetobserver.com), BDW Publishing, Colorado Springs.
Mortara, L., and Fowler, A. 1981, in *Solid State Imagers for Astronomy*, SPIE Conf. Proc. 290, Society for Photo-Optical Instrumentation Engineers, Bellingham, WA, 28.
NASA/IPAC. 2021, NASA/IPAC Infrared Science Archive, Galactic Dust Reddening and Extinction (https://irsa.ipac.caltech.edu/applications/DUST/).
Nelson, R. H. 2009, WDwint56a: Astronomy Software by Bob Nelson (https://www.variablestarssouth.org/bob-nelson).
Pecaut, M. J., and Mamajek, E. E. 2013, *Astrophys. J., Suppl. Ser.*, **208**, 9.
Prša, A., and Zwitter, T. 2005, *Astrophys. J.*, **628**, 426.
Qian, S. 2001, *Mon. Not. Roy. Astron. Soc.*, **328**, 635.
Qian, S. 2003, *Mon. Not. Roy. Astron. Soc.*, **342**, 1260.
Ruciński, S. M. 1969, *Acta Astron.*, **19**, 245.
Schlafly, E. F., and Finkbeiner, D. P. 2011, *Astrophys. J.*, **737** 103.
Shappee, B. J., *et al.* 2014, *Astrophys. J.*, **788**, 48.
Smith, T. C., Henden, A. A., and Starkey, D. R. 2011, in *The Society for Astronomical Sciences 30th Annual Symposium on Telescope Science*, Society for Astronomical Sciences, Rancho Cucamonga, CA, 121.
Software Bisque. 2019, TheSkyX Professional Edition 10.5.0 (https://www.bisque.com).
Terrell, D., and Wilson, R. E. 2005, *Astrophys. Space Sci.*, **296**, 221.
van Hamme, W. 1993, *Astron. J.*, **106**, 2096.
Wilson, R. E. 1979, *Astrophys. J.*, **234**, 1054.
Wilson, R. E. 1990, *Astrophys. J.*, **356**, 613.
Wilson, R. E., and Devinney, E. J. 1971, *Astrophys. J.*, **166**, 605.
Woźniak, P. R., *et al.* 2004, *Astron. J.*, **127**, 2436.

A Photometric Study of the Eclipsing Binary LO Ursae Majoris

Edward J. Michaels
Waffelow Creek Observatory, 10780 FM 1878, Nacogdoches, TX 75961; astroed@ejmj.net

Received October 15, 2021; revised November 23, 2021; accepted November 24, 2021

Abstract Multicolor photometric observations of the eclipsing binary LO UMa are presented. Photometric models were determined simultaneously from four sets of light curves using the Wilson-Devinney program. The results indicate LO UMa is a semidetached Algol type binary with a mass ratio of q = 0.62 and primary and secondary star spectral types of F9 and K8, respectively. Based on available times of minimum light, the O – C curves revealed a sinusoidal oscillation with a period of about 16.4 years and an amplitude of 0.0238 day. Two possible causes of the period variation were considered, changes in the quadrupole moment of the secondary star caused by magnetic activity (Applegate mechanism) and the light-time effect of a third body orbiting the binary. It was found that the most plausible explanation for the period oscillation is an unseen body, with a mass of no less than 1.55 $M_\odot$, orbiting the binary. A main sequence star of this mass would be the dominant light source in the system. However, spectra and observed color do not support a star of this mass, nor did the photometric solution find any indication of third light. A massive non-radiating third body suggests a possible neutron star candidate.

1. Introduction

The variability of LO UMa (GSC 03002-00454) was discovered from two images taken with the 25-cm astrograph at Indiana University's Goethe Link Observatory (Williams 2001). Using Harvard College Observatory patrol plates, combined with visual and CCD observations, it was evident this variable was an Algol-type eclipsing binary with a deep primary eclipse and an orbital period of 1.856 days (Baldwin *et al.* 2001). The ASAS-SN Variable Star Database gives a mean visual magnitude of 12.86 with a primary eclipse amplitude of 1.87 (Jayasinghe *et al.* 2019; Shappee *et al.* 2014). The LAMOST DR5 catalog gives an effective temperature of 6018 K (Luo *et al.* 2015). There are several minima times available (48), but no precision multiband photometric observations have been published for this system.

In this paper, a photometric study of LO UMa is presented. The photometric observations and data reduction methods are presented in section 2. A period analysis is presented in section 3. Analysis of the light curves using the Wilson-Devinney (WD) model is presented in section 4. Discussion of the results is presented in section 5 and conclusions are presented in section 6.

2. Photometric observations

Multicolor photometric observations were acquired with a 0.36-m Ritchey-Chrétien robotic telescope located at the Waffelow Creek Observatory, Nacogdoches, Texas (https://obs.ejmj.net). A SBIG-STXL camera with a cooled KAF-6303E CCD (–20° C, 9 µm pixels) was used for imaging. Each night, images were obtained in four passbands: Johnson V and Sloan g', r', and i'. The observation dates and number of images acquired are shown in the Table 1 observation log. The images were calibrated using bias, dark, and flat frames. MIRA software (Mirametrics 2015) was used for image calibration and the ensemble differential aperture photometry of the light images. The locations of the comparison and check stars are shown in Figure 1, and Table 2 gives their coordinates and standard magnitudes. The standard magnitudes were taken from the AAVSO Photometric All-Sky Survey database (APASS; Henden *et al.* 2015). The instrumental magnitudes were converted to standard magnitudes using the APASS comparison star magnitudes. The Heliocentric Julian Date (HJD) of each observation was converted to orbital phase (φ) using the following epoch and orbital period: T_0 = 2459292.6600 and

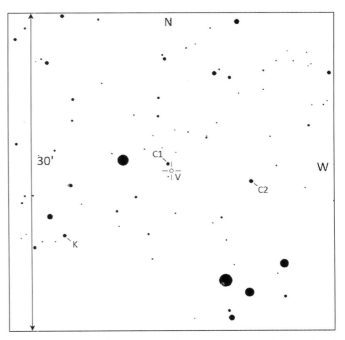

Figure 1. Finder chart for LO UMa (V), comparison stars (C1 and C2), and check (K) stars. This chart was generated by the AAVSO Variable Star Plotter (VSP; https://www.aavso.org/apps/vsp/).

Table 1. Observation log.

Filter	Dates	No. Nights	No. Images
V, g', r', i'	2021 Feb 22	1	66
V, g', r', i'	2021 Mar 18, 19, 23, 28, 31	5	280
V, g', r', i'	2021 Apr 1, 11, 18, 20, 21, 24	6	219
V, g', r', i'	2021 May 2, 4, 5, 6, 13	5	181

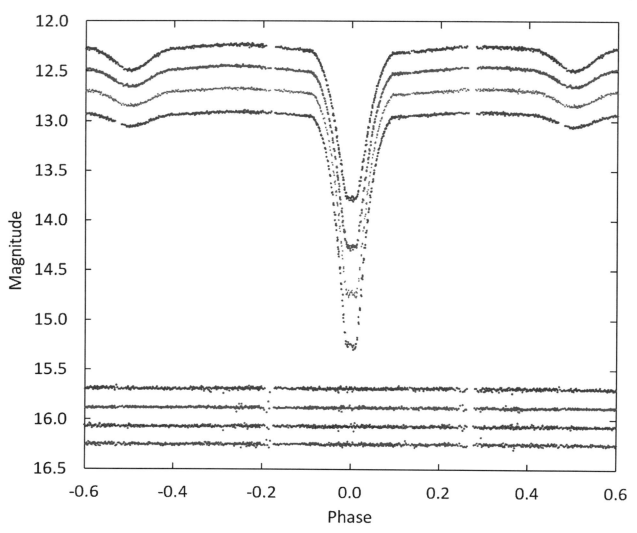

Figure 2. The folded CCD light curves in standard magnitudes. From top to bottom the passbands are i', r', V, and g'. In the same order, the bottom curves are the check-star magnitudes with offsets of +3.15, +3.15, +3.10, and +3.00 magnitudes, respectively. Error bars were omitted from the plotted points for clarity.

Table 2. APASS comparison and check star magnitudes.

System	R.A. (2000) h	Dec (2000) °	V	g'	r'	i'
LO UMa	10.497760	+39.94108				
GSC 03002-00277 (C1)	10.498258	+39.95172	13.227	13.365	13.105	12.999
GSC 03002-00389 (C2)	10.487341	+39.92442	12.810	13.149	12.501	12.240
GSC 03002-00145 (K)	10.511830	+39.83925	12.971	13.289	12.733	12.509
Standard deviation of K-star magnitudes			± 0.009	± 0.010	± 0.008	± 0.016

Table 3. Average light curve properties.

	Min I Mag.	Min II Mag.	Δ Mag. Min II – Min I	Max I Mag.	Max II Mag.	Δ Mag. Max II – Max I	Mag. Range Max II – Min I
V	14.731 ± 0.012	12.844 ± 0.003	−1.887 ± 0.012	12.675 ± 0.003	12.672 ± 0.003	−0.003 ± 0.004	2.059 ± 0.012
g'	15.255 ± 0.008	13.055 ± 0.006	−2.200 ± 0.010	12.910 ± 0.012	12.907 ± 0.004	−0.004 ± 0.013	2.348 ± 0.009
r'	14.264 ± 0.003	12.652 ± 0.002	−1.613 ± 0.003	12.464 ± 0.002	12.451 ± 0.001	−0.014 ± 0.003	1.814 ± 0.003
i'	13.775 ± 0.003	12.492 ± 0.002	−1.283 ± 0.003	12.253 ± 0.004	12.237 ± 0.002	−0.016 ± 0.005	1.539 ± 0.003

Note: Primary total eclipse duration: ~54 minutes.

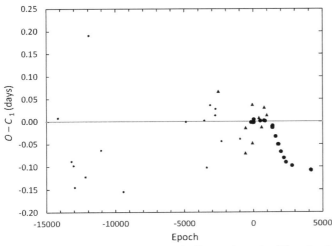

Figure 3. The residuals calculated from the linear ephemeris of Equation 1. The dots are the photovisual minima, the triangles the visual, and the filled circles the CCD.

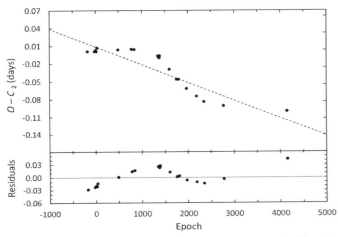

Figure 4. The top panel shows the residuals (filled circles) calculated from the linear ephemeris of Equation 1 using the CCD minima times from 1999–2021. The dashed line is the best-fit linear line from Equation 2. The bottom panel shows the residuals from the linear fit of Equation 2.

P = 1.8558690. Figure 2 shows the folded light curves plotted from orbital phase −0.6 to 0.6, with negative phase defined as (φ−1). The nearly complete light curves required over two months of observations. The error of a single observation ranged from 7 to 23 mmag. The check star magnitudes were plotted and inspected each night, but no significant variability was found (see bottom of Figure 2). The standard deviations for all check star observations are listed in Table 2. The minimum light at primary eclipse for each passband was briefly constant, which confirms the total eclipse reported by Baldwin et al. (2001). The light curve properties for each passband are given in Table 3 (Min I, Min II, Max I, Max II, Δm, and total eclipse duration). The observations can be accessed from the AAVSO International Database (Kafka 2017).

3. Period study

A literature search located 42 minimum timings for this period study. From the current observations two new primary

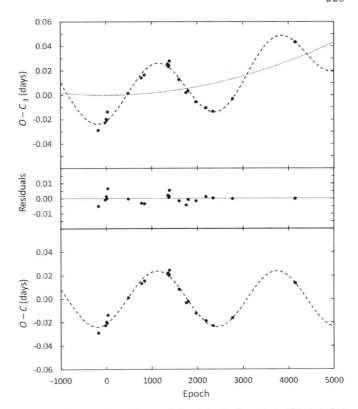

Figure 5. The Levenberg–Marquardt fit of the O–C residuals (filled circles) calculated from the updated linear ephemeris (Equation 2) using only the CCD minima times. In the top panel the dashed line shows the fit for a circular orbit (e = 0) for a supposed third body and the dotted (blue) line gives the quadratic fit from the residuals. The middle panel displays the total residuals after subtraction of both the upward parabolic change and the cyclic variation. The bottom panel shows the model fit after subtracting out the quadratic component.

minima were found using the Kwee and van Woerden (1956) method. In addition, SuperWASP data were identified with sufficient cadence and light curve quality, from which another four minima were determined. All the minima times have been collected in Table 4. The difference between the observed and predicted eclipse timings, the $O - C_1$ residuals in Table 4, were calculated using Baldwin's (2001) linear ephemeris:

$$\text{HJD Min I} = 2451603.7691 + 1.8559010 \text{ E}. \quad (1)$$

These residuals are shown in Figure 3. Compared to the CCD minima, the photovisual and visual minima show a large amount of scatter (small dots and triangles in Figure 3). This is not unexpected given that each photovisual minima was determined from a single plate. These minima times occur at some point during the eclipse but not at mid-eclipse (Cycle Numbers: −14178 to −554, −63, −57). In addition, four of the visual minima covered only the ingress or the egress of a primary eclipse. The precision of the photovisual and most of the visual minima is unknown, since the standard errors were not provided.

A first attempt at a period analysis utilized only the CCD minima. Those observations are of higher accuracy and span 21 years (1999–2021). A least-squares solution to the residuals of Equation 1 gives the following new linear ephemeris:

$$\text{HJD Min I} = 2451603.7777\,(12) + 1.8558710\,(8)\,E. \quad (2)$$

This ephemeris should be useful in predicting the times of future primary eclipses. The results of the linear fit are displayed in the $O-C_2$ diagram of Figure 4 (top panel). The residuals plotted in the bottom panel provide information on any orbital period changes that may have occurred since 1999. Visual inspection of the residuals reveals possible sinusoidal and linear changes in the orbital period. A long-term linear change causes the $O-C$ residuals to take on a parabolic shape, which is often attributed to mass transfer or angular momentum loss caused by magnetic braking. The cyclic variation may be the result of a third body orbiting the binary or magnetic activity of the stars. Each of these will be investigated in turn.

The motion of the binary around the barycenter of a tertiary system causes an apparent periodic change in the binary's orbital period. This results from the changing light travel time between Earth and the binary (Light-Time Effect or LITE). The period of the LITE oscillations corresponds to the orbital period of the binary and the tertiary component about their barycenter. An initial attempt to investigate both the parabolic and sinusoidal variations in the orbital period used the following equation:

$$\text{HJD Min I} = \text{HJD}_0 + PE + QE^2 + A\sin(\omega E + \varphi). \quad (3)$$

The computed result of the first three terms, $\text{HJD}_0 + PE + QE^2$, is the quadratic ephemeris where Q measures the long-term period change of the binary. The fourth term in Equation 3 is the time difference due to the binary's orbital motion about the barycenter. In this model, the periodic oscillation should appear symmetrical, and the orbit of the tertiary component is circular (e = 0). The parameter values HJD_0, P, Q, A, ω, and φ were determined using the Levenberg-Marquardt algorithm. The results (LITE-1) are displayed in Figure 5, and the calculated parameters are listed in column 2 of Table 5. The $O-C_3$ diagram in Figure 5 shows a possible long-term increase in the orbital period of the binary (dotted line). The calculated quadradic coefficient, $Q = 1.7\,(2) \times 10^{-9}$ d, measures this long-term change. The rate of period change since 1999 was calculated using the following equation:

$$\frac{dP}{dt} = \frac{2Q}{P} \cdot 365.24. \quad (4)$$

The orbital period appears to be increasing at a rate of $7(1) \times 10^{-7}$ d yr^{-1} or about 6 seconds per century. The coefficient of the sine term, $A = 0.0238 \pm 0.0002$ d, is the semi-amplitude of the oscillation. The period of oscillation was calculated using the following equation:

$$P_3 = \frac{2\pi P}{\omega}, \quad (5)$$

where ω is the angular frequency and P the binary orbital period in days. The oscillation period, $P_3 = 13.36 \pm 0.09$ yr, is the orbital period of the binary and tertiary component about the barycenter. There are no additional periodic variations seen in the residuals (see center panel of Figure 5).

To analyze the possibility of a non-circular orbit, the sine term in Equation 3 was replaced with Irwin's (1959) formula:

$$\text{HJD (Min)} = \text{HJD}_0 + PE + QE^2 + \frac{a_{12}\sin i_3}{c}$$
$$\left[\frac{1-e^2}{1+e\cos v}\sin(v+\omega) + e\sin\omega\right]. \quad (6)$$

Table 4. Times of minima and O–C residuals.

Method	Epoch HJD 2400000+	Error	Cycle No.	$(O-C)_1$	Ref.
pg	25290.8130[a]	—	–14178.0	0.00828	1
pg	27092.7970	—	–13207.0	–0.08759	1
pg	27374.8840	—	–13055.0	–0.09755	1
pg	27532.5880	—	–12970.0	–0.14513	1
pg	28961.6550	—	–12200.0	–0.12190	1
pg	29429.6550[a]	—	–11948.0	0.19105	1
pg	31084.8640	—	–11056.0	–0.06364	1
pg	34072.7740[a]	—	–9446.0	–0.15425	1
pg	42485.7270	—	–4913.0	–0.00049	1
pg	44996.7640	—	–3560.0	0.00246	1
pg	45289.8920[a]	—	–3402.0	–0.10190	1
pg	45757.7170	—	–3150.0	0.03605	1
pg	46438.8100	—	–2783.0	0.01338	1
pg	46492.6460	—	–2754.0	0.02825	1
pg	46878.7120	—	–2546.0	0.06685	1
pg	47264.6290[a]	—	–2338.0	–0.04356	1
pg	49801.6510	—	–971.0	–0.03823	1
vis	50545.8360	—	–570.0	–0.06953	2
vis	50573.7300	—	–555.0	–0.01404	2
vis	50575.5860	—	–554.0	–0.01395	2
ccd	51273.4170	—	–178.0	–0.00172	3
vis	51486.8850	—	–63.0	0.03766	2
vis	51497.9350	—	–57.0	–0.04774	2
ccd	51551.8020	0.0030	–28.0	–0.00187	2
ccd	51603.7691	0.0001	0.0	0.00000	2
ccd	51629.7502	0.0002	14.0	–0.00151	2
ccd	51656.6670	0.0020	28.5	0.00472	2
vis	52368.4080	0.005	412.0	0.00769	4
ccd	52500.1707	—	483.0	0.00142	5
vis	52691.3150	—	586.0	–0.01209	6
ccd	53038.3828	0.0002	773.0	0.00223	7
vis	53049.5470	0.0090	779.0	0.03102	8
ccd	53157.1601	0.0010	837.0	0.00186	7
vis	53409.5750	0.0070	973.0	0.01423	9
ccd	54103.6577[b]	0.0001	1347.0	–0.01008	10
ccd	54142.6287[b]	0.0001	1368.0	–0.01299	10
ccd	54155.6204[b]	0.0001	1375.0	–0.01261	10
ccd	54170.4710[b]	0.0001	1383.0	–0.00921	10
ccd	54562.0423	—	1594.0	–0.03299	11
ccd	54860.8250	0.0008	1755.0	–0.05035	12
ccd	54942.4844	0.0004	1799.0	–0.05060	13
ccd	55259.8273	0.0010	1970.0	–0.06677	14
ccd	55660.6883	0.0003	2186.0	–0.08039	15
ccd	55953.9111	0.0005	2344.0	–0.08994	16
ccd	56038.7184[a]	0.0029	2389.5	0.27386	16
ccd	56744.5178	0.0003	2770.0	–0.09707	17
ccd	59292.6600	0.0003	4143.0	–0.10693	18
ccd	59305.6511	0.0002	4150.0	–0.10720	18

(a) CCD outlier not used in the period analysis. (b) Minima determined from SuperWASP data. References: (1) Williams (2001); (2) Baldwin et al. (2001); (3) Paschke and Brat (2021); (4) Locher et al. (2002); (5) Kreiner (2004); (6) Diethelm (2003); (7) Krajci (2005); (8) Diethelm (2004); (9) Locher (2005); (10) Butters et al. (2010); (11) Nagai (2009); (12) Diethelm (2009); (13) Hübscher et al. (2010); (14) Diethelm (2010); (15) Diethelm (2011); (16) Diethelm (2012); (17) Hübscher (2015); (18) This paper.

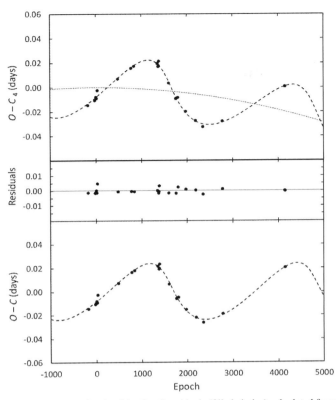

Figure 6. The simplex fit of the O − C residuals (filled circles) calculated from the updated linear ephemeris (Equation 2) using only CCD minima times. In the top panel the dashed line shows the fit for an elliptical orbit (e = 0.42) for a supposed third body and the dotted (blue) line defines the quadratic fit from the residuals. The middle panel displays the total residuals after subtraction of both the upward parabolic change and the cyclic variation. The bottom panel shows the model fit after subtracting out the quadratic component.

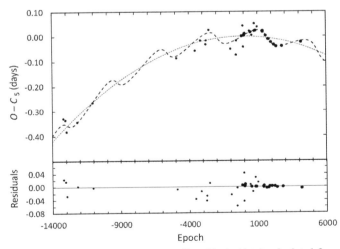

Figure 7. The simplex fit of the O − C residuals (dots) calculated from the Equation 1 linear ephemeris using all available minima times (CCD, photovisual, and visual). In the top panel the dashed line shows the fit for an elliptical orbit (e = 0.47) for a supposed third body and the dotted (blue) line defines the quadratic fit from the residuals. The bottom panel displays the total residuals remaining after LITE analysis.

Table 5. Parameters of the tertiary component.

Parameter	CCD minima only LITE 1	CCD minima only LITE 2	All minima LITE 3
JD_0 [HJD]	2451603.7887(5)	2451603.777(7)	2451603.763(5)
P [day]	1.8558600(8)	1.855873(4)	1.855882(2)
P_3 [yr]	13.36(9)	16.4(3)	18.6(3)
T_0 [HJD]	—	2454529(81)	2454771(403)
ω [°]	—	159(5)	177(19)
e	0	0.42(4)	0.47(33)
A_3 [day]	0.0238(2)	0.0238(8)	0.0273(5)
$a_{12} \sin i_3$ [a.u.]	4.4(1)	4.5(1)	5.4(9)
$f(M_3)$ [$M_\odot$]	0.39(1)	0.33(3)	0.45(7)
M_3 (i = 90°) [$M_\odot$]	1.68(3)	1.55(6)	1.8(2)
M_3 (i = 60°) [$M_\odot$]	2.09(4)	1.91(7)	2.2(2)
M_3 (i = 30°) [$M_\odot$]	5.5(9)	4.9(2)	6.0(7)
Q [day] [10^{-9}]	1.734(2)	−1.1518(8)	−2.1414(3)
dP/dt [10^{-7} d/y]	7(1)	−4.534(3)	−8.429(1)
Sum Res²	0.00154	0.00063	—

Table 6. Results derived from light-curve modeling.

Parameter	No Spots	Spots
i (°)	85.96 ± 0.12	86.02 ± 0.09
T_1 (K)	6018[1]	6018[1]
T_2 (K)	3975 ± 9	3980 ± 4
Ω_1	4.963 ± 0.017	4.959 ± 0.013
Ω_2	3.107[2]	3.091[2]
q (M_2 / M_1)	0.624 ± 0.003	0.615 ± 0.002
$L_1 / (L_1 + L_2)$ (V)	0.8135 ± 0.0009	0.8130 ± 0.0006
$L_1 / (L_1 + L_2)$ (g')	0.8555 ± 0.0007	0.8550 ± 0.0007
$L_1 / (L_1 + L_2)$ (r')	0.7718 ± 0.0010	0.7714 ± 0.0006
$L_1 / (L_1 + L_2)$ (i')	0.7062 ± 0.0013	0.7060 ± 0.0009
r_1 side	0.2255 ± 0.0008	0.2280 ± 0.0007
r_2 side	0.3330 ± 0.0004	0.3304 ± 0.0003
Residuals	0.00070	0.00040
Star 1		Hot Spot
co-latitude (°)	—	85 ± 12
longitude (°)	—	24 ± 2
spot radius (°)	—	10 ± 5
temp. factor	—	1.06 ± 0.06
Star 1		Cool Spot
co-latitude (°)	—	120 ± 6
longitude (°)	—	288 ± 3
spot radius (°)	—	15 ± 9
temp. factor	—	0.92 ± 0.11
Star 2		Hot Spot
co-latitude (°)	—	76 ± 3
longitude (°)	—	13 ± 2
spot radius (°)	—	12 ± 4
temp. factor	—	1.16 ± 0.06

Note: The errors in the stellar parameters result from the least–squares fit to the model. The actual uncertainties are considerably larger. The subscripts 1 and 2 refer to the star being eclipsed at primary and secondary minimum, respectively.
[1] Assumed. [2] Calculated.

Table 7. Provisional absolute parameters.

Parameter	Symbol	Value
Stellar masses	M_1 ($M_\odot$)	1.10 ± 0.11
	M_2 ($M_\odot$)	0.68 ± 0.07
Semi–major axis	a ($R_\odot$)	7.7 ± 0.2
Mean stellar radii	R_1 ($R_\odot$)	1.78 ± 0.08
	R_2 ($R_\odot$)	2.60 ± 0.09
Bolometric magnitude	$M_{bol,1}$	3.3 ± 0.1
	$M_{bol,2}$	4.3 ± 0.4
Stellar luminosity	L_1 ($L_\odot$)	3.7 ± 0.4
	L_2 ($L_\odot$)	1.5 ± 0.4
Surface gravity	$\log g_1$ (cgs)	3.98 ± 0.05
	$\log g_2$ (cgs)	3.44 ± 0.05

Note: The calculated values in this table are provisional. Radial velocity observations are necessary for direct determination of M_1, M_2, and a.

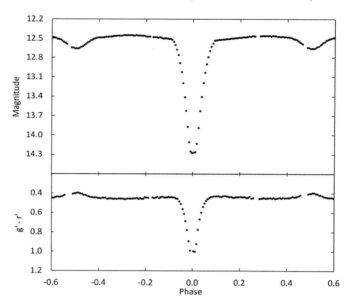

Figure 8. Light curve of the binned Sloan r' passband observations in standard magnitudes (top panel). The observations were binned with a phase width of 0.0067. The errors for each binned point are about the size of the plotted points. The (g' – r') colors (bottom panel) were calculated by subtracting the linearly interpolated binned g' and r' magnitudes.

The tertiary component's associated orbital parameters for this term include the orbital period P_3, inclination i_3, orbital eccentricity e, amplitude $A = a_{12} \sin i_3$, argument of periastron ω, and time of periastron passage T_0. A simplex optimization was used to solve for the parameters using the MATLAB code written by Zasche *et al.* (2009). The initial parameter values were taken from the LITE-1 solution. The results (LITE-2) are listed in column 3 of Table 5 and are displayed in Figure 6. This solution gave a better fit, with a 41% reduction in residuals, compared to a circular orbit solution (LITE-1). To include information from the photovisual and visual minima times dating back to 1933, a third LITE solution (LITE-3) was attempted. This solution utilized most of the minima timings in Table 4 (photovisual, visual, and CCD), with only a few outliers excluded (Cycles −14178, −11948, −9446, −3402, and −2338). An arbitrary weighting scheme was applied, with w = 10 for CCD and w = 1 for photovisual and visual minima. The initial parameter values were taken from the LITE-2 solution. The results are tabulated in column 4 of Table 5 and displayed in Figure 7. The tertiary component masses listed in Table 5 were derived for each LITE solution using the mass function of the third body and the fitted parameters $A = a_{12} \sin i_3$ and P_3. The mass function is given by:

$$f(M_3) = \frac{(M_3 \sin i_3)^3}{(M_1 + M_2 + M_3)^2} = \frac{4\pi^2}{GP_3^2}(a_{12} \sin i_3)^3, \quad (7)$$

where G is the gravitational constant, $M_1 = 1.10 \pm 0.11\,M_\odot$, and $M_2 = 0.68 \pm 0.07\,M_\odot$ (see section 5 for binary component masses). The minimum mass occurs when the orbit of the tertiary component is co-planar with the binary's orbit ($i_3 = 90°$). For each LITE solution, Table 5 lists the values for the mass function f(m), the semimajor axis of the binary's orbit about the barycenter ($a_{12} \sin i_3$), and the tertiary masses for inclinations of 30°, 60°, and 90°. The tertiary component's minimum mass ranged from $1.6\,M_\odot$ to $1.8\,M_\odot$ for the three LITE solutions. Main sequence stars in this mass range would have approximate luminosities of $7-9\,L_\odot$ and temperatures from 7200 K to 7600 K. The observed color and the LAMOST spectra do not support a star of this temperature in the system. A tertiary component of this luminosity would also greatly reduce the eclipse depths and would result in large third-light values (l_3) during Roche modeling. The results of the LITE solutions will be discussed further in section 5.

Alternate explanations for a modulated orbital period include magnetic cycles in late-type stars and apsidal motion. Algol binaries with short orbital periods (<6 days) have circular orbits and are tidally locked, thus making apsidal motion unlikely as the cause of period modulation (Qian *et al.* 2018). The period changes may be caused by the Applegate mechanism, which postulates a change in the gravitational quadrupole moment of the binary's magnetically active secondary star (Applegate 1992; Lanza and Rodonò 1999; Völschow *et al.* 2016). This change is caused by the redistribution of angular momentum within the star due to the magnetic activity. To drive a period oscillation, a certain amount of energy is required to build a strong magnetic field. Eventually this field is dissipated, only to be built and dissipated again in a hydromagnetic dynamo cycle. A detailed investigation of the energetics ($\Delta E / E_{sec}$) was done by Völschow *et al.* (2016). The ratio $\Delta E / E_{sec}$ gives the energy necessary to drive the Applegate mechanism over the available energy produced by the magnetically active secondary star. This quantity determines the feasibility of the Applegate mechanism for LO UMa. The assessment of this mechanism for driving the period variations used Völschow *et al.*'s (2016) analytical two-zone model with different densities for the secondary's core and its convective shell. The $\Delta E / E_{sec}$ value was calculated using the "Eclipsing Time Variation Calculator" web module (http://theory-starformation-group.cl/applegate/index.php; Völschow *et al.* 2016). The module requires the following measured quantities for the calculation: the secondary star's mass (M_{sec}), radius (R_{sec}), and temperature (T_{sec}); the semimajor axis of the binary (a_{bin}); and $\Delta P / P_{bin}$, which is given by:

$$\frac{\Delta P}{P_{bin}} = 2\pi \frac{A_{o-c}}{P_{mod}}. \quad (8)$$

The calculations for this approximation used parameter values from each LITE solution and stellar parameters from Table 7 (see section 5). The resulting $\Delta E/E_{sec}$ values for LITE-1, LITE-2, and LITE-3 were 3.7, 1.9, and 1.7, respectively. In each case the relative threshold energy is greater than unity, indicating the energy necessary to drive the period oscillations is greater than the total energy generated by the secondary star. This implies the period modulation cannot be explained by the secondary star's magnetic activity. The period of modulation (P_{mod}) can also be estimated using the empirical relationship derived by Lanza and Rodonò (1999):

$$\log P_{mod} = -0.36(\pm 0.10)\log \Omega + 0.018, \qquad (9)$$

where $\Omega = 2\pi/P$, P_{mod} is in years, and P is in seconds. Equation 9 predicts a modulation period of about 40 years, which is much longer than the values found in the LITE analysis (13.4–18.6 years). This result also indicates magnetic activity is unlikely the cause of the period modulation.

4. Light curve analysis

4.1. Color, temperature, spectral type, absolute magnitude, and luminosity

For measuring color change and Roche modeling, the large number of photometric observations was binned in both phase and magnitude. This resulted in 150 points for each color with a phase width of 0.0067. The phases and magnitudes of the observations in each bin were averaged. For color index, the binned r' magnitudes were then subtracted from the linearly interpolated g' magnitudes. The binned points of the r' light curve and the (g'–r') color index are shown in Figure 8. The large color change during primary eclipse indicates a significant temperature difference between the primary and secondary stars. The average observed color over the entire phase range is (g'–r') = 0.479 ± 0.010. The color excess for this system, E(g'–r') = 0.020 ± 0.015, was determined from dust maps based on Pan-STARRS1 and 2MASS photometry and Gaia parallaxes (Green et al. 2018). Subtracting the color excess from the average observed color gives an intrinsic color of $(g'-r')_o = 0.46 \pm 0.02$.

The LAMOST spectral survey DR5 catalog gives an effective temperature of $T_{eff} = 6018 \pm 34$ K for LO UMa's primary star (Luo et al. 2015). The LAMOST pipeline measures the spectra as single stars even though in the case of Algol binaries, there are two stars of different temperatures. There are subtle differences between the spectra of Algol binaries and single stars (Qian et al. 2018). This results in a small systematic bias of less than 200 K in the effective temperature. For stars with temperature differences larger than 1000 K, as is the case for LO UMa, the systematic biases are even smaller. The effective temperature's error was set to ± 100 K to account for this bias. The observed color index can also be used to estimate the effective temperature. The dereddened color at orbital phase $\varphi = 0.5$ is (g'–r') = 0.379 ± 0.016. At this orbital phase, the secondary star's contribution to the total light is at a minimum. The effective temperature for this color, $T_{eff} = 6055 \pm 106$ K, was interpolated from Table 5 of Pecaut and Mamajek (2013). The effective temperatures from both methods are consistent, giving a spectral type of F9 for the primary star.

The absolute visual magnitude at quadrature ($\varphi = 0.75$), $M_v = 2.94 \pm 0.06$, was calculated using the Gaia distance and the apparent visual magnitude corrected for extinction. Using the bolometric correction for the effective temperature gives the combined luminosity of both stars, $L_{12} = 5.5 \pm 0.3 \, L_\odot$ (Pecaut and Mamajek 2013).

4.2. Synthetic light curve modeling

Simultaneous four-color light curve solutions were obtained using the 2015 version of the Wilson-Devinney (WD) program (Wilson and Devinney 1971; van Hamme and Wilson 1998). The input data consisted of 150 normal points for each color (see section 4.1). The normal points were converted from magnitudes to flux, with each point assigned a weight equal to the number of observations forming that point.

The light curves (see Figure 2) display a deep primary minimum that is briefly total, a shallow secondary minimum, and small brightness changes outside of eclipses. This light curve morphology is typical of an Algol binary where there are large temperature differences between the component stars. Algols are binaries that are often detached with spherical or slightly elliptical components, but some are semidetached with one star filling its Roche lobe. Not knowing the configuration of this system, the WD program was initially configured to Mode-2 for detached binaries. The primary star's effective temperature was fixed at $T_1 = 6018$ K (see section 4.1). The subscripts 1 and 2 refer to the hotter and cooler components, respectively. With both component temperatures less than 7500 K, internal energy transfer to the surface is due to convection rather than radiative transfer. Standard convective parameters were used: gravity brightening, $g_1 = g_2 = 0.32$ (Lucy 1968) and bolometric albedo, $A_1 = A_2 = 0.5$ (Ruciński 1969). Logarithmic limb-darkening coefficients were calculated by the program from tabulated values using the method of van Hamme (1993). The adjustable parameters include the inclination (i), mass ratio ($q = M_2/M_1$), potentials (Ω_1, Ω_2), temperature of the secondary star (T_2), band-specific luminosity for each wavelength (L), and third light (l). Given the evidence for a possible tertiary component (see section 3), third light was included from the beginning and throughout the solution process.

Preliminary fits to each light curve were made using the BINARY MAKER 3.0 program (BM3; Bradstreet and Steelman 2002). The primary star's temperature was set to 6018 K, standard convective parameters were used, and limb-darkening coefficients were taken from van Hamme's (1993) tabular values. The other parameters—inclination, mass ratio, potentials, and secondary star temperature—were adjusted in sequence until a good fit was obtained between the synthetic light curves and the observations for each passband. The parameters from the BM3 synthetic light curve fits were then averaged and used as the inputs for the computation of simultaneous four-color light curve solutions with the WD program. The Mode-2 iterations quickly converged to a semidetached configuration. Subsequent runs and solutions used Mode-5, in which the secondary potential (Ω_2) was no longer adjustable. A preliminary WD solution was completed using the Kurucz (2002) stellar atmosphere radiation

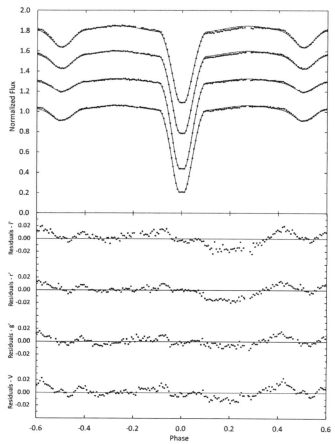

Figure 9. Comparison between the WD spotless best-fit model (solid curve) and the observed normalized flux curve. From top to bottom, the passbands are i', r', g', and V. Each light curve is offset by 0.25 for this combined plot. The residuals are shown in the bottom panel. Error bars are omitted from the points for clarity.

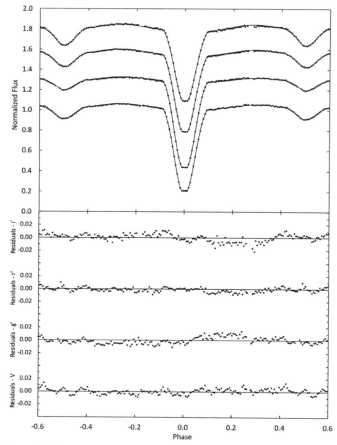

Figure 10. Comparison between the WD spotted best-fit model (solid curve) and the observed normalized flux curve. From top to bottom, the passbands are i', r', g', and V. Each light curve is offset by 0.25 for this combined plot. The residuals are shown in the bottom panel. Error bars are omitted from the points for clarity.

formulas, but this solution resulted in poor fits to the g'- and V-band observations. The final solution iterations were performed using blackbody radiation formulas, which resulted in better fits in all four passbands. The best-fit final solution parameters are shown in column 2 of Table 6. Figure 9 displays the normalized light curves overlaid by the synthetic solution curves (solid line), with the residuals in the bottom panel. Spectroscopic observations are not available to verify the mass ratio (q) found in this solution, but the total primary eclipses provide the necessary constraints for a reliable value (Wilson 1978; Terrell and Wilson 2005). Throughout the solution iteration process, the third-light corrections were negligibly small and often negative.

4.3. Spot model

The light curve asymmetries seen in Figure 9 are usually attributed to magnetic activity that causes cool spots or hot regions (faculae) in the star's photosphere. In Algol systems, a gas stream from the donor star can also form a hot spot on its companion from impact heating. The residuals in Figure 9 show the same asymmetries in all four colors: a small loss of light between orbital phases 0.05 and 0.30 and two small peaks of excess light at phases 0.40 and 0.60. To model these asymmetries, several different spot configurations were modeled using the BM3 program. The spot parameters, latitude, longitude, spot size, and temperature were adjusted until asymmetries were minimized. The process was repeated several times using different numbers of spots (1 to 3) and spot configurations until the asymmetries and residuals were minimized. The best-fit parameter values were then incorporated into a new WD model. The final spotted model resulted in a much-improved fit, with a 57% reduction in residuals compared to the spotless model. This model is not definitive; other spot configurations may give equal or better results. It does indicate that the light curve asymmetries are likely caused by star spots and that the stars are magnetically active. The final spotted solution parameters are shown in column 3 of Table 6. Figure 10 displays the spotted model fit (solid lines) overlaid on the observed light curves. Figure 11 shows a graphical representation of LO UMa created using BM3 (Bradstreet and Steelman 2002).

5. Discussion

The provisional absolute orbital and stellar parameters for each star can be determined with knowledge of one of the star's masses and the mass ratio. There are no spectroscopic observations currently available to directly determine the stellar masses, but the primaries in Algol systems are typically main

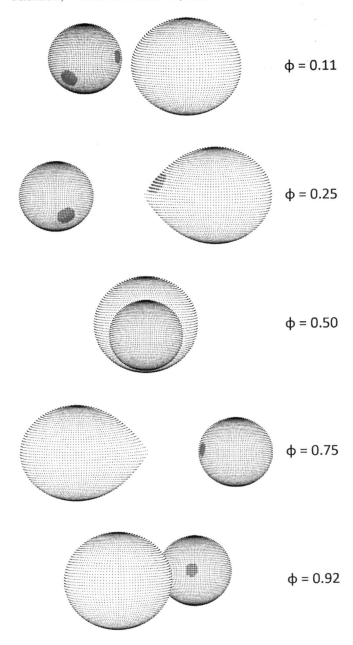

Figure 11. Roche Lobe surfaces of the best–fit WD spot model showing spot locations. The orbital phase is shown next to each diagram.

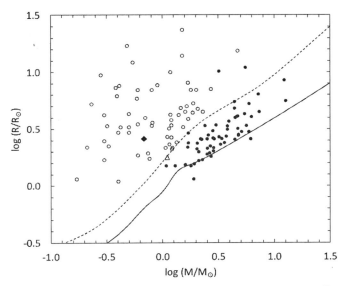

Figure 12. Positions of both components of LO UMa on the Mass-Radius diagram of 62 semidetached Algol systems with well-determined parameters. Closed circles are the primary stars and open circles the secondary stars. The triangle and the diamond are the primary and the secondary of LO UMa, respectively. Solid and dotted lines refer to ZAMS and TAMS, respectively (Tout *et al.* 1996).

sequence stars. The mass of those stars can be estimated from their spectral type. The primary's mass, $M_1 = 1.10 \pm 0.11\,M_\odot$, was interpolated from Table 5 of Pecaut and Mamajek (2013) using its effective temperature. This mass, combined with the spotted WD solution mass ratio, gives a secondary mass of $M_2 = 0.68 \pm 0.07\,M_\odot$. Applying Kepler's Third Law gives the distance between the mass centers as $7.7 \pm 0.2\,R_\odot$. The bolometric magnitudes, radii, and surface gravities of the stars were calculated by the WD light curve program (LC). The stellar luminosities, $L_1 = 3.7 \pm 0.4\,L_\odot$ and $L_2 = 1.5 \pm 0.4\,L_\odot$, were computed using the LC bolometric magnitudes in the following equation:

$$M_{bol} = 4.74 - 2.5 \log\left(\frac{L}{L_\odot}\right). \quad (10)$$

The total system luminosity, $L_{12} = 5.2 \pm 0.7\,L_\odot$, is in good agreement with the value calculated in section 4.1 using observed quantities, $5.5 \pm 0.3\,L_\odot$. All the provisional stellar parameter values are collected in Table 7. The distance modulus gives a distance of 853 ± 145 pc, which is consistent with the Gaia distance of 863^{14}_{18} pc (Bailer-Jones *et al.* 2021; Gaia 2016, 2018). In Figure 12, the provisional radii and masses of LO UMa are compared with the values from 62 semi-detached systems with well-determined absolute parameters (Ibanoğlu *et al.* 2006). The zero-age main sequence lines (ZAMS) and the terminal-age main sequence (TAMS) lines are displayed in Figure 12 as well. The primary component of LO UMa (triangle point) has one of the lowest masses of this group and a larger radius compared to a ZAMS star of the same mass. The secondary component (diamond point) is located above the TAMS line indicating an evolved star.

In the period study (see section 2), the parameter values of the three LITE solutions are comparable, but there was one significant difference. The LITE-2 solution gave an orbital eccentricity of $e = 0.42 \pm 0.04$ and a period of $P_3 = 16.4 \pm 0.03$ y for the tertiary component. LITE-3 gave similar values, with $e = 0.47 \pm 0.33$ and $P_3 = 18.6 \pm 0.3$ y. These eccentricity values differ by only 11%, but the error in LITE-3 is very large, as is the error for the time of periastron passage ($T_0 = 2454771 \pm 403$ HJD). The large errors are likely the result of the sparse coverage, large data gaps, and the lower accuracy of the minima timings from the years 1939–1999. The main difference between the LITE solutions concerns the long-term period change given by the quadratic coefficient term (Q). Its value was positive for the LITE-1 solution but negative for LITE-2 and LITE-3. In a close semidetached Algol binary, conservative mass exchange from the less massive Roche-lobe-filling component to the more massive star always causes an increase in orbital period (Q>0). Matter transferred through the inner Lagrangian point may hit the primary star, causing impact heating, or miss the primary and form a gaseous disk around the star. The distance separating the component stars (A_{orb}) and the radius of the primary determine which one of these configurations occurs. The minimum primary-star radius required for the formation of a gaseous disk was calculated from an empirical relationship derived by Nanouris et al. (2015):

$$R_{min} = (0.04930 + 0.03387 \log q + 0.05915 (\log q)^2) A_{orb}, \quad (11)$$

where q is the mass ratio. The resulting value, $R_{min} = 0.345 \pm 0.002\ R_\odot$, is much smaller than the estimated radius for this star (1.78 $R_\odot$). This means the matter stream would collide with the primary star. The hot spot modeled on the side of the primary star facing the secondary suggests mass transfer is presently active. The downward parabolic O–C diagrams found in the LITE-2 and LITE-3 solutions suggest just the opposite of LITE-1: the orbital period is decreasing (Q<0; see Figures 6 and 7). This implies a nonconservative mass-loss process, which is usually attributed to magnetic braking caused by a coupling between the magnetic field and stellar winds in low-mass stars. The spots found in the light curve solution support current magnetic activity on both stars. This non-conservative mass loss would remove orbital angular momentum from the system, leading to a downward parabolic O–C curve and a long-term decrease in the orbital period. In a comprehensive study on the efficiency of O–C diagrams for diagnosing long-term period changes, it was found that a combination of the mass transfer process and wind-driven mass loss may be at work in close binaries (Nanouris et al. 2011, 2015; Erdem and Öztürk 2014). In semidetached systems, these two mechanisms may be strongly competitive. Based on the LITE-2 solution, the period of LO UMa is decreasing at a rate of 4.5×10^{-7} d yr^{-1}, or 4 seconds per century. The LITE-3 solution, with its much longer temporal base (82 years), also supports a decreasing orbital period.

The period modulation found in the O–C residuals was presented as evidence for a third body orbiting the system's barycenter. Cyclical orbital-period variation in binary systems is common; it is observed in 49% of Algols and 64% of W-UMa systems (Liao and Qian 2010). The sinusoidal-like behavior found in the period analysis of LO UMa is mostly supported by CCD minima timings collected over the past 21 years. This time interval is less than two cycles of the proposed orbital period (P_3). This third-body hypothesis should therefore be considered preliminary. Another 10–15 years of precision minimum times will be necessary to confirm that the period modulation is continuing as predicted and thus to prove the existence of the tertiary component. Those future observations could also confirm that the binary's orbital period is decreasing and reduce the errors in the orbital parameters. The LITE solutions indicate the tertiary component has a minimum mass between 1.6 $M_\odot$ and 1.8 $M_\odot$, yet the light curve solutions found no evidence of excess luminosity ($l_3=0$). A main sequence star of this mass would have a spectral type of F0, but the LAMOST spectra gives an F9 spectral type that is consistent with the observed color. A massive orbiting object not emitting normal stellar radiation would suggest that it is a noninteracting compact stellar object. With the minimum mass above the Chandrasekhar limit of $\simeq 1.4\ M_\odot$, the tertiary component would most likely be a neutron star. If the orbit has a high inclination i_3, a black hole is also possible. The LITE-2 solution gives periastron distance between this object and the binary of 5.6 ± 0.4 AU ($i_3 = 90°$) and an apastron distance of 13.7 ± 0.4 AU. The barycenter is almost equally distant between the binary and the tertiary object. A compact object of this mass and distance would play a significant role in the evolution of this system.

6. Conclusions

Multiband CCD photometric observations collected in V, g', r', and i' bands resulted in the first precision light curves for LO UMa and two new minimum times for primary eclipse. The light curves displayed deep total primary eclipse and shallow secondary eclipse. Light curve modeling with the WD program found the binary configuration to be semidetached, with primary and secondary stars of spectral types F9 and K8, respectively. Three spots were included in the final Roche model to address light curve asymmetries: a cool spot and a hot spot on the primary star and a single hot spot on the secondary star. This spotting is an indication of magnetically active stars. The linear ephemeris was updated using CCD minima timing observations from the years 1999–2021. A detailed analysis of the O–C diagram found the orbital period of LO UMa is undergoing a sinusoidal variation superimposed on a downward parabolic change. The downward parabolic change indicates that the orbital period of the binary is decreasing due to a combination of magnetic braking and mass transfer. Two possible causes for the period modulation were investigated: the existence of an object of significant mass that is gravitationally bound to the binary and the Applegate effect. The Applegate mechanism requires a certain amount of energy to build a strong magnetic field to drive a dynamo cycle that causes the period variations. Calculations showed the energy available from the secondary star was insufficient to drive the Applegate mechanism. The results of the LITE analysis showed that an object orbiting in either a circular or an elliptical revolution would explain the period modulation. The best-fit LITE solution gave the third body's orbital eccentricity as 0.42 and its minimum mass as 1.6 $M_\odot$. Given this mass, the object is hypothesized to be a neutron star.

LO UMa is an interesting system worthy of additional study. A spectroscopic study would be particularly useful to gain a better understanding of this binary system. Radial velocity measurements are needed to pin down the individual masses and separation distance of the binary stars. Velocity changes in the binary's barycenter could provide supporting evidence for the unseen companion. Spectroscopy could also check the metallicity of the binary stars for possible contamination from a supernova which would have formed the neutron star. Precision CCD minima timings over many years will be very important in confirming the third body and the decreasing orbital period of the binary.

7. Acknowledgements

This research was made possible through use of the AAVSO Photometric All-Sky Survey (APASS), funded by the Robert Martin Ayers Sciences Fund. This research has made use of the SIMBAD database and the VizieR catalog access tool, operated at CDS, Strasbourg, France. This work has made use of data from the European Space Agency (ESA) mission Gaia (https://www.cosmos.esa.int/gaia), processed by the Gaia Data Processing and Analysis Consortium (DPAC, https://www.cosmos.esa.int/web/gaia/dpac/consortium). Funding for DPAC has been provided by national institutions, in particular the institutions participating in the Gaia Multilateral Agreement.

References

Applegate, J. H. 1992, *Astrophys. J.*, **385**, 621.
Bailer-Jones, C. A. L., Rybizki, J., Fouesneau, M., Demleitner, M., and Andrae, R. 2021, *Astron. J.*, **161**, 147.
Baldwin, M. E., Guilbault, P. R., Henden, A. A., Kaiser, D. H., Lubcke, G. C., Samolyk, G., and Williams, D. B. 2001, *J. Amer. Assoc. Var. Star Obs.*, **29**, 89.
Bradstreet, D. H., and Steelman, D. P. 2002, *Bull. Amer. Astron. Soc.*, **34**, 1224.
Butters, O. W., et al. 2010, *Astron. Astrophys.*, **520**, L10 (SuperWASP, https://wasp.cerit-sc.cz/form).
Diethelm, R. 2003, *Inf. Bull. Var. Stars*, No. 5438, 1.
Diethelm, R. 2004, *Inf. Bull. Var. Stars*, No. 5543, 1.
Diethelm, R. 2009, *Inf. Bull. Var. Stars*, No. 5894, 1.
Diethelm, R. 2010, *Inf. Bull. Var. Stars*, No. 5945, 1.
Diethelm, R. 2011, *Inf. Bull. Var. Stars*, No. 5992, 1.
Diethelm, R. 2012, *Inf. Bull. Var. Stars*, No. 6029, 1.
Erdem, A., and Öztürk, O. 2014, *Mon. Not. Roy. Astron. Soc.*, **441**, 1166.
Gaia Collaboration, et al. 2016, *Astron. Astrophys.*, **595A**, 1.
Gaia Collaboration, et al. 2018, *Astron. Astrophys.*, **616A**, 1.
Green, G. M., et al. 2018, *Mon. Not. Roy. Astron. Soc.*, **478**, 651.
Henden, A. A., et al. 2015, AAVSO Photometric All-Sky Survey, data release 9, (https://www.aavso.org/apass).
Hübscher, J., and Lehmann, P. B. 2015, *Inf. Bull. Var. Stars*, No. 6149, 1.
Hübscher, J., Lehmann, P. B., Monninger, G., Steinbach, H.-M., and Walter, F. 2010, *Inf. Bull. Var. Stars*, No. 5918, 1.
Ibanoğlu, C., Soydugan, F., Soydugan, E., and Dervişoğlu, A. 2006, *Mon. Not. Roy. Astron. Soc.*, **373**, 435.
Irwin, J. 1959, *Astron. J.*, **64**, 149.
Jayasinghe, T., et al. 2019, *Mon. Not. Roy. Astron. Soc.*, **486**, 1907.
Kafka, S. 2017, variable star observations from the AAVSO International Database (https://www.aavso.org/aavso-international-database).
Krajci, T. 2005, *Inf. Bull. Var. Stars*, No. 5592, 1.
Kreiner, J. M. 2004, *Acta Astron.*, **54**, 207.
Kurucz, R. L. 2002, *Baltic Astron.*, **11**, 101.
Kwee, K. K., and van Woerden, H. 1956, *Bull. Astron. Inst. Netherlands*, **12**, 327.
Lanza, A. F., and Rodonò, M. 1999, *Astron. Astrophys.*, **349**, 887.
Liao, W.-P., and Qian S.-B. 2010, *Mon. Not. Roy. Astron. Soc.*, **405**, 1930.
Locher, K. 2005, *Open Eur. J. Var. Stars*, **3**, 1.
Locher, K., Blättler, E., and Diethelm, R. 2002, *BBSAG Bull.*, No. 128, 1 (http://www.variables.ch/observations_BBSAG.html).
Lucy, L. B. 1968, *Astrophys. J.*, **151**, 1123.
Luo, A-Li., et al. 2015, *Res. Astron. Astrophys.*, **15**, 1095.
Mirametrics. 2015, Image Processing, Visualization, Data Analysis, (https://www.mirametrics.com).
Nagai, K. 2009, *Bull. Var. Star Obs. League Japan*, **48**, 1.
Nanouris, N., Kalimeris, A., Antonopoulou, E., and Rovithis-Livaniou, H. 2011, *Astron. Astrophys.*, **535A**, 126.
Nanouris, N., Kalimeris, A., Antonopoulou, E., and Rovithis-Livaniou, H. 2015, *Astron. Astrophys.*, **575A**, 64.
Paschke, A. and Brat, B. 2021, O–C Gateway (http://var2.astro.cz/ocgate/).
Pecaut, M. J., and Mamajek, E. E. 2013, *Astrophys. J., Suppl. Ser.*, **208**, 9 (http://www.pas.rochester.edu/~emamajek/EEM_dwarf_UBVIJHK_colors_Teff.txt).
Qian, S.-B., Zhang, J., He, J.-J., Zhu, L.-Y., Zhao, E.-G., Shi, X.-D., Zhou, X., and Han, Z.-T. 2018, *Astrophys. J. Suppl. Ser.*, **235**, 5.
Ruciński, S. M. 1969, *Acta Astron.*, **19**, 245.
Shappee, B. J., et al. 2014, *Astrophys. J.*, **788**, 48.
Terrell, D., and Wilson, R. E. 2005, *Astrophys. Space Sci.*, **296**, 221.
Tout, C. A., Pols, O. R., Eggleton, P. P., and Han, Z. 1996, *Mon. Not. Roy. Astron. Soc.*, **281**, 257.
van Hamme, W. 1993, *Astron. J.*, **106**, 2096.
van Hamme, W. V., and Wilson, R. E. 1998, *Bull. Amer. Astron. Soc.*, **30**, 1402.
Völschow, M., Schleicher, D. R. G., Perdelwitz, V., and Banerjee, R. 2016, *Astron. Astrophys.*, **587A**, 34.
Williams, D. B. 2001, *Inf. Bull. Var. Stars*, No. 5084, 1.
Wilson, R. E. 1978, *Astrophys. J.*, **224**, 885.
Wilson, R. E., and Devinney, E. J. 1971, *Astrophys. J.*, **166**, 605.
Zasche, P., Liakos, A., Niarchos, P., Wolf, M., Manimanis, V., and Gazeas, K. 2009, *New Astron.*, **14**, 121.

Photometric Observations of the Dwarf Nova AH Herculis

Corrado Spogli
Via Palazzolo 21 Frazione Spada 06020 Gubbio (PG), Italy; corradospogli@yahoo.it

Gianni Rocchi
Via Achille Grandi 14, 06038 Spello (PG), Italy; giannirocchi2@gmail.com

Stefano Ciprini
Space Science Data Center, Agenzia Spaziale Italiana (SSDC-ASI), I-00133, Roma, Italy, and Istituto Nazionale di Fisica Nucleare (INFN), Sezione di Roma Tor Vergata, I-00133, Roma, Italy; stefano.ciprini.asdc@gmail.com

Dario Vergari
Via Cantalmaggi 24, 06024 Gubbio (PG), Italy

Jacopo Rosati
Via XVIII Maggio N°4 06024 Gubbio (PG), Italy; jacopo.rosati@studenti.unipg.it

Received May 24, 2021; revised August 24, September 20, 2021; accepted September 20, 2021

Abstract We present the results of 274 nights of observations of the dwarf nova AH Herculis made in the years 2012, 2014, 2017, and 2018 for a total of 725 photometric data points. Observations were made in the B, V, R_c, and I_c Johnson-Cousins photometric bands. In 2012 AH Her was observed for 49 nights, in 2014 for 21 nights, and in 2017 and 2018 for 102 nights each year. Overall, we obtained 186 data points with the photometric filter B, 270 observations with the V filter, 165 with the R_c filter, and 104 with the I_c filter. The variable was well sampled in 2017 and 2018 and was observed on almost all clear nights; comments are missing in some filters due to technical problems with the filter wheel. The observations were all made at Gianni Rocchi's private observatory. In 2017 and 2018 we observed several outbursts of AH Her and in 2017 a standstill of short duration. In this work we present the observational data, the light curves obtained in 2012, 2017, and 2018, a study of the color indices, and the temporal characteristics of the outbursts of this dwarf nova.

1. Introduction

Cataclysmic variables (CVs) are binary stars containing a white dwarf that is accreting material from a red dwarf secondary or main sequence or subgiant companion (see Warner 1995 for a comprehensive review). An important subclass is the dwarf novae. Based on their photometric behavior, we distinguish a few subclasses of dwarf novae: U Gem, showing more or less similar outbursts; SU UMa, characterized by the so-called superoutbursts in addition to normal outbursts; Z Cam, with outbursts interrupted by irregular standstill (activity suspensions) intervals of constant brightness.

The U Gem variables have explosions that raise their brightness by 2 to 6 magnitudes and last for one or two days. In the following days the system returns to its usual brightness. SS Cyg variables are also called after their alternative prototype, SS Cygni, which periodically exhibits the brightest events of this subtype of variables.

The SU UMa sub-class is characterized by two very distinct outbursts types: short ones (lasting a few days) and superoutbursts which can last two weeks or longer in their rather bright "plateau" phase. Normal explosions are similar to those that occur in U Gem variables, while superoutbursts are two magnitudes brighter, last five times longer, and are three times less frequent. Typical superoutburst cycle lengths of these "ordinary" SU UMa stars range from 100 to 500 days. SU UMa systems generally have an orbital period $P_{orb} < 2$ hours and brighter superoutbursts occurring every few months, while U Gem and Z Cam systems have $P_{orb} > 3$ hours and normal outbursts. Within the SU UMa class there is an additional distinction, from the most to the least active ones: ER UMa-type, pure SU UMa-type, and WZ Sge-type stars (see Hellier 2001 and Warner 2003 for a detailed overview). ER UMa stars have very short (much less than 100 d) regular supercycles, very short recurrence times of normal outbursts, and long duty cycles (for a review, see Kato *et al.* 1999) while WZ Sge-type dwarf novae are considered to be objects at the terminal stage of the cataclysmic variable (CV) evolution. WZ Sge stars are characterized by the large amplitude and long duration of superoutbursts which are accompanied by "early superhumps" in the early terms of the superoutbursts (see Kato 2015).

The main characteristics of the Z Cam subclass are: the short duration of minimum; the irregularity of the light curve, described as rare for U Gem types and almost the norm for Z Cams; the lesser amplitudes of variation compared to U Gems; and a "curious and very special feature" wherein the variable remains nearly constant at a magnitude in between the maximum and minimum: this peculiarity is called "standstill" and it is the most significant characteristic of assigning membership to the Z Cam classification of dwarf novae. The Z Cams are not very numerous; about 30 are known, and only 17 of the 19 bona fide Z Cams have orbital periods in the literature. All have periods from 3.048 hours (0.127 d) to 8.4 hours (0.38 d), the average being 5.272 hours (0.2196 d). Z Cams are very active systems.

Most have outburst cycles (the time between successive maxima) between 10 and 30 days. Their normal cycles between maxima and minima look very much like U Gem stars but they spend very little time at minimum.

Outburst amplitudes of Z Cam stars range from 2.3 to 4.9 magnitudes in V. The average amplitude is 3.7 V magnitudes. This is identical to the range of amplitudes seen in U Gem stars, so it cannot be used to distinguish them from these more common dwarf novae. Z Cam systems that show "standstill" in their light curves are thought to be on the boundary between nova-like variable stars with their hot stable discs, and dwarf novae with their unstable discs (Smak 1983). The duration of standstills has a wide range, from tens of days to several years.

AH Her is a dwarf nova, Z Cam subtype, that varies in magnitude from V = 14.3 in quiescence to V = 11.3 during outburst, with outbursts lasting 4 to 18 days and recurring at intervals of 7 to 27 days (Ritter and Kolb 1998). AH Her is a very active dwarf nova. Spectroscopic observations were made by Williams (1983) that published a spectrum of the variable at minimum, giving the equivalent width of some lines of the Balmer series. Through spectroscopic observations, Horne et al. (1986) determined an orbital period for AH Her equal to P = 0.258116 day (6.19 hours). They found a M2/M1 mass ratio of 0.80 with M1 = 0.95 and M2 = 0.70 solar masses; they calculated the inclination of the orbital plane and found i = 46° with respect to the secondary star of spectral type K. AH Her was detected in the ROSAT all-sky survey at a low rate (Verbunt et al. 1997). Simultaneus optical and UV (IUE) observations show that the UV flux follows the optical flux during an outburst (Verbunt et al. 1984). Wils et al. (2010) reports that the variable distance is 450 parsecs, while Ramsey et al. (2017), through Gaia satellite estimations, report a distance of this variable of 325.0 ± 47.2 parsecs. Further spectroscopic observations made by Echevarria et al. (2021), during a deep quiescent state, indicate that K5 is the most likely spectral type of the secondary and that the orbital period is $P = 0.25812 \pm 0.00032$ d, a value consistent with those determined by Horne et al. (1986).

Dwarf novae can have type A or type B outbursts. In type A with fast optical rise, the system brightens at longer wavelenghths first, with shorter wavelengths delayed progressively. In type B with a slower rise, the rise is almost simultaneous at all wavelenghts with at most only a small delay between optical and UV. This affects the light curve, which can be asymmetrical (type A) or symmetrical (type B). In a separate section we will deal with this issue and in the case of AH Her we will see that this dwarf nova can have both type A and type B outbursts.

2. Photometric observations and light curve

All the observations were obtained with a 0.12-m f/7 apochromatic refractor telescope by Skywatcher Esprit trademark, equipped with an Orion G3 CCD camera (Sony $I_c \times 419$ all), R_c, I_c Schuler filters, and U, B, V Baader filters. The exposure time was 240 sec. Our photometric system has been carefully tested by observing the M67 sequence (Chevalier and Ilovaisky 1991). The CCD frames were first corrected for de-biasing and flat fielding, then processed for aperture photometry. All the B, V, R_c data were obtained via differential photometry using the photometric comparison stars 1, 2, 3 reported by Misselt (1996). To estimate the observations of AH Her made with the I_c filter we used the values $I_c(1) = 12.07 \pm 0.03$, $I_c(2) = 14.22 \pm 0.05$, $I_c(3) = 13.40 \pm 0.04$ reported by Spogli et al. (2001). Magnitude errors were evaluated as standard deviations of the mean. All observational data relating to the years 2012, 2014, 2017, and 2018 are shown in Appendix A after the references. A finding chart for AH Her is shown in Figure 1.

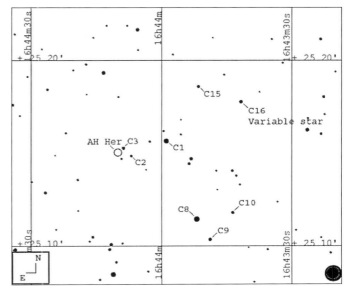

Figure 1. Finding chart for AH Her field.

2.1. Observations made in 2012

AH Her was observed sporadically in the V and R_c filters from 10 July 2012 to 7 November 2012 for a total of 49 nights, 42 of which were for observations in V and 7 in R_c. The star seems to maintain an average level of luminosity equal to $V = 12.58 \pm 0.11$ magnitudes and $R_c = 12.56 \pm 0.13$ mag; however, the star oscillates in the V band between magnitude 12.88 and 12.31 and in the R_c band between 12.71 and 12.35, even if in the latter case the photometric data are few. In Figure 2 we present our light curve from 2012.

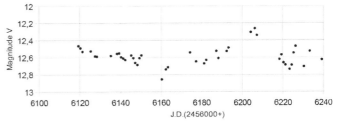

Figure 2. AH Her V band light curve in 2012.

2.2. Observations made in 2014

For the year 2014, we obtained sporadic observations in the four photometric filters: 5 observations in B, 18 in V, 9 in R_c, and 9 in I_c, over a total of 21 nights. The low numbers of observations are due to technical problems at Gianni Rocchi's telescope. We can, however, make a rough estimate of the color indices; at a minimum, the average color indices are:

(B–V) = 0.54 ± 0.08, (V-R) = 0.49 ± 0.06, (V–I) = 0.73 ± 0.14, while in the phase of maximum light we only have a single estimate of the color index that is: (B–V) = –0.02.

2.3. Observations made in 2017

In 2017 AH Her was observed for 102 nights, from 22 April 2017 to 11 November 2017 in three photometric filters: B, V, R_c. There were few observations in the I_c filter due to technical problems with the filter wheel. The photometric data obtained were 314, divided as follows: 97 in B, 109 in V, 94 in R_c, and only 14 in I_c. Based on these data we have built the light curve in V that is presented in Figure 3, while in Figure 4 we present the light curve of AH Her in all four photometric filters.

From the analysis of the light curve of the variable in the V band, we can see that nine maximum brightnesses of AH Her and one standstill were observed.

The temporal distance between two consecutive maxima of the star is on average 20.5 days, while the duration of the standstill phase was almost 21 days. In Figure 5 we have represented the light curve of the variable during the standstill phase.

During the standstill, the average brightness values of AH Her in the different photometric bands were as follows: B = 12.6 ± 0.3, V = 12.5 ± 0.2, R_c = 12.3 ± 0.2, and I_c = 12.1 ± 0.1. After the standstill the star has a maximum brightness and it suggests that AH Her may belong to the IW And subclass of the Z Cam stars (Kato 2019).

Table 1 shows the main characteristics of our observational data for 2017.

2.4. Observations made in 2018

In 2018 AH Her was observed in four photometric filters; for three of the four filters data were obtained for 102 nights. We collected 331 photometric data divided as follows in the various filters: 84 data in B, 101 in V, 55 in R_c, and 81 in I_c. We occasionally had problems with the R_c filter, so we did not always manage to use it. In fact, in our observations there are missing data in particular from 02 July to 01 September 2018. Table 2 shows the main characteristics of our observational data for 2018. Figure 6 shows the light curve of AH Her in the V band for the year 2018, while Figure 7 shows the light curve of AH Her in all four photometric filters. Figure 8 shows the maximum brightness values reached during the various AH Her outbursts in 2018.

You may notice a slight decrease in brightness, and this is due to the worsening of the weather conditions in the months of October and November and to the fact that the outbursts were no longer observed continuously, hence the fragmented data. From a check of the observational data for October and November, the possible influence of the air mass on our observations does not emerge, since the variable at the time it was observed was high on above the horizon, and, also, the difference between the instrumental magnitudes of the comparison stars C1 and C3 always remained constant in the various observational bands and for the entire period of time in which AH Her was observed.

In Figure 9 we report the observational values of AH Her in B, V, and I_c, when the star was in the phase of minimum light. We can see an oscillating trend in the light curve.

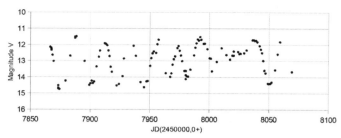

Figure 3. The V-band light curve of AH Her in 2017.

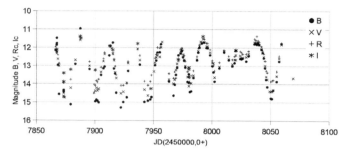

Figure 4. The 2017 light curve of AH Her in all four photometric filters (B, V, R_c, I_c).

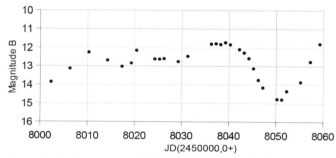

Figure 5. The light curve of AH Her during the standstill ranging from JD 2458010 to 2458030. Note that after the standstill phase, the variable has a new maximum and then a slow descent to the minimum brightness typical after a normal outburst.

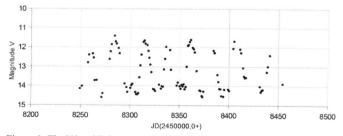

Figure 6. The V band light curve of AH Her in 2018; eight maximum brightness values corresponding to eight outbursts are clearly evident.

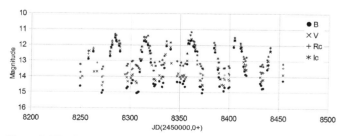

Figure 7. The 2018 light curve of AH Her in the four photometric bands (B, V, R_c, I_c).

Table 1. Summary data of AH Her 2017.

	B	B Error	V	V Error	R_c	R_c Error	I_c	I_c Error
Maximum Values	10.96	0.04	11.49	0.05	11.37	0.03	11.38	0.03
Minimum Values	15.29	0.12	14.74	0.04	14.39	0.05	13.88	0.04
Mean Values at Minimum	14.66	0.31	14.25	0.30	13.78	0.25	—	—
Mean Values at Maximum	11.83	0.24	11.84	0.20	11.75	0.22	—	—

Table 2. Summary data of AH Her 2018.

	B	B Error	V	V Error	Rc	Rc Error	Ic	Ic Error
Maximum Values	11.55	0.02	11.41	0.02	11.43	0.02	11.36	0.07
Minimum Values	15.06	0.02	14.56	0.03	14.09	0.02	13.63	0.11
Mean Values at Minimum	14.64	0.25	14.19	0.22	13.75	0.21	13.28	0.16
Mean Values at Maximum	11.89	0.19	11.87	0.19	11.84	0.22	11.66	0.24

Table 3. The mean values of color index in 2018. The errors on the color indices were calculated as standard deviation from the mean.

	B–V	B–V Error	V–R	V–R Error	R–I	R–I Error	V–I	V–I Error
Mean Values at Maximum	0.04	0.05	0.11	0.06	0.12	0.08	0.28	0.10
Mean values at Minimum	0.45	0.11	0.43	0.09	0.51	0.09	0.93	0.14

Note: The errors on the color indices were calculated as standard deviation from the mean.

In Figure 10 we report the observed values of AH Her in Ic during the minimum light phase. We can see that the star oscillates between I_c magnitudes 13.6 and 12.9. The observational data relating to the star in the minimum luminous phase were selected by selecting a posteriori, from the analysis of the light curve, the days in which the star appeared faintly luminous.

3. A study of color indices

Bruch (1984) reported that the color index B-V varies from 0.04 to 0.13 in the maximum of an outburst, while in the minimum B–V varies from 0.24 to 0.55. In the years in which AH Her was better monitored, i.e. in 2017 and 2018, the color indices had different values depending on the state of the star. In 2017, in the first three outbursts observed, the B–V color index is strongly negative, as we can see from Figure 15, in which we have represented the color index B–V as a function of time, with values of B–V oscillating between –0.3 and –0.8, something that no longer occurred in subsequent outbursts. Excluding these first observational data, in the following outbursts during the maximum phase, the B–V color index assumed values between 0.09 and –0.05, with an average value equal to B–V = 0.03. Table 3 shows the mean values of the B–V, V–R, R–I, and V–I color indices calculated for AH Her in 2018, in phases of minimum and maximum brightness. Figure 15 shows a comparison between the light curve of AH Her in 2017 and the corresponding trend of the B–V color index.

As for the V–R color index, it varies between 0.01 and 0.19, with an average value of 0.12. Considering the limited data available, the mean value of V–I_c is equal to 0.18 but this result is not significant. We also calculated the color indices of AH Her in the minimum light phase: B–V varies between 0.3 and 0.9, with an average value of 0.58, while V–R varies between 0.14 and 0.73, with an average value of 0.41—in excellent agreement with the value found by Spogli et al. (2001). As for the color indices estimated in 2018, they do not differ much from those calculated in 2017. At the minimum brightness of the star, the B–V varies between 0.22 and 0.69, while V–R varies between 0.22 and 0.67, R-I varies between 0.25 and 0.67, and V–I has values between 0.70 and 1.11.

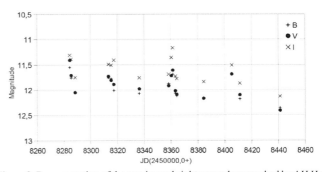

Figure 8. Representation of the maximum brightness values reached by AH Her in the various outbursts of 2018.

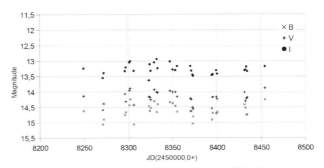

Figure 9. AH Her 2018 BVI$_c$ light curve in the minimum light phase, constructed excluding the phases concerning the outbursts.

Clearly, the variable star has a variation of its spectral type when it reaches the maximum of the outburst from the phase of minimum light: AH Her tends to become bluer, and from type K it changes to type A, according to the Harvard classification; this is what can be deduced from the variation of the color indices.

During the various outbursts of the variable in 2018, in the maximum phase the B–V color index varies from –0.04 to 0.15, the V–R from –0.02 to 0.16, the R–I between 0.1 and 0.27, and the V–I between 0.1 and 0.48.

Figure 11 shows how the B–V color index varies as a function of the R_c magnitude: it can be seen that in the maximum phase the points accumulate around the value B–V = 0 or are negative, while in the phase of minimum light the values of B–V are included between 0.3 and 1.

Figure 12 shows how the $V–R_c$ color index varies as a function of the R_c magnitude in the year 2017: in the phase of maximum $V–R_c$ has values between 0 and 0.2 while at minimum brightness $V–R_c$ has values between 0.2 and 0.8.

Figure 13 shows how the $V–R_c$ color index varies as a function of time in the year 2017: we can see, comparing this graph with the light curve of AH Her, how V–R at the maximum brightness of the outbursts has values between 0 and 0.2, with some negative data, while in the minimum phase the values oscillate between 0.4 and 0.6, with peaks up to 0.8.

Figure 14 shows how the B–V color index varies as a function of time in the observations made in 2017: B–V is sharply negative in the rising phases preceding the outburst of the dwarf nova, assumes values between 0 and 0.1 at maximum, and values between 0.4 and 1.0 at minimum light.

Figure 15 shows a comparison between the AH Her light curve in 2017 and the corresponding change in the B–V color index. The color index assumes negative values during the maximum brightness of the outburst and positive values during the minimum brightness phase. During the standstill the B–V color index fluctuates around the value of zero.

In Figure 16 we can see how the $V–I_c$ color index varies as a function of the V magnitude in the observations made in 2018. It may be noted that at the minimum the $V–I_c$ values are between 0.6 and 1.2, while at the maximum $V–I_c$ oscillates between 0.1 and 0.4. The overall trend of the points draws an arc of a parabola.

In Figure 17 we can see how the $V–I_c$ color index varies as a function of time in the observations made in 2018. Clearly, during the numerous outbursts the $V–I_c$ color index varies between 0.4 and 0.1, while in the phase of minimum light it oscillates between 0.7 and 1.1.

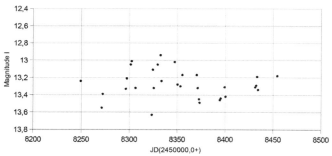

Figure 10. AH Her light curve in the minimum light phase, in the I_c band.

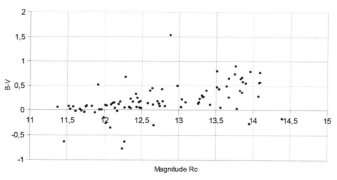

Figure 11. The B–V color index as a function of the magnitude R_c in 2017.

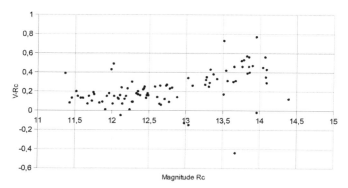

Figure 12. The $V–R_c$ color index as a function of the magnitude R_c in 2017.

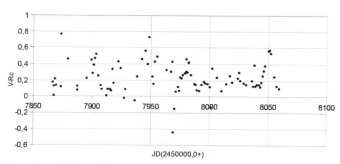

Figure 13. Variation of the $V–R_c$ color index as a function of time in 2017.

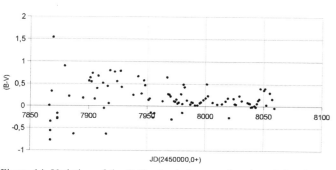

Figure 14. Variation of the B-V color index as a function of time in the observations made in 2017.

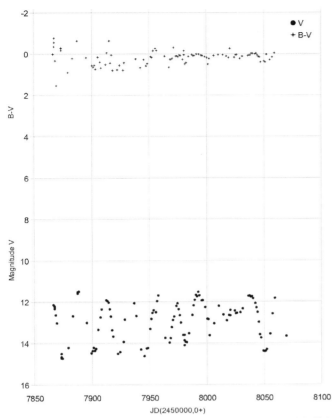

Figure 15. Variation of the B–V color index in relation to the trend of the light curve of AH Her in 2017.

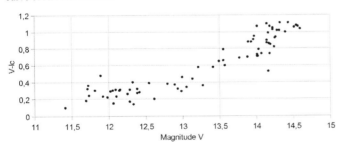

Figure 16. The V–I$_c$ color index as a function of the V magnitude in the 2018 observations.

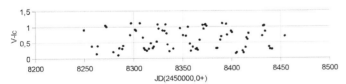

Figure 17. Variations of the V–I$_c$ color index with time in the 2018 observations.

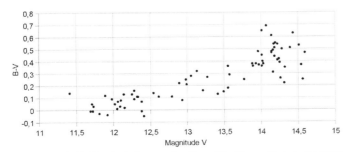

Figure 18. The B–V color index as a function of the V magnitude in the 2018 observations.

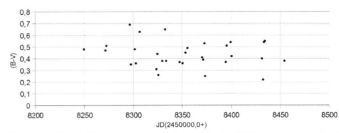

Figure 19. The B–V color index at a minimum in the 2018 observations: B–V values fluctuate between 0.2 and 0.7.

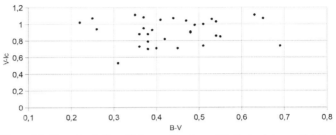

Figure 20. The color indices V–I$_c$ as a function of B–V at minimum light in the 2018 observations.

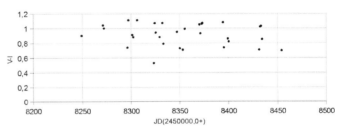

Figure 21. Distribution of the V–I$_c$ color index at a minimum as a function of time in 2018.

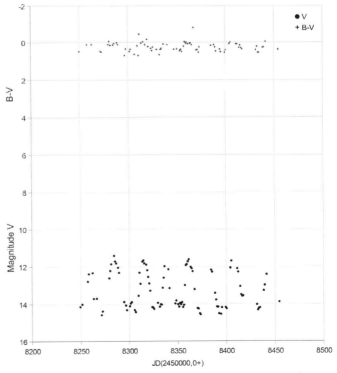

Figure 22. This figure shows the comparison between the light curve of AH Her in 2018 and the corresponding variation of the B–V color index referred to the same times and the same values of V magnitude.

In Figure 18 we can see how the color B–V index varies as a function of the V magnitude for the observations made in 2018: in the minimum phase, B–V is between 0.3 and 0.7, while at maximum, during the outburst B–V varies between 0.2 and –0.1. A negative B–V color index indicates that the star emits more in the blue than in the visible.

Figures 19 and 21 show the variations of the B–V and V–I_c color indices in the phase of minimum light and as a function of time. These changes relate to AH Her observations made in 2018.

Figure 20 shows the distribution of the V–I_c color index as a function of the B–V color index in the minimum light phase for the observations relating to 2018; we have that V–I_c is between 0.6 and 1.2, while B–V oscillates between 0.2 and 0.7.

Figure 22 shows a comparison between the trend of the variation of the B–V color index and the light curve of AH Her in 2018. We can note that in the phase of the maximum of the various outbursts, B–V tends to assume negative values or close to zero, while in the phase of minimum light B–V assumes positive values.

4. Typology of outbursts

The outbursts of dwarf novae have long been known to originate in the accretion disk surrounding the white dwarf (Smak 1971; Osaki 1974) due to a mechanism identified by Meyer and Meyer-Hofmeister (1981). This instability occurs when the temperature is low enough in the accretion disk that hydrogen recombines. The steep dependence of the opacity with temperature in this regime triggers a thermal and a viscous instability that leads the disk to cycle through two states. In the eruptive state, the disk has a high temperature > 10,000 K, hydrogen is highly ionized, and the mass accretion rate d_M/d_t ($\dot{M}$) from the disk on the white dwarf is higher than the mass transfer rate $\dot{M}_t$ from the companion star on the disk. In the quiescent state, the disk has a temperature < 3000 K, hydrogen is mostly neutral, and $\dot{M} < \dot{M}_t$. The disk instability model (DIM) aims at exploring the consequences of this instability on disk accretion and explaining the variety of observed light curves (Osaki 1996; Lasota 2001). Dwarf novae can have outbursts that are classified as type A or type B.

In type A outbursts, an outburst begins with the heating up and brightening of the outer parts of the disk; at the same time the viscosity increases, causing the material to flow inward and thus preventing an excessive heating of those outer parts. As the instability propagates, the inner parts become hotter and begin to contribute to the integrated luminosity. The type A outburst corresponds to higher levels of the mass-transfer rate. The outburst light curve has an asymmetrical profile. This asymmetric trend for AH Her can be seen in Figures 23 and 24; through linear regressions we have determined very different d_V/d_t between the phases of rise and the phases of decline.

In type B outbursts the instability occurs as a result of redistribution of the surface density in the inner parts of the disc and inward and outward propagation. Hence, the outburst begins almost simultaneously at all wavelengths and the emission is very strong in the U band. The instability of the B type outburst, starting in the inner parts of the disk and propagating outwards

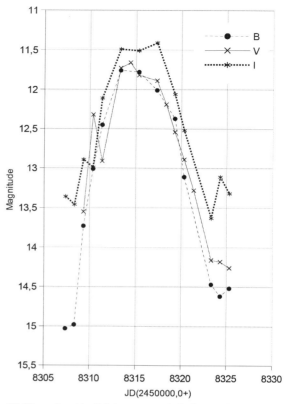

Figure 23. The outburst is slightly asymmetrical, since we have $d_V/d_t = -0.57$ mag/day in the ascent to the bright maximum, while $d_V/d_t = 0.38$ mag/day in the decline phase.

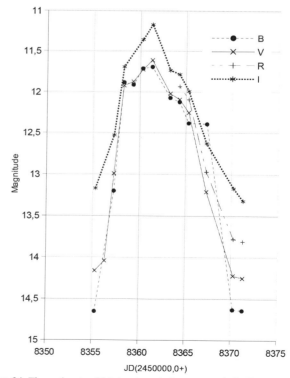

Figure 24. The outburst, which is a type A, is asymmetrical. The star rapidly increases in brightness and after reaching its maximum, it slowly declines. We have $d_V/d_t = -0.68$ mag/day for the maximum rise and $d_V/d_t = 0.30$ mag/day for the decline.

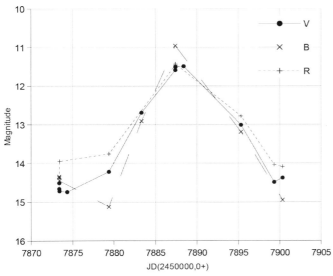

Figure 25. In this first example of a symmetrical outburst from 2017, the rise to maximum brightness expressed by $d_V/d_t = -0.31$ mag/day is almost equal to the decline time of $d_V/d_t = 0.25$ mag/day.

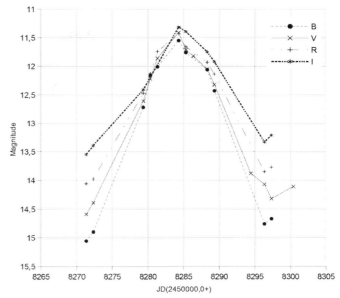

Figure 27. A typical Type B outburst: note that the light curve is almost symmetrical. The rate of climb at maximum light is equal to $d_V/d_t = -0.26$ mag/day, a value almost identical to that of decline which is $d_V/d_t = 0.24$ mag/day.

5. The intra-night time series in V band

We performed intra-night time series observations of AH Her on three nights: 05 May 2017, 27 September 2018, and 13 October 2018. The exposure time for each single observation was 240 seconds. Tables 4, 5, and 6 report the values of the estimated magnitudes for AH Her in the V band for these nights, while Table 7 reports the magnitudes of the star C8 (in the same field as AH Her) in the same bands.

In the observations of 05 May 2017 AH Her was in the phase of minimum light and the magnitude of the star varied by 0.5 magnitude, passing from V = 14.2 to V = 14.7. The star was tracked for about 1.9 hours in V, for a total of 15 photometric points. During this time the average value of AH Her was V = 14.48 ± 0.19 magnitude, while the value of the reference star, C8 was V(C8) = 12.58 ± 0.02 magnitude. Figures 28 and 29 show the trend of AH Her in the phase of minimum light.

In the observations of 27 September 2018, the star was in decline and the brightness of the variable went from V = 13.7 to V = 14.0, decreasing by 0.3 magnitude. In this phase AH Her was followed for 1.32 hours, 21 photometric points in the V band. The mean value of AH Her was V = 13.89 ± 0.09 magnitude, while C8 had an average value equal to V(C8) = 12.53 ± 0.02 magnitude. Figures 30 and 31 show the trend of AH Her in the phase of decline.

On 13 October 2018 AH her was followed in the maximum phase during an outburst for 0.93 hour. Its brightness did not vary, but remained constant around the mean value of V = 11.7 ± 0.03 magnitude. A total of 14 photometric points were obtained. The average value of the star C8 in this third series of observations was V(C8) = 12.51 ± 0.03 magnitude, a value which agrees with the previous data, but differs from the first data by 0.05 magnitude; this difference is within the margin of error. Figure 32 shows the trend of AH Her during the maximum brightness of this outburst.

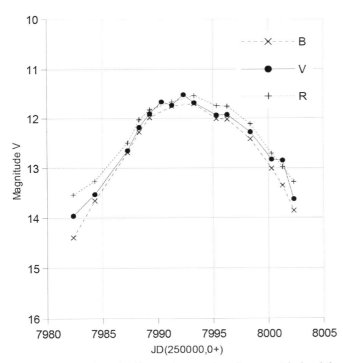

Figure 26. The outburst in this case is approximately symmetrical and the rise time at the maximum is almost equal to the decline time. We have $d_V/d_t = -0.23$ mag/day in the ascent phase and $d_V/d_t = 0.25$ mag/day in the decline.

(inside-out outburst), produces a rather symmetric light curve with a relatively low mass transfer rate (Smak 1984). In Figures 25, 26, and 27 we can see this type of outburst represented for AH Her; through linear regressions we have determined practically identical values of the d_V/d_t both for the phases of rise and for the phases of decline of the outbursts.

So as we can also see from the conformation of the light curves of the following outbursts, AH Her has both type A and type B outbursts.

Table 4. AH Her time series 5/05/2017.

JD	V Magnitude	Error
2457879.379	14.22	0.01
2457879.401	14.19	0.01
2457879.406	14.29	0.03
2457879.408	14.24	0.03
2457879.412	14.22	0.02
2457879.423	14.43	0.02
2457879.427	14.61	0.04
2457879.431	14.51	0.03
2457879.435	14.56	0.02
2457879.437	14.63	0.02
2457879.441	14.59	0.01
2457879.447	14.68	0.02
2457879.451	14.65	0.02
2457879.454	14.71	0.02
2457879.458	14.69	0.02

Table 5. AH Her time series 27/09/2018.

JD	V Magnitude	Error
2458389.281	13.77	0.03
2458389.283	13.86	0.01
2458389.286	13.82	0.02
2458389.289	13.76	0.01
2458389.291	13.77	0.01
2458389.294	13.73	0.03
2458389.297	13.82	0.03
2458389.301	13.84	0.02
2458389.303	13.87	0.02
2458389.305	13.90	0.04
2458389.308	13.82	0.01
2458389.311	13.92	0.03
2458389.314	13.92	0.01
2458389.317	13.92	0.04
2458389.319	13.97	0.05
2458389.322	13.94	0.01
2458389.325	13.98	0.01
2458389.328	14.03	0.04
2458389.331	14.01	0.01
2458389.334	13.97	0.08
2458389.336	13.98	0.03

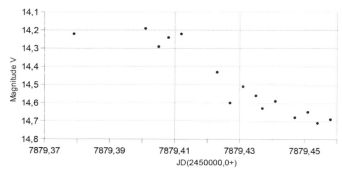

Figure 28. Time series in V from 05 May 2017. AH Her is in the phase of minimum light.

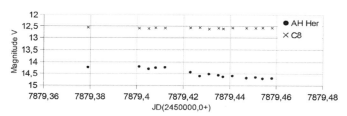

Figure 29. Time series in V from 05 May 2017 and comparison with the star C8 in the AH Her field.

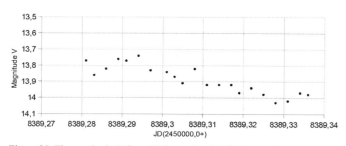

Figure 30. Time series in V from 27 September 2018. AH Her is in the decline phase.

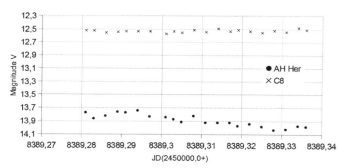

Figure 31. Time series in V from 27 September 2018 and comparison with the star C8 in the AH Her field.

6. Conclusions

We presented B, V, R_c, I_c observations of AH Her, a very active dwarf nova characterized by very frequent outbursts with recurrence times around 20 days. The variable star was systematically observed over the years 2017–2018 whenever the weather conditions allowed it. We can say that AH Her was particularly active and bright in 2017, reaching brightness values never reported before.

All observations were made at Gianni Rocchi's private observatory. The profile of the outbursts, which are both type A and type B, and the presence of a standstill even if of short duration in 2017 and of longer duration in 2012, confirm that this dwarf nova belongs to the Z Camelopardalis subgroup. Analyzing the 2017 AH Her standstill, we see that it does not end with a descent to the minimum as a classic Z Cam should do, but with a maximum rise of an outburst. This anomalous behavior is typical of the IW And subclass of the Z Cams (Kato 2019). This unusual feature was identified for the first time by Wils *et al.* (2011).

The color indices are also typical of a dwarf nova of the Z Cam subgroup and correspond, in substantial agreement, with the color indices determined by other authors in the past years. The observations presented here are part of a project aimed to obtain light curves at different wavelengths of a certain sample of dwarf novae. This is being done in order to increase the information on and the historical database of this subgroup of cataclysmic variables, which can help astrophysicists in the construction of theoretical models closer to reality.

Table 6. AH Her time series 13/10/2018.

JD	V Magnitude	Error
2458405.287	11.68	0.04
2458405.289	11.68	0.04
2458405.293	11.70	0.08
2458405.295	11.70	0.02
2458405.298	11.70	0.02
2458405.301	11.68	0.05
2458405.303	11.70	0.02
2458405.307	11.73	0.02
2458405.311	11.64	0.03
2458405.313	11.73	0.04
2458405.315	11.70	0.02
2458405.318	11.73	0.02
2458405.321	11.69	0.05
2458405.323	11.72	0.07

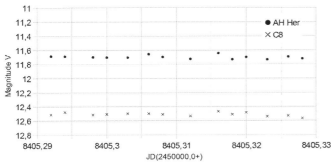

Figure 32. Time series in V from 13 October 2018 during the maximum of an AH Her outburst.

Table 7. Time series observations of the star C8.

JD	V (C8)	Error	JD	V (C8)	Error
2457879.379	12.55	0.12	2458389.308	12.50	0.04
2457879.401	12.59	0.01	2458389.311	12.54	0.01
2457879.406	12.60	0.02	2458389.314	12.50	0.03
2457879.408	12.57	0.02	2458389.317	12.54	0.01
2457879.412	12.58	0.03	2458389.319	12.49	0.04
2457879.423	12.57	0.01	2458389.322	12.53	0.05
2457879.427	12.56	0.02	2458389.325	12.55	0.01
2457879.431	12.62	0.03	2458389.328	12.51	0.03
2457879.435	12.58	0.01	2458389.331	12.53	0.01
2457879.437	12.61	0.01	2458389.334	12.48	0.09
2457879.441	12.57	0.02	2458389.336	12.50	0.03
2457879.447	12.58	0.01	2458405.287	12.52	0.05
2457879.451	12.58	0.02	2458405.289	12.48	0.04
2457879.454	12.57	0.02	2458405.293	12.52	0.08
2457879.458	12.57	0.01	2458405.295	12.51	0.02
2458389.281	12.52	0.02	2458405.298	12.50	0.02
2458389.283	12.52	0.01	2458405.301	12.50	0.05
2458389.286	12.55	0.02	2458405.303	12.51	0.02
2458389.289	12.54	0.01	2458405.307	12.53	0.01
2458389.291	12.53	0.01	2458405.311	12.47	0.03
2458389.294	12.53	0.02	2458405.313	12.51	0.05
2458389.297	12.53	0.02	2458405.315	12.48	0.01
2458389.301	12.57	0.02	2458405.318	12.54	0.02
2458389.303	12.53	0.02	2458405.321	12.53	0.05
2458389.305	12.55	0.03	2458405.323	12.57	0.05

Note: C8 coordinates (2000.0) = R.A. $16^h\ 43^m\ 52^s$ Dec. $+25°\ 11'\ 34''$.

7. Acknowledgements

Special thanks to professor Rosella Piccotti who oversaw the English translation of our paper and warm thanks to the former Dean of the IIS Cassata Gattapone Professor Cecilia Tabarrini and her secretary Giancarlo Cerafischi, to the current Dean David Nadery, and to the Dean of the Liceo G. Mazzatinti of Gubbio, Maria Marinangeli for organizing seminars and conferences in their schools on themes of Astrophysics.

The most sincere thanks from professor Spogli Corrado to the direction of the Don Nicola Mazza College of Padua for having helped him financially during the period of time in which he attended the faculty of Astromomy at the University of Padua where he then graduated.

References

Bruch, A. 1984, *Astron. Astrophys., Suppl. Ser.*, **56**, 441.
Chevalier, C., and Ilovaisky, S. A. 1991, *Astron. Astrophys., Suppl. Ser.*, **90**, 225.
Echevarria, J., et al. 2021, *Mon. Not. Roy. Astron. Soc.*, **501**, 596.
Hellier, C. 2001, *Cataclysmic Variable Stars*, Springer, New York.
Horne, K., Wade, R. A., and Szkody, P. 1986, *Mon. Not. Roy. Astron. Soc.*, **219**, 791.
Kato, T. 2015, *Publ. Astron. Soc. Japan*, **67**, 108.
Kato, T. 2019, *Publ. Astron. Soc. Japan*, **71**, 20.
Kato, T., Nogami, D., Baba, H., Masuda, S., Matsumoto, K., and Kunjaya, C. 1999, in *Disk Instabilities in Close Binary Systems. 25 Years of the Disk-Instability Model*, eds. S. Mineshige, J. C. Wheeler, Frontiers Sci. Ser. 26, Universal Academy Press, Inc., Tokyo, 45.
Lasota, J.-P. 2001, *New Astron. Rev.*, **45**, 449.
Meyer, F., and Meyer-Hofmeister, E. 1981, *Astron. Astrophys.*, **104**, L10.
Misselt, K. A. 1996, *Publ. Astron. Soc. Pacific*, **108**, 146.
Osaki, Y. 1974, *Publ. Astron. Soc. Japan*, **26**, 429.
Osaki, Y. 1996, *Publ. Astron. Soc. Pacific*, **108**, 39.
Ramsay G., Scheiber M., Gensicke B. T., and Wheatley, P. J. 2017, *Astron. Astrophys.*, **604A**, 107.
Ritter, H., and Kolb U. 1998, *Astron. Astropyys., Suppl. Ser.*, **129**, 83.
Smak, J. 1971, *Acta Astron.*, **21**, 15.
Smak, J. 1983, *Astrophys. J.*, **272**, 234.
Smak, J. 1984, *Acta Astron.*, **34**, 161.
Spogli, C., Fiorucci, M., Tosti, G., and Raimondo, G. 2001, *Inf. Bull. Var. Stars*, No. 5147, 1.
Verbunt, F., Bunk, W. H., Ritter, H., and Pfeffermann E. 1997, *Astron. Astrophys.*, **327**, 602.
Verbunt, F., et al. 1984, *Mon. Not. Roy. Astron. Soc.*, **210**, 197.
Warner, B. 1995, *Cataclysmic Variable Stars*, Cambridge Univ. Press, Cambridge.
Warner, B. 2003, *Cataclysmic Variable Stars*, Cambridge Univ. Press, Cambridge (doi:10.1017/CBO9780511586491).
Williams, G. 1983, *Astrophys. J., Suppl. Ser.*, **53**, 523.
Wils, P., Gänsicke, B. T., Drake, A. J., and Southworth, J. 2010, *Mon. Not. Roy. Astron. Soc.*, **402**, 436.
Wils, P., Krajci, T., and Simonsen, M. 2011, *J. Amer. Assoc. Var. Star Obs.*, **39**, 77.

Appendix A: B, V, R_c, I_c observed magnitude data for the dwarf nova AH Her during the years 2012, 2014, 2017, and 2018.

Date	JD(2450000.0+)	B	Error	V	Error	R_c	Error	I_c	Error
10 July 2012	6120.33	—	—	12.49	0.05	—	—	—	—
12 July 2012	6121.33	—	—	12.53	0.02	—	—	—	—
16 July 2012	6125.38	—	—	12.52	0.03	—	—	—	—
18 July 2012	6127.47	—	—	12.58	0.03	—	—	—	—
19 July 2012	6128.34	—	—	12.59	0.03	—	—	—	—
26 July 2012	6135.34	—	—	12.58	0.03	—	—	—	—
29 July 2012	6138.32	—	—	12.56	0.02	—	—	—	—
30 July 2012	6139.32	—	—	12.55	0.02	—	—	—	—
31 July 2012	6140.32	—	—	12.60	0.02	—	—	—	—
01 August 2012	6141.35	—	—	12.61	0.02	—	—	—	—
02 August 2012	6142.33	—	—	12.63	0.02	—	—	—	—
05 August 2012	6145.32	—	—	12.57	0.04	—	—	—	—
06 August 2012	6146.31	—	—	12.61	0.03	—	—	—	—
07 August 2012	6147.23	—	—	12.66	0.03	—	—	—	—
08 August 2012	6148.32	—	—	12.69	0.02	—	—	—	—
09 August 2012	6149.31	—	—	12.61	0.01	—	—	—	—
10 August 2012	6150.32	—	—	12.88	0.04	—	—	—	—
19 August 2012	6159.40	—	—	—	—	12.72	0.02	—	—
20 August 2012	6160.30	—	—	12.86	0.02	—	—	—	—
21 August 2012	6161.33	—	—	—	—	12.69	0.07	—	—
22 August 2012	6162.32	—	—	12.74	0.03	—	—	—	—
23 August 2012	6163.30	—	—	12.71	0.02	—	—	—	—
24 August 2012	6164.30	—	—	—	—	12.61	0.05	—	—
25 August 2012	6165.34	—	—	—	—	12.59	0.02	—	—
03 September 2012	6174.33	—	—	12.54	0.03	—	—	—	—
06 September 2012	6177.28	—	—	12.65	0.01	—	—	—	—
07 September 2012	6178.28	—	—	—	—	12.51	0.02	—	—
08 September 2012	6179.31	—	—	—	—	12.46	0.02	—	—
10 September 2012	6181.27	—	—	12.67	0.06	—	—	—	—
11 September 2012	6182.27	—	—	12.63	0.01	—	—	—	—
15 September 2012	6186.32	—	—	—	—	12.35	0.03	—	—
16 September 2012	6187.27	—	—	12.53	0.04	—	—	—	—
17 September 2012	6188.27	—	—	12.61	0.02	—	—	—	—
21 September 2012	6192.33	—	—	12.53	0.03	—	—	—	—
22 September 2012	6193.27	—	—	12.49	0.03	—	—	—	—
03 October 2012	6204.24	—	—	12.31	0.02	—	—	—	—
05 October 2012	6206.24	—	—	12.26	0.02	—	—	—	—
06 October 2012	6207.23	—	—	12.34	0.04	—	—	—	—
17 October 2012	6218.26	—	—	12.62	0.04	—	—	—	—
18 October 2012	6219.24	—	—	12.57	0.02	—	—	—	—
19 October 2012	6220.23	—	—	12.66	0.02	—	—	—	—
20 October 2012	6221.24	—	—	12.69	0.02	—	—	—	—
22 October 2012	6223.23	—	—	12.74	0.05	—	—	—	—
23 October 2012	6224.27	—	—	12.69	0.01	—	—	—	—
24 October 2012	6225.25	—	—	12.55	0.05	—	—	—	—
25 October 2012	6226.22	—	—	12.47	0.03	—	—	—	—
29 October 2012	6230.28	—	—	12.71	0.02	—	—	—	—
01 November 2012	6233.21	—	—	12.53	0.05	—	—	—	—
07 November 2012	6239.21	—	—	12.62	0.02	—	—	—	—
03 August 2014	6873.36	—	—	14.11	0.04	13.69	0.04	—	—
11 August 2014	6881.31	—	—	12.14	0.04	12.04	0.01	—	—
12 August 2014	6882.32	—	—	—	—	12.26	0.02	12.21	0.01
16 August 2014	6886.40	—	—	14.09	0.03	—	—	—	—
24 August 2014	6894.30	—	—	—	—	12.01	0.02	—	—
29 August 2014	6899.31	—	—	13.09	0.02	—	—	12.57	0.04
06 September 2014	6907.29	—	—	13.59	0.03	—	—	12.91	0.01
14 September 2014	6915.28	11.73	0.02	11.75	0.02	—	—	—	—
22 September 2014	6923.28	14.57	0.08	14.01	0.02	—	—	—	—
26 September 2014	6927.30	—	—	13.92	0.01	—	—	13.14	0.03
27 September 2014	6928.31	—	—	—	—	—	—	12.99	0.01
29 September 2014	6930.30	—	—	12.17	0.02	—	—	11.88	0.05
04 October 2014	6935.33	—	—	13.99	0.02	—	—	13.21	0.05
08 October 2014	6939.27	14.38	0.06	13.93	0.02	13.41	0.02	—	—
09 October 2014	6940.25	14.19	0.02	13.59	0.01	—	—	—	—
11 October 2014	6942.31	—	—	12.32	0.03	—	—	11.84	0.05
18 October 2014	6949.27	—	—	14.03	0.05	—	—	13.15	0.05
22 October 2014	6953.24	—	—	13.49	0.03	12.96	0.01	—	—

Table continued on following pages

Appendix A: B, V, R_c, I_c observed magnitude data for the dwarf nova AH Her during the years 2012, 2014, 2017, and 2018 (cont).

Date	JD(2450000.0+)	B	Error	V	Error	R_c	Error	I_c	Error
25 October 2014	6956.24	—	—	12.26	0.04	12.11	0.03	—	—
27 October 2014	6958.27	—	—	12.74	0.01	12.52	0.02	—	—
29 October 2014	6959.21	13.53	0.03	13.18	0.03	12.94	0.02	—	—
22 April 2017	7866.44	12.15	0.05	12.14	0.02	11.96	0.05	—	—
23 April 2017	7867.46	11.86	0.07	12.21	0.04	12.08	0.02	—	—
23 April 2017	7867.47	11.48	0.05	12.25	0.03	—	—	—	—
23 April 2017	7867.47	11.77	0.08	12.33	0.04	12.24	0.02	12.11	0.03
24 April 2017	7868.40	12.87	0.05	12.64	0.06	12.42	0.05	—	—
25 April 2017	7869.39	14.56	0.03	13.02	0.08	12.88	0.05	—	—
29 April 2017	7873.37	14.34	0.04	14.51	0.07	—	—	—	—
29 April 2017	7973.38	14.37	0.05	14.66	0.03	14.39	0.05	13.88	0.04
29 April 2017	7973.40	14.45	0.03	14.72	0.04	13.95	0.05	13.40	0.06
30 April 2017	7874.33	—	—	14.74	0.04	—	—	—	—
05 May 2017	7879.38	15.12	0.05	14.22	0.02	13.76	0.02	13.21	0.05
09 May 2017	7883.31	12.91	0.05	12.69	0.07	—	—	—	—
13 May 2017	7887.41	10.96	0.04	11.59	0.05	11.46	0.04	11.38	0.05
13 May 2017	7887.41	—	—	11.43	0.02	11.43	0.02	—	—
14 May 2017	7888.41	—	—	11.49	0.07	—	—	—	—
21 May 2017	7895.32	13.19	0.05	13.01	0.08	12.78	0.03	—	—
25 May 2017	7899.31	—	—	14.49	0.07	14.04	0.02	—	—
26 May 2017	7900.32	14.95	0.02	14.38	0.05	14.09	0.04	13.24	0.02
27 May 2017	7901.36	14.86	0.11	14.22	0.02	13.83	0.02	—	—
28 May 2017	7902.33	14.92	0.09	14.37	0.02	13.90	0.05	—	—
29 May 2017	7903.33	15.01	0.03	14.27	0.02	13.75	0.03	—	—
31 May 2017	7905.32	13.51	0.05	13.35	0.02	13.09	0.05	—	—
02 June 2017	7907.35	13.15	0.05	12.75	0.02	12.61	0.03	—	—
03 June 2017	7908.33	13.05	0.01	12.37	0.04	12.28	0.04	—	—
07 June 2017	7912.32	12.44	0.03	11.92	0.05	11.91	0.08	—	—
08 June 2017	7913.33	11.91	0.08	11.96	0.02	11.87	0.02	—	—
09 June 2017	7914.41	—	—	12.03	0.03	—	—	—	—
10 June 2017	7915.38	11.73	0.05	12.36	0.03	12.27	0.04	12.13	0.02
11 June 2017	7916.32	13.16	0.15	12.71	0.16	12.64	0.04	—	—
12 June 2017	7917.32	14.43	0.18	13.37	0.06	13.03	0.07	—	—
13 June 2017	7918.32	14.48	0.05	13.68	0.16	13.51	0.07	—	—
17 June 2017	7922.33	15.29	0.12	14.52	0.07	14.09	0.02	—	—
19 June 2017	7924.33	14.99	0.02	14.43	0.02	14.08	0.05	—	—
22 June 2017	7927.33	14.73	0.21	13.94	0.09	13.96	0.03	—	—
23 June 2017	7928.32	13.29	0.09	13.27	0.17	12.77	0.02	—	—
01 July 2017	7936.44	—	—	12.07	0.03	12.12	0.04	11.78	0.03
04 July 2017	7938.34	12.94	0.09	12.69	0.02	12.44	0.05	—	—
07 July 2017	7942.34	14.98	0.11	14.31	0.02	13.85	0.04	—	—
10 July 2017	7945.32	14.91	0.03	14.63	0.01	14.07	0.04	—	—
12 July 2017	7947.33	14.84	0.09	14.26	0.06	13.86	0.03	—	—
13 July 2017	7948.33	14.71	0.08	14.24	0.02	13.51	0.07	—	—
15 July 2017	7950.35	—	—	13.31	0.03	—	—	—	—
16 July 2017	7951.33	12.97	0.06	12.83	0.04	12.58	0.08	—	—
17 July 2017	7952.32	12.71	0.05	12.54	0.05	12.38	0.05	—	—
18 July 2017	7953.32	12.27	0.06	12.42	0.02	11.99	0.08	—	—
20 July 2017	7955.32	12.25	0.05	12.51	0.02	12.02	0.05	—	—
21 July 2017	7956.32	11.81	0.05	11.97	0.05	—	—	—	—
22 July 2017	7957.34	—	—	11.69	0.04	—	—	—	—
28 July 2017	7963.41	13.86	0.05	13.75	0.04	13.42	0.06	—	—
01 August 2017	7967.31	14.63	0.05	13.98	0.06	13.67	0.05	—	—
02 August 207	7968.32	14.03	0.02	13.76	0.06	13.33	0.03	—	—
03 August 2017	7969.32	13.49	0.06	13.23	0.07	13.67	0.05	—	—
04 August 2017	7970.33	13.11	0.08	12.89	0.02	13.04	0.09	—	—
05 August 2017	7971.40	12.42	0.05	12.72	0.04	12.66	0.05	12.63	0.04
07 August 2017	7973.33	12.32	0.03	12.21	0.03	12.09	0.02	—	—
08 August 2017	7974.31	12.18	0.03	12.08	0.05	12.01	0.02	—	—
09 August 2017	7975.30	12.52	0.12	12.37	0.02	12.13	0.03	—	—
10 August 2017	7976.33	12.72	0.02	12.66	0.02	12.43	0.01	—	—
11 August 2017	7977.31	13.09	0.05	13.01	0.01	12.74	0.01	—	—
13 August 2017	7979.31	13.88	0.06	13.61	0.02	13.32	0.07	—	—
14 August 2017	7980.40	13.46	0.08	13.64	0.04	—	—	13.18	0.03
14 August 2017	7980.42	13.92	0.09	13.87	0.02	13.66	0.04	13.09	0.05
14 August 2017	7980.44	—	—	14.11	0.03	—	—	—	—
15 August 2017	7981.3	14.43	0.03	13.94	0.02	13.62	0.05	—	—
16 August 2017	7982.33	14.39	0.03	13.96	0.02	13.54	0.11	—	—

Table continued on following pages

Appendix A: B, V, R_c, I_c observed magnitude data for the dwarf nova AH Her during the years 2012, 2014, 2017, and 2018 (cont).

Date	JD(2450000.0+)	B	Error	V	Error	R_c	Error	I_c	Error
18 August 2017	7984.29	13.66	0.03	13.53	0.04	13.28	0.13	—	—
21 August 2017	7987.29	12.69	0.02	12.65	0.02	12.49	0.07	—	—
22 August 2017	7988.33	12.27	0.05	12.18	0.03	12.03	0.03	—	—
23 August 2017	7989.29	11.99	0.08	11.91	0.02	11.83	0.04	—	—
24 August 2017	7990.36	—	—	11.67	0.05	—	—	—	—
25 August 2017	7991.30	11.76	0.02	11.74	0.02	11.67	0.02	—	—
26 August 2017	7992.35	—	—	11.52	0.02	—	—	11.38	0.03
27 August 2017	7993.31	11.71	0.03	11.69	0.08	11.54	0.03	—	—
29 August 2017	7995.31	12.01	0.05	11.94	0.04	11.75	0.04	—	—
30 August 2017	7996.28	12.02	0.04	11.93	0.04	11.76	0.02	—	—
01 September 2017	7998.38	12.41	0.03	12.27	0.03	12.11	0.02	—	—
03 September 2017	8000.29	13.01	0.08	12.83	0.08	12.71	0.04	—	—
04 September 2017	8001.27	13.35	0.07	12.85	0.07	12.98	0.05	—	—
05 September 2017	8002.28	13.86	0.08	13.63	0.08	13.28	0.02	—	—
09 September 2017	8006.27	13.13	0.03	13.06	0.03	12.81	0.02	—	—
13 September 2017	8010.27	12.25	0.02	12.21	0.02	12.14	0.03	—	—
17 September 2017	8014.25	12.68	0.03	12.63	0.03	12.47	0.03	—	—
20 September 2017	8017.32	13.02	0.02	12.91	0.02	12.65	0.05	—	—
22 September 2017	8018.27	12.84	0.04	12.66	0.04	12.48	0.02	—	—
23 September 2017	8020.36	—	—	12.67	0.06	—	—	12.13	0.05
23 September 2017	8020.40	12.15	0.05	12.41	0.05	—	—	12.15	0.04
27 September 2017	8024.25	12.61	0.02	12.44	0.05	12.21	0.08	—	—
28 September 2017	8025.26	12.62	0.02	12.57	0.02	12.27	0.09	—	—
29 September 2017	8026.26	12.59	0.04	12.55	0.02	12.36	0.04	—	—
02 October 2017	8029.25	12.75	0.07	12.52	0.02	12.35	0.03	—	—
04 October 2017	8031.29	12.46	0.05	12.34	0.02	12.19	0.03	—	—
09 October 2017	8036.29	11.80	0.05	11.72	0.02	11.52	0.02	—	—
10 October 2017	8037.24	11.78	0.03	11.71	0.02	11.58	0.03	—	—
11 October 2017	8038.22	11.82	0.05	11.76	0.05	11.37	0.03	—	—
12 October 2017	8039.23	11.72	0.05	11.74	0.03	11.61	0.03	—	—
13 October 2017	8040.27	11.83	0.04	11.84	0.02	11.69	0.02	—	—
15 October 2017	8042.22	12.09	0.08	12.08	0.04	11.93	0.08	—	—
16 October 2017	8043.23	12.27	0.15	12.29	0.05	12.17	0.02	—	—
17 October 2017	8044.22	12.57	0.04	12.51	0.02	12.33	0.02	—	—
18 October 2017	8045.22	13.12	0.03	12.99	0.03	12.73	0.05	—	—
19 October 2017	8046.24	13.75	0.02	13.59	0.03	13.27	0.05	—	—
20 October 2017	8047.21	14.15	0.09	13.75	0.02	13.37	0.05	—	—
22 October 2017	8049.22	—	—	14.39	0.04	—	—	—	—
23 October 2017	8050.22	14.77	0.02	14.41	0.04	13.85	0.05	—	—
24 October 2017	8051.22	14.79	0.02	14.39	0.06	13.82	0.04	—	—
25 October 2017	8052.23	14.34	0.17	14.31	0.05	13.78	0.01	—	—
28 October 2017	8055.21	13.86	0.02	13.56	0.03	13.31	0.01	—	—
30 October 2017	8057.24	12.75	0.06	12.59	0.05	12.46	0.02	—	—
01 November 2017	8059.24	11.79	0.07	11.83	0.02	11.73	0.02	—	—
11 November 2017	8069.21	14.01	0.12	13.67	0.07	13.39	0.04	—	—
10 May 2018	8249.34	14.62	0.02	14.14	0.03	13.75	0.08	13.24	0.08
12 May 2018	8251.37	—	—	14.01	0.03	—	—	—	—
18 May 2018	8257.34	12.90	0.02	12.79	0.02	12.58	0.06	12.41	0.05
19 May 2018	8258.37	—	—	12.39	0.02	—	—	—	—
23 May 2018	8262.32	12.44	0.02	12.33	0.04	12.21	0.02	12.19	0.06
24 May 2018	8263.31	12.68	0.02	12.54	0.04	12.31	0.08	12.15	0.05
25 May 2018	8264.32	—	—	13.71	0.06	—	—	—	—
27 May 2018	8266.35	—	—	13.69	0.04	13.14	0.16	—	—
01 June 2018	8271.33	15.06	0.02	14.59	0.04	14.06	0.15	13.65	0.03
02 June 2018	8272.33	14.90	0.02	14.39	0.05	13.98	0.09	13.38	0.05
09 June 2018	8279.33	12..72	0.06	12.61	0.03	12.47	0.04	12.41	0.02
10 June 2018	8280.32	12.16	0.05	12.22	0.03	12.13	0.02	12.11	0.09
11 June 2018	8281.32	12.02	0.05	11.86	0.02	11.74	0.11	11.72	0.08
14 June 2018	8284.32	11.55	0.02	11.41	0.02	11.43	0.02	11.81	0.03
15 June 2018	8285.33	11.76	0.02	11.71	0.04	11.66	0.03	11.39	0.11
16 June 2018	8286.32	—	—	11.82	0.03	—	—	—	—
18 June 2018	8288.32	12.06	0.12	12.05	0.02	11.93	0.05	11.75	0.09
19 June 2018	8289.33	12.43	0.02	12.32	0.03	12.14	0.12	11.92	0.10
24 June 2018	8294.44	—	—	13.88	0.03	—	—	—	—
26 June 2018	8296.33	14.76	0.02	14.02	0.02	13.86	0.02	13.33	0.09
27 June 2018	8297.32	14.67	0.02	14.32	0.07	13.77	0.03	13.21	0.09
30 June 2018	8300.39	—	—	14.11	0.05	13.57	0.05	—	—
01 July 2018	8301.34	14.44	0.02	13.96	0.05	13.51	0.05	13.05	0.11

Table continued on following pages

Appendix A: B, V, R_c, I_c observed magnitude data for the dwarf nova AH Her during the years 2012, 2014, 2017, and 2018 (cont).

Date	JD(2450000.0+)	B	Error	V	Error	R_c	Error	I_c	Error
02 July 2018	8302.34	14.25	0.03	13.89	0.06	13.61	0.05	13.01	0.06
05 July 2018	8305.33	—	—	14.32	0.04	—	—	13.22	0.02
06 July 2018	8306.34	15.06	0.02	14.43	0.02	—	—	13.32	0.03
07 July 2018	8307.35	15.03	0.03	14.44	0.02	—	—	13.36	0.06
08 July 2018	8308.34	14.98	0.06	14.38	0.02	—	—	13.46	0.02
09 July 2018	8309.33	13.73	0.06	13.55	0.02	—	—	12.89	0.02
10 July 2018	8310.36	13.01	0.04	12.93	0.03	—	—	12.61	0.05
11 July 2018	8311.35	12.45	0.04	12.38	0.04	—	—	12.11	0.03
13 July 2018	8313.32	12.16.	0.05	11.73	0.04	—	—	11.49	0.02
14 July 2018	8314.38	—	—	11.66	0.02	—	—	—	—
15 July 2018	8315.33	11.78	0.02	11.82	0.02	—	—	11.51	0.03
17 July 2018	8317.32	12.01	0.03	11.89	0.07	—	—	11.41	0.12
18 July 2018	8318.33	—	—	12.19	0.06	—	—	—	—
19 July 2018	8319.30	12.37	0.05	12.38	0.02	—	—	12.06	0.18
20 July 2018	8320.33	13.11	0.05	12.89	0.02	—	—	12.52	0.05
21 July 2018	8321.38	—	—	13.28	0.02	12.89	0.03	—	—
23 July 2018	8323.32	14.47	0.03	14.16	0.05	—	—	13.63	0.11
24 July 2018	8324.31	14.62	0.02	14.18	0.04	—	—	13.11	0.03
25 July 2018	8325.31	14.52	0.03	14.26	0.09	—	—	13.32	0.11
29 July 2018	8329.31	14.31	0.05	13.93	0.10	—	—	13.05	0.02
31 July 2018	8331.32	14.75	0.02	14.14	0.03	—	—	13.27	0.08
01 August 2018	8332.30	14.66	0.02	14.01	0.11	—	—	12.94	0.07
02 August 2018	8333.32	14.41	0.01	14.03	0.04	—	—	13.24	0.02
03 August 2018	8334.32	13.45	0.02	13.13	0.06	—	—	12.69	0.04
04 August 2018	8335.35	—	—	12.59	0.04	12.47	0.01	—	—
05 August 2018	8336.31	12.07	0.03	11.98	0.01	—	—	11.76	0.03
09 August 2018	8340.34	12.27	0.03	12.14	0.04	—	—	11.84	0.02
10 August 2018	8341.34	—	—	12.28	0.02	12.11	0.02	—	—
16 August 2018	8347.31	14.34	0.04	13.97	0.06	—	—	13.02	0.10
17 August 2018	8348.36	—	—	13.81	0.03	13.41	0.02	—	—
19 August 2018	8350.30	14.37	0.02	14.01	0.04	—	—	13.28	0.02
20 August 2018	8351.30	—	—	14.14	0.06	—	—	13.10	0.05
21 August 2018	8352.33	—	—	13.94	0.02	—	—	—	—
22 August 2018	8353.30	14.46	0.03	14.01	0.03	—	—	13.30	0.06
23 August 2018	8354.30	14.28	0.07	13.90	0.02	—	—	—	—
24 August 2018	8355.32	14.65	0.10	14.16	0.02	—	—	13.17	0.06
25 August 2018	8356.34	—	—	14.04	0.09	13.43	0.04	—	—
26 August 2018	8357.29	14.28	0.07	12.99	0.04	—	—	12.53	0.08
27 August 2018	8358.28	11.88	0.03	11.91	0.03	—	—	11.69	0.05
28 August 2018	8359.28	11.91	0.05	11.87	0.02	—	—	—	—
28 August 2018	8359.30	11.75	0.05	—	—	—	—	—	—
29 August 2018	8360.30	11.71	0.05	11.72	0.03	—	—	11.36	0.07
30 August 2018	8361.34	11.69	0.04	11.61	0.02	—	—	11.17	0.15
01 September 2018	8363.28	12.07	0.04	12.02	0.03	—	—	11.73	0.03
02 September 2018	8364.29	12.12	0.02	12.09	0.02	11.93	0.01	11.78	0.03
03 September 2018	8365.30	12.38	0.02	12.25	0.05	12.09	0.06	11.99	0.06
05 September 2018	8367.27	13.37	0.05	13.21	0.03	12.97	0.02	12.63	0.04
08 September 2018	8370.26	14.63	0.09	14.22	0.02	13.78	0.01	13.17	0.04
09 September 2018	8371.29	14.64	0.01	14.25	0.02	13.81	0.02	13.32	0.06
10 September 2018	8372.27	15.04	0.08	14.51	0.06	14.10	0.01	13.45	0.06
11 September 2018	8373.26	14.81	0.02	14.56	0.03	13.89	0.04	13.49	0.01
12 September 2018	8374.27	14.78	0.02	14.28	0.02	13.86	0.01	13.26	0.02
22 September 2018	8384.26	12.17	0.03	12.17	0.07	11.96	0.05	11.84	0.02
23 September 2018	8385.26	12.44	0.02	12.28	0.02	12.09	0.03	11.97	0.05
26 September 2018	8388.26	13.54	0.05	13.41	0.01	13.07	0.02	12.83	0.04
27 September 2018	8389.28	14.02	0.02	13.77	0.04	13.43	0.02	13.08	0.08
28 September 2018	8390.25	14.61	0.02	14.14	0.03	13.71	0.03	13.04	0.01
29 September 2018	8391.32	—	—	14.15	0.03	—	—	—	—
30 September 2018	8392.26	—	—	14.51	0.03	—	—	—	—
02 October 2018	8394.31	14.91	0.07	14.54	0.02	14.09	0.02	13.46	0.08
03 October 2018	8395.23	14.69	0.01	14.18	0.10	13.88	0.06	13.44	0.08
07 October 2018	8399.27	14.71	0.05	14.17	0.11	13.79	0.02	13.31	0.05
08 October 2018	8400.24	14.66	0.06	14.24	0.06	13.87	0.10	13.42	0.05
12 October 2018	8404.25	12.13	0.03	12.06	0.02	11.96	0.03	11.91	0.04
13 October 2018	8405.23	11.68	0.03	11.69	0.02	11.55	0.02	11.51	0.02
19 October 2018	8411.23	12.18	0.03	12.10	0.04	11.97	0.03	11.87	0.02
20 October 2018	8412.21	13.37	0.03	12.28	0.08	—	—	12.11	0.02.
22 October 2018	8414.24	13.33	0.05	13.05	0.05	12.85	0.06	12.71	0.07

Table continued on next page

Appendix A: B, V, R_c, I_c observed magnitude data for the dwarf nova AH Her during the years 2012, 2014, 2017, and 2018 (cont).

Date	JD(2450000.0+)	B	Error	V	Error	R_c	Error	I_c	Error
23 October 2018	8415.23	13.64	0.05	13.49	0.04	13.19	0.05	12.84	0.04
24 October 2018	8416.24	13.86	0.02	13.57	0.02	13.31	0.03	12.97	0.02
25 October 2018	8417.24	13.91	0.08	13.55	0.04	13.22	0.02	12.76	0.06
08 November 2018	8431.25	14.42	0.04	14.02	0.02	13.68	0.02	13.31	0.03
09 November 2018	8432.26	14.53	0.07	14.31	0.06	13.79	0.02	13.29	0.04
10 November 2018	8433.21	14.76	0.01	14.22	0.04	13.71	0.02	13.19	0.05
11 November 2018	8434.20	14.74	0.09	14.19	0.02	13.74	0.03	13.34	0.04
15 November 2018	8438.23	13.54	0.08	13.27	0.02	13.04	0.02	12.91	0.07
16 November 2018	8439.22	13.23	0.07	12.98	0.05	13.45	0.05	12.69	0.05
17 November 2018	8440.23	12.88	0.04	—	—	—	—	12.42	0.03
18 November 2018	8441.19	12.36	0.10	12.41	0.05	12.17	0.12	12.13	0.12
01 December 2018	8454.19	14.26	0.19	13.88	0.03	13.43	0.05	13.18	0.05.

Photometric Determination of the Distance to the RR Lyrae Star YZ Capricorni

Jamie Lester
Rowan Joignant
Mariel Meier
Oglethorpe University, 4484 Peachtree Road NE, Atlanta, GA 30319; jlester3@oglethorpe.edu; pjoignant@oglethorpe.edu; mmeier@oglethorpe.edu

Received June 7, 2021; revised November 30, 2021; accepted November 30, 2021

Abstract The RR Lyrae YZ Cap was observed using photometric methods to determine its pulsation period and distance. Light curves in the Bessell B and V and SDSS i' and z' filters were used to determine the period, which was found to be 0.274 ± 0.003 day. The distances calculated using a luminosity-metallicity (V filter) and period-luminosity-metallicity relationship (i' filter and z' filter) were: V: 1107 ± 52 pc, i': 1191 ± 126 pc, and z': 1092 ± 128 pc. These are in agreement with the distance measurement from the second Gaia data release, 1144 ± 90 pc. This demonstrates the potential to use ground-based observations in the visible and near-infrared to determine the distance to RR Lyrae variable stars.

1. Introduction

RR Lyrae stars are periodic variable stars resting on the horizontal band of the Hertzprung-Russell diagram. The prototype for this classification of star, RR Lyrae itself, was discovered by Williamina Fleming in 1901 (Pickering *et al.* 1901). In modern astronomy, RR Lyraes are used as standard candles, meaning that properties of the stars, including their period and metallicity, can be related to their absolute magnitude in order to determine their distance. This distance can be used to analyze and find the distance to other stellar bodies and the globular clusters within which these stars are often found. In the visible bands, fluxes scale as T_{eff}^4, whereas in the near-infrared, the dependence goes as $T_{eff}^{1.6}$ (Catelan and Smith 2015). This reduced sensitivity to temperature in the i' and z' bands has allowed for the development of period-luminosity-metallicity relationships, which are more accurate than the luminosity-metallicity relationships that can be applied in the V band. In an effort to better understand these stars that are critical to modern astronomy, an RR Lyrae star was selected for the determination of its distance and the analysis of its period and magnitude, particularly in the near-infrared bands. The data collected from this star were used to further the usage of previously determined period-luminosity-metallicity relations.

In selecting an RR Lyrae star for analysis, YZ Capricorni, hereafter referred to as YZ Cap, was determined to be a good candidate. The star was present in a variety of astronomical surveys, including the Sloan Digital Sky Survey and the AAVSO Photometric All-Sky Survey, as well as being visible from PanSTARRS. A summary of data regarding YZ Cap is presented in Table 1. YZ Cap is an RRc type star, meaning that it pulsates in the first overtone mode and its light curve can be expected to have a sinusoidal shape (Monson *et al.* 2017).

There are several published values of metallicity for YZ Cap, summarized in Table 2. Since these measurements differ from each other and no singular measurement has been deemed the most accurate, an average of these measurements, [Fe/H] = –1.33 ± 0.17, was used for this analysis. This [Fe/H] value was converted first to a metals-to-hydrogen ratio, [M/H], using the scaling relationship (Equation 1) (assuming $f \simeq 10^{0.3}$, e.g. (Catelan *et al.* 2004) and then to log Z using Equation 2 in order to be utilized in the magnitude and distance analysis (Catelan *et al.* 2004). This gave an [M/H] value of –1.112 and a log Z value of –2.877.

$$[M/H] = [Fe/H] + \log(0.638f + 0.362) \quad (1)$$

$$\log Z = [M/H] - 1.765 \quad (2)$$

2. Observations

All observations were made using the Las Cumbres Observatory, which has ten sites around the world with an assortment of 0.4-meter, 1.0-meter, and 2.0-meter telescopes) Brown *et al.* 2013). This distribution of telescopes allows for longer observation times, because when an object is no longer visible to one telescope, the object becomes visible to another.

Table 1. Summary of previous published information about YZ Cap.

R.A. (2000) h m s	Dec. (2000) ° ' "	V Mag.	Distance	Spectral Type	Period
21 19 32.41	–15 07 1.14[1]	11.34[2]	1144 ± 90 pc[3]	F5[4]	0.273 d[5]

Sources: 1. Neeley et al. (2017). 2. Henden et al. (2015). 3. Gaia Collab. et al. (2018). 4. Nesterov et al. (1995). 5. Bono et al. (2020); Neeley et al. (2017).

Table 2. YZ Cap metallicity.

[Fe/H]	Source
–1.25	Cacciari *et al.* (1989)
–1.30	Cacciari *et al.* (1989)
–1.06	Feast *et al.* (2008)
–1.54	Govea *et al.* (2014)
–1.48	Govea *et al.* (2014)
–1.33	Average of available sources

Test images were taken using the 0.4-meter telescopes to calculate proper exposure times. Optimal times for each band were chosen to ensure ≈ 100,000 counts for YZ Cap. The exposure times used were 100 seconds for the Bessell B band, 30 seconds for the Bessell V band, 40 seconds for the SDSS i' band, and 130 seconds for the SDSS z' band.

Observations were made between October 19, 2020, and November 1, 2020. A total of 194 images across the four filters was received after an initial processing pass through the Our Solar Siblings (OSS) Pipeline (Fitzgerald 2018). 53 images in B, 48 in V, 47 in i', and 46 z' were retained.

3. Methods

Images gathered from the Las Cumbres Observatory were passed through the Our Solar Siblings (OSS) Pipeline in order to generate photometry files for analysis. From the 194 images that passed through the OSS Pipeline, only 124 were kept for further analysis. In the case of the 70 images that were discarded, YZ Cap and the surrounding comparison stars were either out of focus, too dim, or too blurred. The OSS Pipeline utilizes a variety of photometry methods in order to analyze the images of the stars, and these methods fall into the two main categories of aperture photometry and point-spread function photometry. In aperture photometry, a digital ring is placed around the image of the star, and the magnitude of the star within the ring is measured. This method works well for images in which the star appears very circular and regularly shaped, because an aperture can be accurately placed around a circular-shaped image of a star. The data produced by the SEXtractor method (Bertin and Arnouts 1996), also known as simple aperture method, was the most consistent and was selected for further analysis.

In order to analyze the remaining 124 images, a PYTHON package called ASTROSOURCE was utilized (Fitzgerald *et al.* 2020). ASTROSOURCE analyzes images and photometry data files of variable stars and identifies comparison stars in the star field for calibration. These calibration comparison stars are identified by having a stable magnitudes and already being present in an accessible database. The catalogues used by ASTROSOURCE for the different filters are APASS for B and V (Henden *et al.* 2016), SDSS for i' (Alam *et al.* 2015), and PanSTARRS for z' (Magnier *et al.* 2020; Flewelling *et al.* 2020). The magnitudes of these comparison stars are used to calibrate the magnitude of the variable star being studied, in this case YZ Cap. A list of the coordinates of the identified comparison stars can be found in Table 3. These comparison stars are circled in Figure 1, with YZ Cap circled in the center. After using the comparison stars to calibrate the magnitude, ASTROSOURCE produced light curves and period measurements for each filter using the Phase Dispersion Minimization Method (Stellingwerf 1978).

4. Results/discussion

Depicted in Figure 2 are the light curves in the B, V, i', and z' filters. The light curves are relatively consistent with each other and clearly exhibit the characteristic sinusoidal shape of an RRc type star. The period was measured from each light curve using the phase dispersion method minimization algorithm developed by Altunin *et al.* (2020) based on the technique developed by Stellingwerf (1978). The period was measured independently in each filter to be 0.274 ± 0.003 days, which is consistent with previously measured values. The apparent magnitude in each filter is presented in Table 4. The apparent magnitudes were determined by averaging all of the magnitudes from the light curve, which provides sufficiently accurate values because the light curve is highly symmetrical.

The luminosity-metallicity equations in the V (Equation 3) (Catelan *et al.* 2004), and period-luminosity-metallicity equations in the i' (Equation 4) and z' (Equation 5) bands (Caceres and Catelan 2008) were used to calculate the absolute magnitude of YZ Cap in each filter (Table 4). In Equations 4 and 5, the period used was first converted into the "fundamentalized" period using Equation 6 (Caceres and Catelan 2008).

$$M_V = 2.288 + 0.882 \log Z + 0.108 (\log Z)^2 \quad (3)$$

$$M_i = 0.908 - 1.035 \log P_f + 0.220 \log Z \quad (4)$$

$$M_z = 0.839 - 1.295 \log P_f + 0.211 \log Z \quad (5)$$

$$\log P_f = \log P + 0.128 \quad (6)$$

These magnitude values were used to determine the distance to YZ Cap (Table 5), which is compared to the parallax distance

Table 3. Coordinates of comparison stars with magnitude in the B filter.

R.A. (2000) h m s	Dec. (2000) ° ' "	B Mag.
21 20 04	–15 03 41	13.69 ± 0.02
21 19 26	–15 01 56	12.68 ± 0.03
21 19 31	–15 01 13	13.36 ± 0.04
21 19 50	–14 53 20	14.31 ± 0.05
21 19 15	–14 55 03	13.83 ± 0.04
21 20 04	–15 11 12	12.41 ± 0.04
21 19 45	–15 01 58	12.41 ± 0.04
21 19 07	–15 07 25	11.81 ± 0.04
21 19 35	–15 01 30	14.63 ± 0.07
21 19 10	–15 16 47	13.92 ± 0.04

Table 4. Apparent and calculated absolute magnitudes.

Filter	m	M
B	11.48 ± 0.01	—
V	11.19 ± 0.01	0.721 ± 0.05
i'	11.17 ± 0.01	0.783 ± 0.161
z'	11.15 ± 0.01	0.850 ± 0.192

Table 5. Comparison of distance measurements.

	Distance (pc)	Error (pc)
Gaia	1144	90
V	1107	52
i'	1191	126
z'	1092	128
mean	1130	59

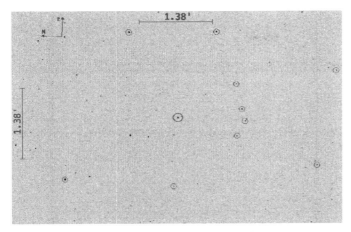

Figure 1. Image taken in the i' filter with circled comparison stars.

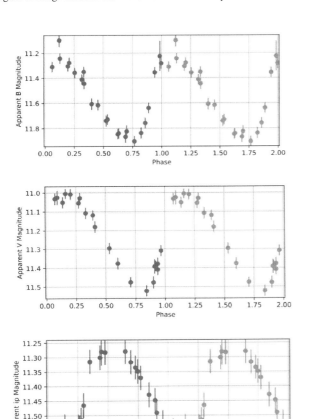

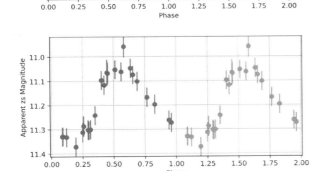

Figure 2. Light curves in the B, V, i', and z' filters, respectively.

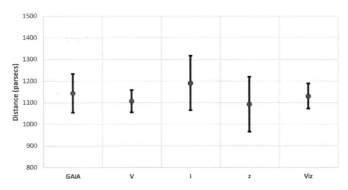

Figure 3. Distance measurements with error after extinction correction. Viz is the mean of all three filters.

measurement published in the second Gaia data release. The interstellar reddening (E(B–V)) of YZ Cap was determined to be 0.14 by minimizing the variance in the values of the true distance moduli for the V (Equation 7), i' (Equation 8), and z' (Equation 9) filters. Extinction values A_m/A_V were calculated using an extinction curve from Cardelli *et al.* (1989), where $R_V = 3.1$ and $A_V = 1$.

$$\mu(V) = V - M_V - A_V \qquad (7)$$

$$\mu(i') = i' - M_i - (A_i / A_V) A_V \qquad (8)$$

$$\mu(z') = z' - M_z - (A_z / A_V) A_V \qquad (9)$$

Calculated error in the reported distances incorporated errors in the apparent magnitude, period, and metallicity. When the average across all three filters is taken, the distance is found with an error smaller than that reported by Gaia. The distance measurements with error for each filter, the mean of the filters, and the Gaia measured distance are summarized in Figure 3.

5. Conclusion

Using the photometric analysis of the infrared and visible images, the period of YZ Cap was determined to be 0.274 ± 0.003 day, which agrees with previous measurements. Using the equations found in Catelan *et al.* (2004) and Caceres and Catelan (2008), the distance and magnitudes of YZ Cap in the V, i', and z' filters were determined. These values were found to agree with the Gaia measured distance of 1144 ± 90 pc. This study expanded the existing pool of data on RR Lyrae stars, and it provides an example of the ability to apply the infrared period-luminosity relations to an individual star to determine the distance to the star.

6. Acknowledgements

We would like to thank Dr. Michael Fitzgerald and the Our Solar Siblings Staff, who provided guidance and support for the use of Astrosource and provided us time to use the Las Cumbres Observatory Global Telescope Network.

This research has made use of the VizieR catalogue access tool, CDS, Strasbourg, France (DOI: 10.26093/cds/vizier).

The original description of the VizieR service was published in A&AS 143, 23.

This work made use of the International Variable Star Index (VSX) operated by AAVSO, Cambridge, MA, USA.

This research was made possible through the use of the AAVSO Photometric All-Sky Survey (APASS), funded by the Robert Martin Ayers Sciences Fund and NSF AST-1412587.

References

Alam, S., *et al.* 2015, *Astrophys. J., Suppl. Ser.*, **219**, 12.

Altunin, I., Caputo, R., and Tock, K., 2020, *Astron. Theory, Obs., Methods (ATOM)*, 1 (doi:https://doi.org/10.32374/atom.2020.1.1).

Bertin, E., and Arnouts, S. 1996, *Astron. Astrophys., Suppl. Ser.*, **117**, 393.

Bono, G., *et al.* 2020, *Astrophys. J., Lett.*, **896**, 15.

Brown, T., *et al.* 2013, *Publ. Astron. Soc. Pacific*, **125**, 1031.

Cacciari, C., Clementini, G., and Buser, R. 1989, *Astron. Astrophys.*, **209**, 154.

Cáceres, C., and Catelan, M. 2008, *Astrophys. J., Suppl. Ser.*, **179**, 242.

Cardelli, J. A., Clayton, G. C., and Mathis, J. S. 1989, *Astrophys. J.*, **345**, 245.

Catelan, M., Pritzl, B. J., and Smith, H. A. 2004, *Astrophys. J., Suppl. Ser.*, **154**, 633.

Catelan, M., and Smith, H. 2015, *Pulsating Stars*, 1st ed., Wiley-VCH, Berlin.

Feast, M. W., Laney, C. D., Kinman, T. D., van Leeuwn, F., and Whitelock, P. A. 2008, *Mon. Not. Roy. Astron. Soc.*, **386**, 2115.

Fitzgerald, M. T. 2018, *Robotic Telesc., Student Res. Education Proc.*, **1**, 343.

Fitzgerald, M., *et al.* 2020, *J. Open Source Software*, in review.

Flewelling, H. A., *et al.* 2020, *Astrophys. J., Suppl. Ser.*, **251**, 7.

Gaia Collaboration, *et al.* 2018, *Astron. Astrophys.*, **616A**, 1.

Govea, J., Gomez, T., Preston, G. W., and Sneden, C. 2014, *Astrophys. J.*, **782**, 59.

Henden, A. A., Levine, S., Terrell, D., and Welch, D. L. 2015, Amer. Astron. Soc. Meeting 225, id.336.16.

Henden *et al.* 2016 AAS VizieR Online Data Catalog.

Magnier, E. *et al.* 2020, *Astrophys. J., Suppl. Ser.*, **24**, 3.

Monson, A. J., *et al.* 2017, *Astron. J.*, **153**, 96.

Neeley, J. R., *et al.* 2017, *Astrophys. J.*, **841**, 64.

Nesterov, V. V., Kuzmin, A. V., Ashimbaeva, N. T., Volchkov, A. A., Röser, S., and Bastian, U. 1995, *Astron. Astrophys, Suppl. Ser.*, **110**, 367.

Pickering, E. C., Colson, H. R., Fleming, W. P., and Wells, L. D. 1901, *Astrophys. J.*, **13**, 226.

Stellingwerf, R. F. 1978, *Astrophys. J.*, **224**, 953 (doi: 10.1086156444).

Southern Eclipsing Binary Minima and Light Elements in 2020

Tom Richards
Pretty Hill Observatory, POB 323, Kangaroo Ground, VIC 3097, Australia; tomprettyhill@gmail.com

Roy A. Axelsen
P.O. Box 706, Kenmore, QLD 4069, Australia; royaxelsen7@gmail.com

Mark Blackford
Congarinni Observatory, Congarinni, NSW 2447, Australia; markgblackford@outlook.com

Robert Jenkins
Theta Observatory, Salisbury, SA 5108, Australia; robynnabara@ozemail.com.au

David J. W. Moriarty
School of Mathematics and Physics, The University of Queensland, QLD 4072, Australia; djwmoriarty@bigpond.com

Received June 13, 2021; revised September 29, 2021; accepted September 29, 2021

Abstract We present 246 times of minima of 77 southern hemisphere eclipsing binary stars acquired in 2020. These observations were acquired and analyzed by the authors, who are members of the Southern Eclipsing Binary group of Variable Stars South (VSS) (http://www.variablestarssouth.org), using CCD detectors. For four of the systems we have derived updated light elements and present those as well as O–C values for the VSS minima. This paper is the sixth in a series by Richards *et al.*

1. Observations

Equipment and software used are set out in Table 1. Observer initials abbreviate the name of an author of this paper, surname last. Instrument refers to the telescope and objective diameter in cm and the camera used. Remaining columns refer to the software used for the purposes listed.

All observers using PERANSO (Paunzen and Vanmunster 2016) employed polynomial fitting for minima estimation. Minima25 (Nelson 2019) estimates the time of an eclipse minimum (with standard deviation) by six methods: parabolic fit, tracing paper, bisectors of chords, the Kwee-van Woerden method (Kwee and Van Woerden 1956), Fourier fit, and sliding integrations. Then the mean of the individual minimum estimates is calculated, weighted by the standard deviation of each, which is returned together with its standard error as the best estimate.

Online information and sources for the software in Table 1 are as follows. In all cases the current versions of the software were used.

AstroImageJ (Collins *et al.* 2017)
  https://www.astro.louisville.edu/software/astroimagej/

Astrophotography Tool (Incanus Ltd. 2009–2021)
  https://www.astrophotography.app/

IRIS (Buil 1999–2018)
  http://www.astrosurf.com/buil/iris-software.html

MaxIm DL (Diffraction Limited 2012)
  http://www.cyanogen.com

Minima25 (Nelson 2019)
  http://www.variablestarssouth.org

Muniwin (Motl 2011)
  http://c-munipack.sourceforge.net

PERANSO (Vanmunster 2013)
  http://www.peranso.com

TheSkyX (Software Bisque 2020)
  http://www.bisque.com

VStar (Benn 2013)
  https://www.aavso.org/vstar

Table 1. Observers, equipment, and software.

Observer	Instrument	Imaging	Calibration	Photometry	Minima
TR	41 cm R-C + SBIG STXL-6303e	MaxIm-DL	Muniwin	Muniwin	PERANSO
RA	12-cm refractor + ZWO ASI1600MM CMOS	Astrophotography Tool	AstroImageJ	AstroImageJ	VStar
MB	8-cm refractor + Atik One 6.0	TheSkyX Professional	MaxIm-DL	MaxIm-DL	PERANSO
MB	35-cm R-C + SBIG STT-3200	TheSkyX Professional	MaxIm-DL	MaxIm-DL	PERANSO
RJ	25 cm GSO RCA + QSI-583 CCD.	MaxIm-DL	MaxIm-DL	MaxIm-DL	Minima25e
DM	36-cm S-C + Moravian G3-6303 CCD	MaxIm-DL	MaxIm-DL	MaxIm-DL	PERANSO

Observers: TR, T. Richards; RA, R. A. Axelsen; MB, M. Blackford; RJ, R. Jenkins; DM, D. J. W. Moriarty.

Image sets were obtained in hours-long runs. Each observer analyzed their own image sets as follows:

1. Calibrated them using bias frames, dark frames, and flat field frames.

2. Executed differential aperture photometric measurements on the calibrated sets.

3. Performed minima estimation on the photometric data.

2. Results

Appendix A lists the minima estimates. Columns 1 and 2 list the GCVS designation of the target stars in lexical order of constellation abbreviation, and GCVS variability type as listed in (Samus *et al.* 2017). In some cases, more recent work may propose different variability types. Columns 3 and 4 record the heliocentric Julian dates of minima and the uncertainty (in days) as reported by the algorithm used in the photometry software. Column 5 lists the minimum type, primary (P) or secondary (S). We define the primary minimum as the deeper one in our observations where that can be determined, otherwise we assume the epoch recorded in the AAVSO Variable Star Index (Watson *et al.* 2006)—hereafter referred to as VSX—is of a primary minimum. Column 6 gives the filter used: B and V are Johnson B and V, R, and I are Cousins R and I, and SR is Sloan r'. Column 7 gives the initials of the observer.

3. Analysis

Following our practice in these series of papers we list revised light elements (aka ephemerides) for binaries in Appendix A for which we have derived four or more primary minima in this year (2020) and earlier years, spread over at least three observing seasons. The purpose of this is to see if there's good evidence for period change.

This year there are four such binaries. Their earlier minima estimates are listed in Table 2. Its last column cites the year of publication of the paper in this series (Richards *et al.* 2018, 2019, 2020) in which the estimate is recorded. Binaries with revised light elements reported in earlier papers in this series are excluded from the present paper.

Table 3 contains the resulting linear light elements for the systems we analyzed. These were derived by ordinary least squares regression. The regression used all the VSS primary minima times and the VSX epoch time as minima data. The VSX epoch and period were used as reference ephemeris to obtain an orbital cycle count for the minima data. By regressing the (HJD) minima times against cycles we obtained a best-fit period from the slope of the regression line, and a new zero epoch as the y-intercept at the earliest VSS minimum.

The first five columns list the system, the epoch and its standard error, the period and its standard error. The next column records the standard deviation of the residuals of the minima from the regression prediction. The smaller the number, the better are the minima data fitted to a linear fit (constant period). No. Obs. is the number of VSS primary minima estimates used in the regression, and Interval is the time interval in days covered by them.

For each system in Table 2, we present in Figure 1 plots of the residuals of the observed minima from the calculated regression (O–C values). The regression is represented by the horizontal line at O–C = 0. The left-hand panel in each pair for a star shows (by the left edge) the VSX minimum, together with (near the right edge) the VSS minima. The latter are zoomed into in the right-hand panel to exhibit any structure in the residuals which may indicate variation in the period. The error bars are those reported for the time of minimum by the software used for minima estimation.

The primary interest in the Table 3 light elements, and the Figure 1 residual plots, lies in indication of period change. In all four cases, the linear periods in Table 3 are inadequate fits to the data, and hence so are the VSX periods.

V901 Cen may have a decreasing period, but the minima estimates show significant scatter, probably indicative of RS CVn-type chromospheric disturbances shifting minima estimates slightly. An inspection of its photometric light curves shows that the occultation of one star is sometimes deeper, sometimes shallower than for the other star.

For *YY Eri*, while the VSS data are closely linear, the extreme slope is likely spurious, indicating a cycle count jump (Richards 2021) rather than a shorter period. An ephemeris using just the VSS data would plainly give the VSX data point an enormous residual, so such an ephemeris must be wrong.

Table 2. Minima estimates from earlier years used to construct the revised linear ephemerides in Table 3.

Identifier	HJD of Min.	Error	Paper
V901 Cen	2457806.0622	0.0002	2018
V901 Cen	2457849.9859	0.0002	2018
V901 Cen	2458195.96908	0.0013	2019
YY Eri	2458462.16282	0.00010	2019
YY Eri	2458846.02929	0.0013	2020
V Gru	2458362.9813	0.0038	2019
V Gru	2458376.9984	0.0026	2019
V Gru	2458678.20098	0.00031	2020
V Gru	2458727.99620	0.00010	2020
BS Mus	2458188.09221	0.00027	2019
BS Mus	2458204.22829	0.00020	2019
BS Mus	2458221.12957	0.00020	2019
BS Mus	2458228.04311	0.00019	2019
BS Mus	2458231.11474	0.00014	2019
BS Mus	2458576.07301	0.00018	2020

Table 3. Revised linear light elements for systems with four or more VSS primary minima estimates, regressed from the VSX light elements.

Identifier	E_0	E_{0err}	P	P_{err}	SD_{resid}	No. Obs.	Interval
V901 Cen	2452443.50036	0.0010	0.35423330	6.7E-08	0.0010	5	1442
YY Eri	2458462.15812	0.0018	0.3215000	3.0E-07	0.0039	4	730
V Gru	2458362.97649	0.006	0.4834461	5.4E-07	0.0147	5	715
BS Mus	2458188.10910	0.005	0.7682422	3.7E-07	0.0152	11	816

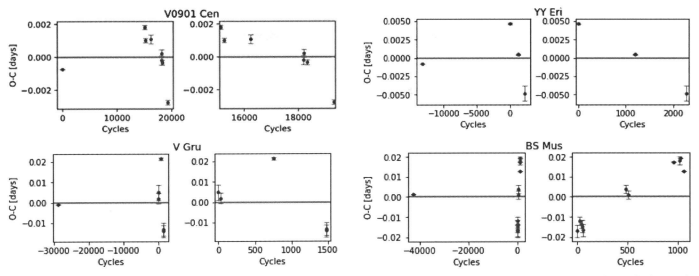

Figure 1. Residual (O–C) plots of the minima estimates against the light elements listed in Table 3. Left-hand panels for each binary system show VSX and VSS minima, right-hand panels the VSS minima only. Orbital cycle numbers count from zero at the first (regressed) VSS minimum, incrementing by the regressed period.

Regarding the data here for *V Gru*, it is diffficult to explain its O–C behavior. One possibility is an oscillating period.

BS Mus shows a very steep upward slope on the VSS data, and as with YY Eri an ephemeris based on VSS data could not accommodate the VSX data point. A cycle count jump is the likely explanation.

4. Conclusion

We have presented 246 minima estimates of 77 southern eclipsing binaries made by the authors in 2020. We have included a preliminary period analysis for four of them where we have acquired four or more minima in this year (2020) and earlier years, spread over at least three observing seasons. In none of these four cases is there a clear indication of a linear period, but nor can the type of any non-linear period (linear period change or period oscillation) be specified, let alone quantified. Because of that they are interesting candidates for further research into period behavior, which, in addition to future minima estimates, requires collation of online sources of photometric data to bridge the gap between early (discovery and VSX) data and the VSS data sets.

For a more extensive study of some of these systems, spectroscopic data are also needed. For example, the contact binary in V1084 Sco is likely a member of a quadruple system with a detached binary (Rucinski and Duerbeck 2006). Most contact and close binary systems may have tertiary or higher order components (Pribulla and Rucinski 2006). Period changes in close binary systems that are part of triple or multiple systems would be expected due to interactions of the various components. Among the binary systems we report on here, preliminary radial velocity analyses with the broadening function show that TY Cap, FQ CMa, and V Gru are triple systems and BS Mus and GZ Pup are triple or quadruple systems (Moriarty 2021).

5. Acknowledgements

This research has made much use of the International Variable Star Index (VSX) database, operated at AAVSO, Cambridge, Massachusetts, USA (Watson *et al.* 2006).

We acknowledge grants from the Edward Corbould Research Fund of the Astronomical Association of Queensland to DJWM for assistance with the purchase of equipment and software for photometry.

References

Benn, D. 2013, VSTAR data analysis software
(https://www.aavso.org/vstar-overview).

Buil, C. 1999–2018, IRIS astronomical images processing software (http://www.astrosurf.com/buil/iris-software.html).

Collins, K. A., Kielkopf, J. F., Stassun1, K. G., and Hessman, F. V. 2017, *Astron. J.*, **153**, 77
(https://www.astro.louisville.edu/software/astroimagej).

Diffraction Limited. 2012, MAXIMDL image processing software (http://www.cyanogen.com).

Incanus Ltd. 2009–2021, ASTROPHOTOGRAPHY TOOL (https://www.astrophotography.app).

Kwee, K.K., and van Woerden, H. 1956, *Bull. Astron. Inst. Neth.*, **12**, 327.

Moriarty, D. J. W. 2021, personal communication.

Motl, D. 2011, C-Munipack
(http://c-munipack.sourceforge.net).

Nelson, R. H. 2019, Software by Bob Nelson
(https://www.variablestarssouth.org).

Paunzen, E., and Vanmunster, T. 2016, *Astron. Nachr.*, **337**, 239.

Pribulla, T., and Rucinski, S. M. 2006. *Astron. J.*, **131**, 2986.

Richards, T. 2021, *Var. Stars South Newsl. 2021–2* (April 2021), No. 14 (https://www.variablestarssouth.org).

Richards, T., Blackford, M., Bohlsen, T., Butterworth, N., Lowther, S., Jenkins, R., and Powles, J. 2018, *Open Eur. J. Var. Stars*, **189**, 1.

Richards, T., Blackford, M., Butterworth, N., Crawford, G., and Jenkins, R. 2019, *Open Eur. J. Var. Stars*, **198**, 1.

Richards, T., Blackford, M., Butterworth, N., and Jenkins, R. 2020, *J. Amer. Assoc. Var. Stars*, **48**, 250.

Rucinski, S. M., and Duerbeck, H. W. 2006, *Astron. J.*, **132**, 1539.

Samus, N. N., Kazarovets, E. V., Durlevich, O. V., Kireeva, N. N., and Pastukhova, E. N. 2017, *Astron. Rep.*, **61**, 80, *General Catalogue of Variable Stars: Version GCVS 5.1* (http://www.sai.msu.su/gcvs/gcvs/index.htm).

Software Bisque. 2020, THESKYX professional (http://www.bisque.com).

Vanmunster, T. 2013, light curve and period analysis software, PERANSO v.2.50 (http://www.cbabelgium.com/peranso).

Watson, C. L., Henden, A. A., and Price, A. 2006, in *The Society for Astronomical Sciences 25th Annual Symposium on Telescope Science*, The Society for Astronomical Sciences, Rancho Cucamonga, CA, 47 (http://www.aavso.org/vsx).

Appendix A: Minima Estimates

Identifier	Type	HJD of Min.	Error	Min.	Filter	Obs.	Identifier	Type	HJD of Min.	Error	Min.	Filter	Obs.
YY Aps	EB	2459009.18076	0.00303	P	SR	TR	V Gru	EW/KW	2459078.02910	0.00148	S	I	MB
NT Aps	EW	2459035.06790	0.00094	P	I	MB	V Gru	EW/KW	2459078.26921	0.00295	P	I	MB
NT Aps	EW	2459036.09990	0.00126	S	I	MB	V Gru	EW/KW	2459078.27006	0.00309	P	V	MB
NT Aps	EW	2459036.10054	0.00135	S	V	MB	V Gru	EW/KW	2459090.11505	0.00104	S	V	MB
NT Aps	EW	2459036.10064	0.00128	S	B	MB	RV Gru	EW/EK	2459075.26875	0.00052	P	V	MB
NT Aps	EW	2459036.98395	0.00108	S	B	MB	RV Gru	EW/KW	2459090.97038	0.00089	S	SR	TR
NT Aps	EW	2459036.98436	0.00147	S	V	MB	RV Gru	EW/KW	2459091.10014	0.00060	P	SR	TR
NT Aps	EW	2459036.98495	0.00148	S	I	MB	RV Gru	EW/EK	2459168.05135	0.00057	S	V	MB
NT Aps	EW	2459037.13033	0.00127	P	V	MB	YY Gru	EW	2459107.00823	0.00065	S	SR	TR
NT Aps	EW	2459037.13057	0.00114	P	B	MB	YY Gru	EW	2459108.03245	0.00070	P	SR	TR
NT Aps	EW	2459037.13105	0.00117	P	I	MB	YY Gru	EW	2459108.17896	0.00066	S	SR	TR
V354 Aps	EA	2459039.01091	0.00180	P	SR	TR	BC Gru	EW:/KW:	2459083.04124	0.00409	S	V	RA
DX Aqr	EA	2459128.05282	0.00169	P	V	MB	BC Gru	EW:/KW:	2459083.19399	0.00532	P	V	RA
EE Aqr	EB	2459118.02437	0.00103	P	I	MB	BC Gru	EW:/KW:	2459085.03911	0.00270	P	V	RA
EE Aqr	EB	2459132.02102	0.00241	S	I	MB	BC Gru	EW:/KW:	2459085.19249	0.00342	S	V	RA
V610 Ara	EW	2459025.11863	0.00417	S	V	RA	BC Gru	EW:/KW:	2459085.96148	0.00317	P	V	RA
V610 Ara	EW	2459025.93232	0.00238	P	V	RA	BC Gru	EW:/KW:	2459086.26803	0.00366	P	V	RA
V610 Ara	EW	2459027.01624	0.00234	P	V	RA	BC Gru	EW:/KW:	2459087.03701	0.00240	S	V	RA
V610 Ara	EW	2459032.99357	0.00316	P	B	RA	BC Gru	EW:/KW:	2459087.18929	0.00285	P	V	RA
V610 Ara	EW	2459035.16639	0.00583	P	B	RA	DY Gru	EW/KW	2459092.05019	0.00094	S	V	MB
V610 Ara	EW	2459037.06851	0.00347	S	B	RA	DY Gru	EW/KW	2459128.00538	0.00097	S	V	MB
V870 Ara	EW	2459003.08566	0.00176	P	I	MB	DY Gru	EW/KW	2459136.91075	0.00094	P	V	MB
V870 Ara	EW	2459016.07958	0.00228	S	V	MB	SZ Hor	EB	2459124.05406	0.00002	S	V	RJ
V870 Ara	EW	2459016.07983	0.00250	S	I	MB	SZ Hor	EB	2459195.00065	0.00308	P	SR	TR
V870 Ara	EW	2459016.08041	0.00235	S	B	MB	WZ Hor	EA	2459127.14875	0.00230	P	I	MB
V870 Ara	EW	2459016.27892	0.00232	P	B	MB	WZ Hor	EA	2459132.97972	0.00211	P	V	MB
V870 Ara	EW	2459016.27893	0.00231	P	V	MB	CP Hyi	EW	2459131.96725	0.00312	P	V	MB
V870 Ara	EW	2459016.27916	0.00259	P	I	MB	CP Hyi	EW	2459136.99961	0.00264	S	V	MB
X Cae	EC	2459164.99236	0.00369	P	I	MB	CN Ind	EW	2459061.02373	0.00538	S	V	RA
X Cae	EC	2459165.13360	0.00392	S	I	MB	CN Ind	EW	2459061.25056	0.00639	P	V	RA
TY Cap	EA/SD+DSCT	2459168.97490	0.00166	P	V	MB	CN Ind	EW	2459062.15687	0.00343	P	V	RA
ST Car	EB/SD	2458955.11084	0.00072	P	V	MB	CN Ind	EW	2459066.23980	0.00619	P	B	RA
BH Cen	EB/KE	2458929.01847	0.00081	S	B	MB	CN Ind	EW	2459066.92096	0.00543	S	B	RA
BH Cen	EB/KE	2458929.01874	0.00064	S	V	MB	CN Ind	EW	2459067.14730	0.00402	P	B	RA
BH Cen	EB/KE	2458929.01876	0.00112	S	I	MB	CN Ind	EW	2459078.94261	0.00231	P	V	RA
V676 Cen	EW/KW	2458986.08538	0.00044	S	SR	TR	CN Ind	EW	2459079.16844	0.00425	S	V	RA
V901 Cen	EW/RS	2458892.13909	0.00124	P	SR	TR	CN Ind	EW	2459080.98361	0.00467	S	V	RA
V901 Cen	EW/RS	2458897.09869	0.00135	P	SR	TR	CN Ind	EW	2459081.21017	0.00435	P	V	RA
V901 Cen	EW/RS	2458939.96070	0.00106	P	SR	TR	CO Ind	EB/KE	2458726.19515	0.00132	P	V	TR
V901 Cen	EW/RS	2458940.13776	0.00110	S	SR	TR	CO Ind	EB/KE	2458744.13395	0.00120	S	SR	TR
WY Cet	EA/SD+DSCT	2459160.01400	0.00134	P	I	MB	CR Ind	EW	2458737.27507	0.00103	S	SR	TR
DM Cir	EW	2458979.99346	0.00095	S	V	MB	CR Ind	EW	2459121.05600	0.00093	S	SR	TR
DM Cir	EW	2458979.99355	0.00093	S	B	MB	CU Ind	EW	2459090.15378	0.00123	S	SI	TR
DM Cir	EW	2458979.99383	0.00122	S	I	MB	RR Lep	EB	2459192.04276	0.00167	P	V	MB
DM Cir	EW	2458980.18764	0.00097	P	B	MB	AU Men	EW	2459198.13718	0.00076	P	SR	TR
DM Cir	EW	2458980.18776	0.00115	P	V	MB	XY Men	EB/KE	2459192.05037	0.00013	P	V	RJ
DM Cir	EW	2458980.18783	0.00100	P	I	MB	XY Men	EB/KE	2459213.01845	0.00436	P	SR	TR
DM Cir	EW	2458980.96012	0.00105	P	V	MB	AH Mic	EW/KW	2459089.94253	0.00085	S	V	MB
DM Cir	EW	2458980.96085	0.00128	P	B	MB	AH Mic	EW/KW	2459090.10476	0.00094	P	V	MB
DM Cir	EW	2458980.96121	0.00115	P	I	MB	AH Mic	EW/KW	2459090.91573	0.00088	S	R	MB
DM Cir	EW	2458981.15462	0.00108	S	B	MB	AH Mic	EW/KW	2459091.07620	0.00115	P	R	MB
DM Cir	EW	2458981.15470	0.00122	S	V	MB	AH Mic	EW/KW	2459091.24165	0.00103	S	R	MB
DM Cir	EW	2458981.15541	0.00135	S	I	MB	AH Mic	EW/KW	2459108.91458	0.00112	P	I	MB
FQ CMa	EA+DSCT	2458870.96372	0.00086	P	V	MB	AH Mic	EW/KW	2459109.07917	0.00122	S	I	MB
FQ CMa	EA+DSCT	2458870.96388	0.00084	P	B	MB	TU Mus	EB/KE	2458986.09161	0.00009	S	V	RJ
FQ CMa	EA+DSCT	2458870.96388	0.00083	P	I	MB	TV Mus	EW/KW	2458948.11848	0.00229	S	SR	TR
FQ CMa	EA+DSCT	2458880.02359	0.00222	S	I	MB	TW Mus	EW/KW	2458920.04601	0.00155	P	SR	TR
FQ CMa	EA+DSCT	2458881.11035	0.00082	P	I	MB	TW Mus	EW/KW	2458920.25557	0.00132	S	SR	TR
FQ CMa	EA+DSCT	2458881.11036	0.00068	P	V	MB	DE Mic	EW	2459118.96789	0.00125	P	V	MB
FQ CMa	EA+DSCT	2458881.11038	0.00070	P	B	MB	DE Mic	EW	2459125.94931	0.00115	P	V	MB
eps CrA	EW	2459117.03715	0.00294	P	V	MB	BR Mus	EW/KE	2458985.97818	0.00123	P	SR	TR
eps CrA	EW	2459117.92324	0.00469	S	V	MB	BS Mus	EB/KE	2458924.10257	0.00009	P	V	RJ
RW Dor	EW:KW	2459187.06542	0.00055	S	SR	TR	BS Mus	EB/KE	2458939.09520	0.00004	S	V	RJ
AP Dor	EW	2459168.02296	0.00191	P	V	MB	BS Mus	EB/KE	2458967.12467	0.00207	P	SR	TR
YY Eri	EW/KW	2459191.95781	0.00097	P	I	MB	BS Mus	EB/KE	2458974.04027	0.00018	P	V	RJ
BV Eri	EW	2459137.05952	0.00178	S	I	MB	BS Mus	EB/KE	2459003.99520	0.00008	P	V	RJ
V Gru	EW/KW	2459078.02824	0.00154	S	V	MB	V395 Nor	EW	2459017.13878	0.00144	P	V	MB
V Gru	EW/KW	2459078.02838	0.00163	S	B	MB	EI Oct	EW	2459063.96363	0.00067	P	V	MB

Table continued on next page

Appendix A: Minima Estimates (cont.)

Identifier	Type	HJD of Min.	Error	Min.	Filter	Obs.	Identifier	Type	HJD of Min.	Error	Min.	Filter	Obs.
EI Oct	EW	2459064.13377	0.00068	S	V	MB	V632 Sco	EA	2459066.04093	0.00170	P	B	MB
EI Oct	EW	2459078.01351	0.00116	S	I	MB	V638 Sco	EA/D:	2456114.10783	0.00300	P	V	DM
EI Oct	EW	2459078.01377	0.00081	S	V	MB	V638 Sco	EA/D:	2459002.98610	0.00300	P	B	DM
EI Oct	EW	2459078.01378	0.00122	S	B	MB	V638 Sco	EA/D:	2459007.70376	0.00200	P	B	PE
EI Oct	EW	2459078.18112	0.00093	P	I	MB	V638 Sco	EA/D:	2459014.77873	0.00200	P	B	PE
EI Oct	EW	2459078.18213	0.00108	P	B	MB	V638 Sco	EA/D:	2459036.00349	0.00210	P	V	MB
EI Oct	EW	2459078.18216	0.00106	P	V	MB	V638 Sco	EA/D:	2459036.00361	0.00240	P	I	MB
EI Oct	EW	2459085.96900	0.00074	P	B	MB	V638 Sco	EA/D:	2459036.00405	0.00238	P	B	MB
EI Oct	EW	2459086.13807	0.00074	S	B	MB	V701 Sco	EW/KE	2459032.12467	0.00264	S	V	MB
EI Oct	EW	2459088.00030	0.00060	P	I	MB	V701 Sco	EW/KE	2459032.12494	0.00344	S	I	MB
EI Oct	EW	2459088.16987	0.00070	S	I	MB	V701 Sco	EW/KE	2459062.97966	0.00336	P	I	MB
EZ Oct	EW/KW	2458883.05284	0.00006	S	V	RJ	V701 Sco	EW/KE	2459062.98109	0.00292	P	B	MB
EZ Oct	EW/KW	2458902.06385	0.00005	P	V	RJ	V701 Sco	EW/KE	2459062.98156	0.00293	P	V	MB
EZ Oct	EW/KW	2459017.98761	0.00096	S	SR	TR	V1055 Sco	EW	2459006.98131	0.00150	S	I	MB
EZ Oct	EW/KW	2459018.13052	0.00087	P	SR	TR	V1055 Sco	EW	2459007.16255	0.00158	P	I	MB
EZ Oct	EW/KW	2459018.27309	0.00073	S	SR	TR	V1055 Sco	EW	2459049.89707	0.00234	S	V	MB
EZ Oct	EW/KW	2459028.99389	0.00130	P	V	TR	V1055 Sco	EW	2459049.89727	0.00208	S	B	MB
EZ Oct	EW/KW	2459029.13698	0.00147	S	V	TR	V1055 Sco	EW	2459050.07873	0.00239	P	V	MB
BF Pav	EW	2459060.99834	0.00042	P	R	MB	V1055 Sco	EW	2459050.07908	0.00195	P	B	MB
HY Pav	EW/KW	2459063.01555	0.00061	S	V	MB	V1055 Sco	EW	2459050.98772	0.00166	S	B	MB
HY Pav	EW/KW	2459063.19024	0.00060	P	V	MB	V1055 Sco	EW	2459074.99078	0.00267	S	I	MB
HY Pav	EW/KW	2459109.96189	0.00091	P	B	MB	V1055 Sco	EW	2459074.99117	0.00257	S	B	MB
HY Pav	EW/KW	2459109.96224	0.00100	P	I	MB	V1055 Sco	EW	2459074.99128	0.00283	S	V	MB
HY Pav	EW/KW	2459110.13765	0.00120	S	I	MB	V1084 Sco	EW	2459064.02666	0.00165	P	V	MB
HY Pav	EW/KW	2459110.13820	0.00125	S	B	MB	QW Tel	EW	2459083.07234	0.00431	P	I	MB
V400 Pav	EB	2459067.11142	0.00265	P	V	MB	GN TrA	EA/KE	2459036.94501	0.00121	P	B	MB
V401 Pav	EW	2459045.00019	0.00087	P	SR	TR	GN TrA	EA/KE	2459036.94508	0.00137	P	I	MB
V401 Pav	EW	2459045.16291	0.00079	S	SR	TR	GN TrA	EA/KE	2459036.94556	0.00113	P	V	MB
YZ Phe	EW	2459162.05443	0.00075	P	V	TR	GQ TrA	EA+DSCT	2458973.06288	0.00455	S	I	MB
YZ Phe	EW	2459162.17327	0.00062	S	V	TR	GQ TrA	EA+DSCT	2458974.23171	0.00140	P	I	MB
AD Phe	EW/KW	2459141.01299	0.00007	P	V	RJ	GQ TrA	EA+DSCT	2459006.98418	0.00166	P	I	MB
AD Phe	EW/KW	2459167.98836	0.00006	P	V	RJ	V336 TrA	EW	2459001.07265	0.00068	P	B	MB
AD Phe	EW/KW	2459186.03242	0.00008	S	V	RJ	V336 TrA	EW	2459001.07268	0.00064	P	I	MB
AD Phe	EW/KE	2459186.98473	0.00004	P	V	RJ	V336 TrA	EW	2459001.07281	0.00048	P	V	MB
AU Phe	EW	2459171.08637	0.00083	S	SR	TR	V336 TrA	EW	2459008.94256	0.00089	S	B	MB
BQ Phe	EW	2459179.01475	0.00002	S	V	RJ	V336 TrA	EW	2459008.94256	0.00079	S	V	MB
GY Pup	EW/KW	2458874.03641	0.00141	P	SR	TR	V336 TrA	EW	2459008.94259	0.00084	S	I	MB
GZ Pup	EW/KW	2458850.04166	0.00082	S	V	TR	V336 TrA	EW	2459009.07572	0.00074	P	I	MB
GZ Pup	EW/KW	2458861.09059	0.00115	P	V	TR	V336 TrA	EW	2459009.07577	0.00063	P	B	MB
GZ Pup	EW/KW	2458879.02495	0.00127	P	V	TR	V336 TrA	EW	2459009.07580	0.00060	P	V	MB
GZ Pup	EW/KW	2458883.98942	0.00097	S	V	TR	V336 TrA	EW	2459016.01171	0.00071	P	I	MB
GZ Pup	EW/KW	2458884.14925	0.00104	P	V	TR	V336 TrA	EW	2459016.01174	0.00073	P	B	MB
GZ Pup	EW/KW	2459194.00698	0.00091	S	V	TR	V336 TrA	EW	2459016.01188	0.00058	P	V	MB
GZ Pup	EW/KW	2459194.16665	0.00101	P	V	TR	V336 TrA	EW	2459016.14512	0.00059	S	B	MB
HI Pup	EW/KW	2458851.06664	0.00143	S	SR	TR	V336 TrA	EW	2459016.14537	0.00066	S	V	MB
V653 Pup	EW	2458486.14701	0.00145	P	SR	TR	V336 TrA	EW	2459016.14543	0.00067	S	I	MB
RT Scl	EB	2459133.97002	0.00105	P	V	MB	AQ Tuc	EW	2458855.01766	0.00024	S	V	RJ
RT Scl	EB	2459136.01645	0.00105	P	V	MB	BU Vel	EW	2458919.03724	0.00191	S	SI	TR
UY Scl	EW	2459133.93005	0.00095	P	V	MB	DU Vel	EA	2458925.06656	0.00502	S	V	TR
UY Scl	EW	2459135.93265	0.00084	S	V	MB	FM Vel	EW/KW	2458896.02208	0.00102	S	SR	TR
UY Scl	EW	2459136.11661	0.00084	P	V	MB	FM Vel	EW/KW	2458896.21661	0.00109	P	SR	TR
BB Scl	EA	2459132.17127	0.00105	P	V	MB	FM Vel	EW/KW	2458917.05622	0.00096	S	SR	TR
V462 Sco	EW	2459050.00781	0.00134	P	R	MB	FM Vel	EW/KW	2458917.25124	0.00130	P	SR	TR
V632 Sco	EA	2459032.22726	0.00180	P	I	MB	V362 Vel	EW	2458952.13686	0.00235	P	I	MB
V632 Sco	EA	2459032.22767	0.00177	P	V	MB	W Vol	EA/AR	2458921.12717	0.00192	P	SR	TR

Observers: TR, T. Richards; RA, R. A. Axelsen; MB, M. Blackford; RJ, R. Jenkins; DM, D. J. W. Moriarty.

The Photometric Period of V1674 Herculis (Nova Her 2021)

Richard E. Schmidt
Burleith Observatory, 1810 35th Street NW, Washington, DC 20007

Sergei Yu. Shugarov
Astronomical Institute of Slovak Academy of Sciences, Tatranská Lomnica, Slovakia; and Sternberg Astronomical Institute, M. V. Lomonosov Moscow State University, Moscow, Russia

Marina D. Afonina
Sternberg Astronomical Institute, M. V. Lomonosov Moscow State Univ., Moscow, Russia; and Faculty of Physics, M. V. Lomonosov Moscow State University, Moscow, Russia

Received July 21, 2021; revised August 18, 29, 2021; accepted September 3, 2021

Abstract A photometric study of the fast galactic nova V1674 Herculis (Nova Her 2021, TCP J18573095+1653396) was undertaken at the Burleith Observatory in Washington, DC, and supplemented with photometry from the Crimean Laboratory of the Sternberg Astronomical Institute and the Astronomical Institute of the Slovak Academy of Science. A total of 979 CCD observations were obtained over a time span of 53.8 days, yielding an orbital period: 0.152934 d ± 0.000034 d, epoch (HJD) of minimum light 2459408.74789.

1. Introduction

V1674 Herculis, (Nova Her 2021, TCP J18573095+1653396), R.A. = $18^h 57^m 30.98^s$, Dec. = $+16° 53' 39.6"$ (2000), was discovered by Seiji Ueda, Kushiro, Hokkaido, Japan, on 2021 June 12.537 UT (Ueda 2021). Spectral observations by Munari, Valisa, and Dallaporta identified the source as a classical nova on June 12.84 (Munari *et al.* 2021). Within three days the nova reached magnitude 5.23 I (6.76 V) (Romanov 2021), then faded by two magnitudes in 1.18 days—making this the fastest nova on record! (Quimby *et al.* 2021). Mroz *et al.* (2021) reported finding an 8.357-min. white dwarf spin period from r-band and g-band images in the Zwicky Transient Facility archives for the period 2018 March 26 to 2021 June 14. The field of V1674 Her is shown in Figure 1.

Schmidt (2021) reported a preliminary photometric period of 0.07115 d ± 0.000044 to the Central Bureau for Astronomical Telegrams on 23 July 2021. However, on 9 Aug. 2021, Shugarov and Afonina (2021) obtained an orbital period of 0.15290(3) d. Shugarov showed that the 0.07-day period was a half-day alias of the true orbital period; that is, with $P_1 = 0.1529$ d, the alias P_2 is found by $1/P_2 = 1/P_1 + 1$ and $P_2/2 \approx 0.07$ d. Shugarov's orbital period was confirmed three days later by (Patterson *et al.* 2021). By late July 2021 the nova faded beyond the limits of Burleith Observatory.

2. Observations

At Burleith Observatory, Washington, DC, CCD observations were obtained with a 0.32-m PlaneWave CDK astrograph and SBIG STL-1001E CCD camera with an Astrodon Cousins I_c filter. Pixel size was 1.95 arc-seconds, yielding on average 2-pixel FWHM. Exposure times ranged from 30 to 300 seconds. The observatory computer was synchronized to USNO NTP before each observing session. Nova Her 2021 was a particularly challenging object, fading on average

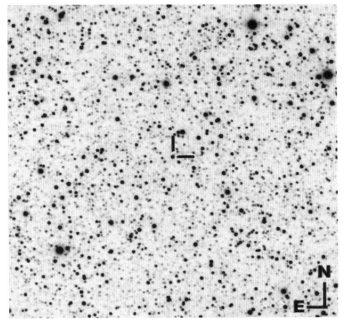

Figure 1. 15 arc-min field of V1674 Her, 13 July 2021.

0.1 mag./day. It would have been desirable to obtain a longer baseline of observations in Washington, but as the nova began its rapid fading, the dense smoke from Western state forest fires greatly hampered photometry by adding significant noise to the sky background.

Shugarov and Afonina (2021) observed with a 0.60-m f/12.5 telescope with FLI ML3041 at the Astronomical Institute of the Slovak Academy of Sciences at Stara Lesna, Slovakia, and with a 0.60-m f/12.5 telescope and FLI-39000 CCD at the Crimean Laboratory of the Sternberg Astronomical Institute, M. V. Lomonosov Moscow State University, Moscow, Russia. Their earliest observations were made with a 0.06-m Zeiss Sonnar T* 2.8/180-mm lens and SBIG ST-10XME CCD with

Cousins B, R, I filters. Their Cousins I band observations were used in this study.

During the span of observations the maximum amplitude of I_c magnitudes increased in a nearly linear manner, as seen in Figures 2 and 3.

3. Reductions

At Burleith Observatory, synthetic aperture photometry was performed using C-Munipack 2.1.29 (Motl 2021), with an aperture of radius 3.6 pixels. Heliocentric corrections were applied to dates of observation. Comparison stars (Table 1) were selected to avoid CCD saturation. Cousins I-band differential ensemble photometry was performed using the comparison stars in Table 1, from AAVSO chart sequence X26663ABW.

At CL Sternberg and at Stara Lesna multicolor CCD photometry was performed using apertures of ~8–10 arcsec. The single standard comparison star 000-BMD-913 was used for the 0.6-m observations. For the wider field of the 0.06-m Zeiss Sonnar observations the standard comparison star was 000-BCD-834.

Table 2 and Figure 4 provide nightly mean times of observation (HJD – 2400000), observed mean magnitude I_c, mean error of the magnitudes, and instrument used. (The slight zero-point offset between observatories is of no consequence, as nightly mean magnitudes were removed.) An example night's observation is shown in Figure 5.

4. Analysis

Prior to Fourier analysis, each nightly observation set from all observers was pre-processed by subtracting nightly average brightness and removing nightly linear trends. Period analysis was performed using PERANSO 2.60 software (Paunzen and Vanmunster 2016), computing an ANOVA spectrum of 100,000 steps over the frequency range 0.3–16 c/d. Figures 6 and 7 show the spectral window with its 1-day alias and the ANOVA frequency of V1674 Her, 6.53875 cycles/day with various aliases. Pre-whitening removing this frequency plus aliases 6.54, 6.03, and 5.55 c/d revealed no other significant periods.

A folded double-phase plot of the most prominent period is shown in Figure 8. The solid curve shown is a 50-point averaging with spline interpolation.

The period error estimate (in parentheses) in the following summary Table 3 are computed by Peranso to provide a 1-sigma confidence level on the period P equal to the line width at the Mean Noise Power Level at P, using the method in section 4.4 of (Schwarzenberg-Czerny 1991); the epoch of extremum is found from a 7-degree polynomial fit to the observations.

5. Conclusion

The fast nova V1674 Her (Nova Her 2021) has been a particularly difficult object for urban photometry, because of its relatively fast fading during a period of coast-to-coast smoke obscuration in the United States. A preliminary Lomb-Scargle solution at Burleith Observatory was based on insufficient observations in the later stages when the double-humped magnetic polar light curve could be observed. Early on, the orbital period was apparently obscured by its bright disk material. The observations from Slovakia and Russia were invaluable in finding the true orbital period. The light curve of V1674 Her shows double humps at phase 0.5, a characteristic of magnetic cataclysmic variables, such as the polars AM Her and TZ Vir.

Table 1. Photometry comparison stars.

AUID	R.A. (2000) h m s	Dec. (2000) deg ' "	Ic	Mag. Error	
000-BMD-913	18 57 41.50	+16 57 29.3	11.250	(0.175)	
000-BMD-915	18 57 38.32	+16 57 29.8	11.908	(0.158)	
000-BCD-758	18 58 11.66	+16 45 10.0	11.944	(0.041)	
000-BMD-912	18 57 42.72	+17 01 13.9	10.838	(0.173)	(check star)
000-BCD-834	18 58 56.83	+16 51 07.3	8.932	(0.087)	

Table 2. Nightly mean magnitudes I_c.

HJD	Mag. I_c	Error	Instrument
59379.38359	6.188	0.018	6-cm Stara Lesna
59380.43751	7.031	0.063	6-cm Stara Lesna
59382.45018	8.144	0.045	6-cm Stara Lesna
59383.72617	8.859	0.003	32-cm Burleith
59384.46286	8.908	0.063	6-cm Stara Lesna
59385.49030	9.213	0.060	6-cm Stara Lesna
59386.48081	9.497	0.068	6-cm Stara Lesna
59388.46603	10.033	0.042	6-cm Stara Lesna
59394.43554	11.345	0.013	6-cm Stara Lesna
59395.65032	11.549	0.004	32-cm Burleith
59398.65690	12.007	0.010	32-cm Burleith
59401.67120	12.413	0.012	32-cm Burleith
59402.62504	12.543	0.014	32-cm Burleith
59405.58067	12.865	0.019	32-cm Burleith
59407.39159	13.109	0.048	60-cm Stara Lesna
59407.58429	13.051	0.030	32-cm Burleith
59408.73693	13.161	0.013	32-cm Burleith
59409.45503	13.371	0.023	60-cm Stara Lesna
59415.67780	13.671	0.034	32-cm Burleith
59416.67040	13.715	0.041	32-cm Burleith
59417.42635	13.893	0.049	60-cm Stara Lesna
59419.54054	13.999	0.034	60-cm Stara Lesna
59424.32052	14.186	0.055	60-cm CL Sternberg
59425.31605	14.234	0.033	60-cm CL Sternberg
59426.33177	14.248	0.045	60-cm CL Sternberg
59427.37227	14.268	0.063	60-cm CL Sternberg
59428.32811	14.267	0.046	60-cm CL Sternberg
59429.35967	14.302	0.037	60-cm CL Sternberg
59430.40949	14.354	0.054	60-cm CL Sternberg
59431.36096	14.367	0.044	60-cm CL Sternberg
59432.35535	14.428	0.054	60-cm CL Sternberg
59433.36201	14.482	0.056	60-cm CL Sternberg
59437.44959	14.753	0.070	60-cm CL Sternberg

Table 3. Period estimate parameters for V1674 Her.

Parameter	Value
Period (d)	0.152934 (0.000034)
Period (h)	3.6704 (0.0008)
Amplitude (mean curve) (mag. I_c)	0.0488
Number of observations	979
Time span (d)	53.80
Epoch of minimum	2459408.74789 (0.0031)

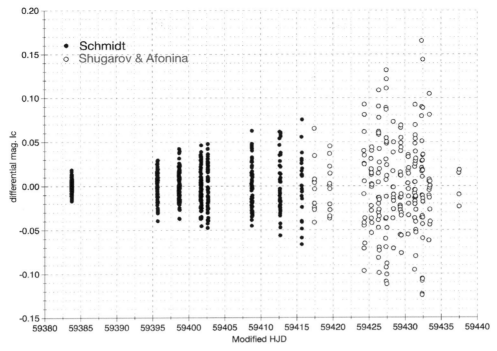

Figure 2. Observations of V1674 Her.

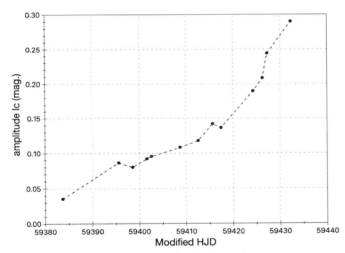

Figure 3. Observed maximum amplitude of I_c observations.

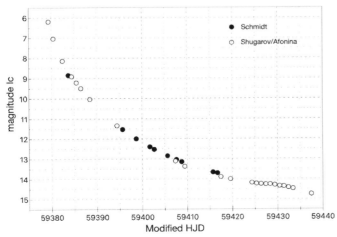

Figure 4. Nightly mean I_c magnitudes.

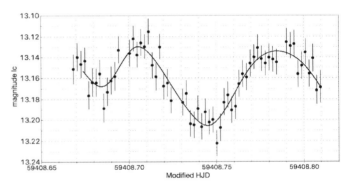

Figure 5. Example observation, Burleith Observatory, 13 July 2021.

The humps result from domination by the radiation of cyclotron emission of electron cooling in shock-treated columns of gas following the white-dwarf's magnetic field lines to impact with its hot surface (Gänsicke *et al.* 2001).

6. Acknowledgements

Schmidt wishes to thank Tonny Vanmunster, author of Peranso, and James A. DeYoung, NRL/USNO (ret.) for helpful comments. Special thanks to the AAVSO for providing photometric standards from the AAVSO Comparison Star Database via its Variable Star Plotter utility. The work of Shugarov was supported by grants from the Slovak Academy VEGA 2/0030/21, APVV-20-0148, with additional support from the M. V. Lomonosov Moscow State University Program "Leading scientific schools", project "Physics of stars, relativistic objects and galaxies." The work of Burleith Observatory on the roof of Schmidt's house was enthusiastically supported by his wife, Margaret.

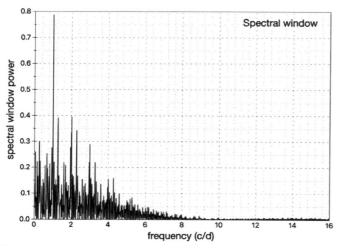

Figure 6. Observing window power spectrum.

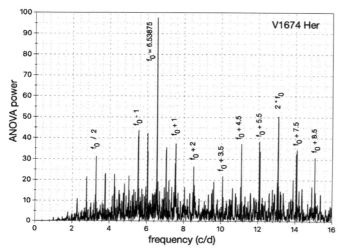

Figure 7. ANOVA periodogram of V1674 Her.

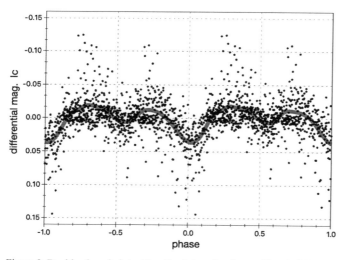

Figure 8. Double phased plot with spline interpolated smoothing (solid line).

References

Gänsicke, B. T., Fischer, A., Sivotti, R., and de Martino, D. 2001, *Astron. Astrophys.*, **372**, 557.

Motl, D. 2021, C-Munipack software utilities (http://c-munipack.sourceforge.net).

Mroz, P, Burdge, K., van Roestel, J., Prince, T., Kong, A. K. H., and Li, K.-L. 2021, *Astron. Telegram*, No. 14720, 1 (15 June 2021).

Munari, U., Valisa, P., and Dallaporta, S. 2021, *Astron. Telegram*, No. 14704, 1 (12 June 2021).

Patterson, J., Epstein-Martin, M., Vanmunster, T, and Kemp, J. 2021, *Astron. Telegram*, No. 14856, 1 (12 August 2021).

Paunzen, E., and Vanmunster, T. 2016, *Astron. Nachr.*, **337**, 239.

Quimby, R. M., Shafter, A. W., and Corbett, H. 2021, arXiv:2107.05763 (http://export.arxiv.org/abs/2107.05763).

Romanov, F. 2021, "Multicolor photometry of V1674 Her", AAVSO Forums/Variable Stars/Novae (15 June 2021).

Schmidt, R. E. 2021, *Cent. Bur. Astron. Telegrams*, No. 5001, 1 (16 July 2021).

Schwarzenberg-Czerny, A. 1991, *Mon. Not. Roy. Astron. Soc.*, **253**, 198.

Shugarov, S., and Afonina, M. 2021, *Astron. Telegram*, No. 14835, 1 (9 August 2021).

Ueda, S. 2021, *Cent. Bur. Astron. Telegrams*, CBAT Transient Object Followup Reports, TCP J18573095+1653396 (http://www.cbat.eps.harvard.edu/unconf/followups/J18573095+1653396.html).

The Photometric Period of V606 Vulpeculae (Nova Vul 2021)

Richard E. Schmidt
Burleith Observatory, 1810 35th Street NW, Washington, DC 20007; schmidt.rich@gmail.com

Received October 4, 2021; revised October 16, 2021; accepted October 18, 2021

Abstract A photometric study of the galactic nova V606 Vulpeculae (Nova Vul 2021, TCP J20210770+2914093) was undertaken at the Burleith Observatory in Washington, DC. A total of 3,511 CCD observations were obtained over a time span of 57.1 days, yielding an observed period 0.133697 d ± 0.000064 d, of amplitude 0.012 magnitude I_c. The epoch (HJD) of minimum light was 2459432.6287 (0.0004). A new δ Scuti (DSCT) variable, GSC 02167-00712, of period 62.526 minutes, was discovered in the field of the nova.

1. Introduction

This is the sixth in a series of reports on the discovery of photometric periods of recent classical novae (Schmidt 2020a, 2020b, 2021a, 2021b; Schmidt et al. 2021). These *JAAVSO* reports serve a dual purpose: adding to the relatively few known orbital periods of novae, and hopefully inspiring urban astronomers to participate in nova research. The reddened color of galactic novae and their typically long period of outbursts lend them well to CCD observation—even in heavily light-polluted cities—when observing in the near infrared, as with the Cousins I_c filter and a monochromatic camera with sensitivity in the 700–900 nm region.

V606 Vulpeculae, (Nova Vulpeculae 2021, TCP J20210770+2914093), R.A. 20^h 21^m 07.70^s, Dec. +29° 14' 09.1" (2000), was discovered by Koichi Itagaki, Yamagata, Japan, on 2021 July 16.475 UT (Itagaki 2021). Spectroscopy by Munari *et al.* on 2021 July 17, 18, and 28 confirmed its type as an Fe-II-type nova (Munari *et al.* 2021). Schmidt reported a preliminary photometric period of 3.096 h to the Central Bureau for Astronomical Telegrams on 8 August 2021 (Schmidt 2021c). The 15 arc-minute field of V606 Vul on 2021 September 24 is shown in Figure 1.

2. Observations

At Burleith Observatory a total of 3,511 CCD observations of V606 Vul were obtained between 2021 July 17.25 and October 2.05 UT with a 0.32-m PlaneWave CDK astrograph and SBIG STL-1001E CCD camera with an Astrodon Cousins I_c filter. Pixel size was 1.95 arc-seconds, yielding on average 2-pixel FWHM, and the field of view was 33 arc-minutes square. The observatory computer was synchronized to USNO NTP before each observing session. Images were de-darked and flat-fielded in real time. Exposure times ranged from 30 to 90 seconds.

3. Reductions

Cousins I-band differential ensemble photometry was performed using the comparison stars in Table 1, which are numbered as in Figure 1. Synthetic aperture photometry was performed using C-MUNIPACK 2.1.29 (Motl 2021). Heliocentric corrections were applied to dates of observation. Data from poor nights and large outliers were filtered out, leaving 2,963 images for analysis.

Table 2 and Figure 2 provide nightly mean times of observation (HJD–2400000), observed mean magnitudes I_c, mean error of the magnitudes, maximum airmass, and duration of nightly observing sessions.

Detrending nightly observations removes linear magnitude changes. For example, Figure 3 shows five hours of observations from 25 August 2021, during which the nova was brightening by 0.1 magnitude. Figure 4 shows the same observations after subtracting a linear solution, y = –0.356*x + 3564.5, from the observed magnitudes.

Table 1. Photometry comparison stars from AAVSO chart sequence X26761AJ.

No.	AUID	R.A. (2000) h m s	Dec. (2000) ° ' "	Mag. I_c	Mag. Err.
1	000-BPB-795	20 20 52.08	+29 14 50.5	11.497	0.206
2	000-BPB-796	20 20 34.83	+29 14 10.1	12.029	0.156
3	000-BPB-797	20 21 02.11	+29 17 36.8	12.470	0.206
4	000-BPB-798	20 20 46.15	+29 20 26.5	12.845	0.222 (check star)

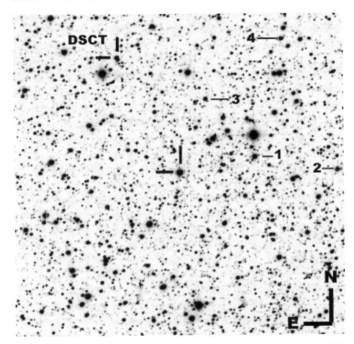

Figure 1. 15 × 15 arc-min field of V606 Vul (center) and DSCT GSC 02167-00712.

Table 2. Nightly mean magnitudes I_c.

HJD	Mag. I_c	Error	Max. Airmass	Duration (hours)
9412.7533	11.341	0.004	1.022	0.77
9419.6657	10.847	0.005	1.500	2.47
9421.6032	10.538	0.007	1.281	0.48
9423.6386	10.002	0.009	1.243	1.98
9429.5902	8.788	0.010	1.375	1.93
9432.6251	9.716	0.009	1.335	3.49
9435.6025	10.023	0.009	1.181	1.05
9439.6253	10.253	0.007	1.110	1.71
9451.6376	10.849	0.005	1.201	5.14
9456.7071	10.396	0.006	1.248	1.76
9460.6180	10.510	0.005	1.178	4.91
9461.5615	10.146	0.007	1.146	2.03
9462.5818	9.787	0.008	1.158	3.35
9464.5698	9.802	0.007	1.112	2.16
9465.5931	9.950	0.005	1.087	3.09
9466.5500	10.013	0.009	1.108	1.50
9467.6611	9.603	0.008	1.225	2.30
9468.5912	9.382	0.007	1.110	3.84
9469.5795	9.313	0.006	1.103	3.15
9471.5688	9.287	0.010	1.107	3.03
9476.5887	9.277	0.007	1.172	3.89
9477.5562	9.620	0.007	1.062	2.19
9482.5801	10.304	0.006	1.204	4.14
9487.5691	10.906	0.004	1.156	3.21
9489.5524	10.795	0.005	1.125	3.24

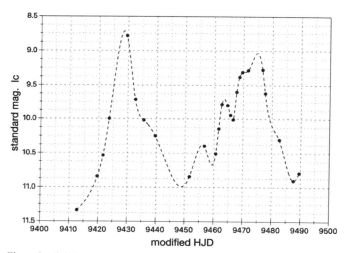

Figure 2. Nightly mean I_c magnitudes.

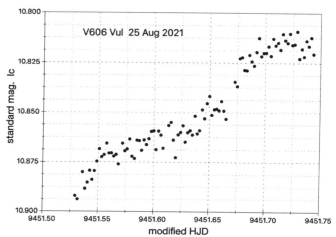

Figure 3. 25 Aug. 2021 as observed.

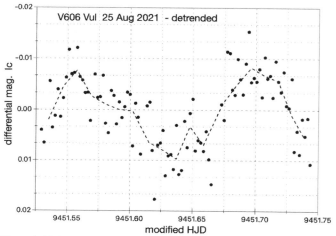

Figure 4. 25 Aug. 2021 detrended.

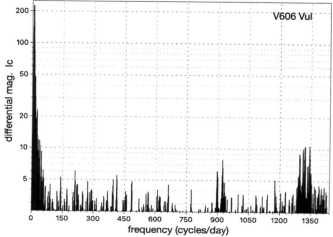

Figure 5. Wide Lomb-Scargle periodogram of V606 Vul.

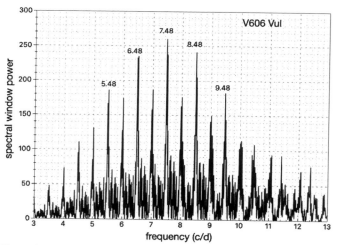

Figure 6. Lomb-Scargle periodogram of V606 Vul.

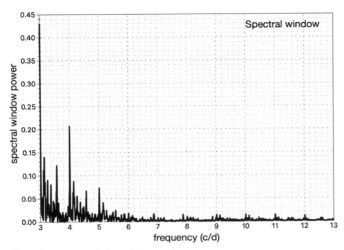

Figure 7. Spectral window of observations.

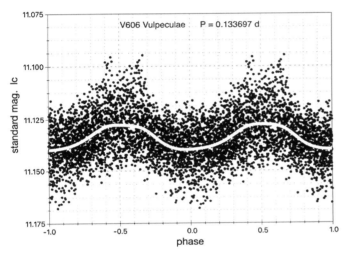

Figure 8. Double phased plot with spline interpolated smoothing (solid line).

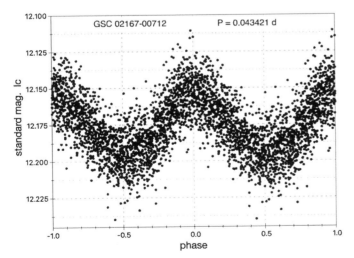

Figure 9. Phase plot of GSC 02167-00712.

Table 3. Observation summary V606 Vulpeculae.

Parameter	Value
Period (d)	0.133697 (0.000064)
Period (h)	3.2087 (0.0015)
Amplitude (mean curve) (mag. Ic)	0.012
Number of observations used	2963
Time span (d)	57.07
Epoch of minimum	2459432.6287 (0.0004)

4. Analysis

Prior to Fourier analysis, each nightly observation set was pre-processed by subtracting nightly average brightness and removing nightly linear trends. Period analysis was performed using PERANSO 2.60 software (Paunzen and Vanmunster 2016), computing a Lomb-Scargle spectra of the observations. Because of the low signal-to-noise ratio of the nova light curve, a resolution of 100,000 steps was computed. Figure 5 shows a Lomb-Scargle periodogram over the range 0–1440 cycles/day (the range of possible photometric periods for novae). On the right of this logarithmic plot we see frequencies due to the observing cadences of 65 and 95 seconds.

Figure 6 shows a Lomb-Scargle periodogram over the frequency range 3–13 cycles/day, which peaks at 7.4796 cycles/day (3.2087 hours). Various aliases appear at ½-day and 1-day intervals due to the diurnal nature of night observing. Pre-whitening (removal of the main period) revealed no other significant periods that were not its aliases.

Figure 7 shows the spectral window for these observations, which displays artifacts caused by the cadence of observations. The absence of a peak at the observed frequency 7.47960 cycles/day shows that this frequency is not an artifact of the observing window.

A folded double-phase plot of the most prominent period is shown in Figure 8. The solid curve shown is a 512-point average with spline interpolation. The magnitude range (I_c) shown is from 01 October 2021.

Table 3 summarizes observed data for V606 Vul, with errors in parentheses. The period error estimate was computed by PERANSO as the 1-σ confidence level on the period P which equals the line width of its Mean Noise Power Level, using the method in section 4.4 of (Schwarzenberg-Czerny 1991). The epoch of extremum is found from a 7-degree polynomial fit to the observations.

5. Discovery of a field δ Scuti variable

Tens of thousands of stars appear in each 33 × 33 arc-minute field taken with this telescope and CCD. Photometry reduction programs such as C-MUNIWIN can generate plots of magnitude vs. standard deviation for each field object. Each object with heightened magnitude standard deviation is checked for possible periodicity. Examination of the field of V606 Vul revealed a previously unknown variable, GSC 02167-00712, which has been assigned AAVSO AUID 000-BPC-988, of magnitude range 12.15–12.20 (I_c), at R.A. $20^h 21^m 20.38^s$, Dec. +29° 19' 39.8" (2000) ("DSCT" in Figure 1). ANOVA analysis of its light curve yields the period 0.043421 d (0.000009) (62.526 min), with epoch of maximum JD 2459467.6147. The fast period and low amplitude is typical of δ Scuti variables. A double phase plot is shown below (Figure 9).

6. Conclusion

The photometric variability of V606 Vulpeculae, though of low amplitude (0.012 magnitude I_c), is readily detected with a Lomb-Scargle spectral analysis based on a large number of observations. The observed period is in agreement with the orbital periods of novae found in the catalogue of galactic novae of (Özdönmez *et al.* 2018). In spite of its location in a heavily light-polluted city, the small telescope of Burleith Observatory yields Cousins I-band photometric measurements with a mean error of 0.007 magnitude. Observations from dark sky sites would naturally yield much less noisy results. The serendipitous discovery of a δ Scuti variable adds to the enjoyment of CCD field photometry.

7. Acknowledgements

Schmidt wishes to thank Sebastián Otero, AAVSO; James A. DeYoung, NRL/USNO (ret.); and Sergei Yu. Shugarov, Sternberg Astronomical Institute, M. V. Lomonosov Moscow State University for helpful comments. Special thanks to the AAVSO for providing photometric standards from the AAVSO Comparison Star Database via its Variable Star Plotter utility. This work from the rooftop observatory of the author's house was enthusiastically supported by his wife, Margaret.

References

Itagaki, K. 2021, *Cent. Bur. Astron. Telegrams*, No. 5007, 1.
Motl, D. 2021, C-Munipack software utilities (http://c-munipack.sourceforge.net).
Munari, U., Moretti, S., Maitan, A., and Andreoli, V. 2021, *Astron. Telegram*, No. 14816, 1.
Özdönmez, A., Ege, E., Güver, T., and Ak, T. 2018, *Mon. Not. Roy. Astron. Soc.*, **476**, 4162.
Paunzen, E., and Vanmunster, T. 2016, *Astron. Nachr.*, **337**, 239.
Schmidt, R. E. 2020a, *J. Amer. Assoc. Var. Star Obs.*, **48**, 13.
Schmidt, R. E. 2020b, *J. Amer. Assoc. Var. Star Obs.*, **48**, 53.
Schmidt, R. E. 2021a, *J. Amer. Assoc. Var. Star Obs.*, **49**, 95.
Schmidt, R. E. 2021b, *J. Amer. Assoc. Var. Star Obs.*, **49**, 99.
Schmidt, R. E. 2021c, *Cent. Bur. Astron. Telegrams*, No. 5012, 1.
Schmidt, R. E., Shugarov, S. Y., and Afonina, M. D. 2021, *J. Amer. Assoc. Var. Star Obs.*, **49**, 246.
Schwarzenberg-Czerny, A. 1991, *Mon. Not. Roy. Astron. Soc.*, **253**, 198.

Recent Minima of 218 Eclipsing Binary Stars

Gerard Samolyk
P.O. Box 20677, Greenfield, WI 53220; gsamolyk@wi.rr.com

Received September 23, 2021; accepted September 23, 2021

Abstract   This paper continues the publication of times of minima for eclipsing binary stars. Times of minima determined from CCD observations of 218 eclipsing binaries received by the AAVSO Eclipsing Binaries section from February 2021 through July 2021 are presented.

1. Recent observations

The accompanying list (Table 1) contains times of minima for 218 variables calculated from recent CCD observations made by participants in the AAVSO's eclipsing binary program. These observations were reduced by the observers or the writer using the method of Kwee and van Worden (1956).

The linear elements in the *General Catalogue of Variable Stars* (GCVS; Kholopov *et al.* 1985) were used to compute the O–C values for most stars. For a few exceptions where the GCVS elements are missing or are in significant error, light elements from another source are used: AC CMi (Samolyk 2008), DV Cep (Frank and Lichtenknecker 1987), Z Dra (Danielkiewicz-Krosniak *et al.* 1996), DF Hya (Samolyk 1992), DK Hya (Samolyk 1990), GU Ori (Samolyk 1985).

The light elements used for V376 And, EK Aqr, FS Aqr, V602 Aql, V688 Aql, V719 Aql, UZ CMi, BH CMi, CZ CMi, V776 Cas, VY Cet, AS CrB, GW Gem, V728 Her, WZ Leo, FS Leo, V351 Peg, CP Psc, DS Psc, DV Psc, DZ Psc, GR Psc, DK Sct, and V1223 Tau are from (Kreiner 2021).

The light elements used for DD Aqr, AW CrB, BD CrB, V470 Hya, V502 Oph, VZ Psc, and ET Psc are from (Paschke 2014).

The light elements used for V775 Cas and HO Psc are from (Nelson 2014).

The standard error is included when available. Column F indicates the filter used. A "C" indicates a clear filter.

This list will be web-archived and made available through the AAVSO ftp site at

ftp://ftp.aavso.org/public/datasets/gsamj492eb218.txt.

This list, along with the eclipsing binary data from earlier AAVSO publications, is also included in the Lichtenknecker database administrated by the Bundesdeutsche Arbeitsgemeinschaft für Veränderliche Sterne e. V. (BAV) at: http://www.bav-astro.de/LkDB/index.php?lang=en.

References

Danielkiewicz-Krosniak, E, Kurpińska-Winiarska, M., eds. 1996, *Rocznik Astron.* (SAC 68), **68**, 1.

Frank, P., and Lichtenknecker, D. 1987, *BAV Mitt.*, No. 47, 1.

Kholopov, P. N., *et al.* 1985, *General Catalogue of Variable Stars*, 4th ed., Moscow.

Kreiner, J. M. 2004, *Acta Astron.*, **54**, 207 (http://www.as.up.krakow.pl/ephem/).

Kwee, K. K., and van Woerden, H. 1956, *Bull. Astron. Inst. Netherlands*, **12**, 327.

Nelson, R. 2014, Eclipsing Binary O–C Files (http://www.aavso.org/bob-nelsons-o-c-files).

Paschke, A. 2014, "O–C Gateway" (http://var.astro.cz/ocgate/).

Samolyk, G. 1985, *J. Amer. Assoc. Var. Star Obs.*, **14**, 12.

Samolyk, G. 1990, *J. Amer. Assoc. Var. Star Obs.*, **19**, 5.

Samolyk, G. 1992, *J. Amer. Assoc. Var. Star Obs.*, **21**, 111.

Samolyk, G. 2008, *J. Amer. Assoc. Var. Star Obs.*, **36**, 171.

Table 1. Recent times of minima of stars in the AAVSO eclipsing binary program.

Star	JD (min) Hel. 2400000+	Cycle	O–C (day)	F	Observer	Standard Error (day)	Star	JD (min) Hel. 2400000+	Cycle	O–C (day)	F	Observer	Standard Error (day)
RT And	59380.8318	29000	–0.0131	V	G. Samolyk	0.0001	BI Cas	59154.6752	3556	0.0384	V	L. Hazel	0.0003
XZ And	59426.7872	26118	0.2089	V	G. Samolyk	0.0001	IR Cas	59426.8127	25067	0.0178	V	G. Samolyk	0.0001
AB And	59413.8323	70216.5	–0.0521	V	G. Samolyk	0.0001	V775 Cas	59183.3693	536.5	0.8551	V	J. Coliac	0.0005
BD And	59413.8344	52822	0.0091	V	G. Samolyk	0.0002	V776 Cas	59184.3613	15176.5	–0.0107	V	J. Coliac	0.0009
V376 And	59177.4216	8360	0.0086	V	J. Coliac	0.0005	SU Cep	59370.8365	36660	0.0072	V	G. Samolyk	0.0001
SU Aqr	59137.6644	22780	–0.0323	C	G. Frey	0.0001	SU Cep	59426.7234	36722	0.0072	V	K. Menzies	0.0004
CX Aqr	59116.6833	40762	0.0188	C	G. Frey	0.0001	WW Cep	59397.8362	22410	0.3652	V	G. Samolyk	0.0001
DD Aqr	59142.6551	15695	–0.0047	C	G. Frey	0.0001	WZ Cep	59378.7090	75071.5	–0.2270	V	G. Samolyk	0.0002
EK Aqr	59132.6822	21638	0.0316	C	G. Frey	0.0001	XX Cep	59382.6858	6222	0.0375	V	L. Hazel	0.0003
FS Aqr	59109.6847	25222	–0.0005	C	G. Frey	0.0002	DV Cep	59398.6611	10874	–0.0059	V	G. Samolyk	0.0003
KO Aql	59418.4595	6121	0.1064	V	T. Arranz	0.0001	EG Cep	59398.6950	30855	0.0059	V	G. Samolyk	0.0001
OO Aql	59368.8245	40955	0.0801	V	G. Samolyk	0.0001	TT Cet	59158.6642	54764	–0.0876	C	G. Frey	0.0001
OO Aql	59416.4623	41049	0.0798	V	T. Arranz	0.0001	TX Cet	59173.6964	21720	0.0119	C	G. Frey	0.0001
V342 Aql	59409.4675	5925	–0.0893	V	T. Arranz	0.0002	VY Cet	59191.6563	19634	–0.0167	C	G. Frey	0.0001
V343 Aql	59371.8247	16767	–0.0508	V	G. Samolyk	0.0001	RW Com	59295.6336	81203	0.0182	V	L. Hazel	0.0003
V343 Aql	59410.5608	16788	–0.0514	V	T. Arranz	0.0001	RW Com	59295.7515	81203.5	0.0174	V	L. Hazel	0.0002
V346 Aql	59387.8401	15790	–0.0157	V	G. Samolyk	0.0001	RW Com	59296.8212	81208	0.0191	V	K. Menzies	0.0001
V602 Aql	59400.6986	2290	–0.0316	V	L. Hazel	0.0009	RW Com	59342.3920	81400	0.0194	V	T. Arranz	0.0002
V688 Aql	59425.4895	1780	–0.0032	V	T. Arranz	0.0004	RW Com	59342.5093	81400.5	0.0181	V	T. Arranz	0.0002
V719 Aql	59423.4482	1025	–0.0021	V	T. Arranz	0.0004	RW Com	59370.6358	81519	0.0191	V	G. Samolyk	0.0001
AP Aur	59280.5874	29571	1.8385	V	K. Menzies	0.0004	RZ Com	59294.8783	72251	0.0586	V	G. Samolyk	0.0001
AP Aur	59291.4066	29590	1.8407	V	T. Arranz	0.0002	RZ Com	59320.4351	72326.5	0.0582	V	T. Arranz	0.0001
AP Aur	59317.5981	29636	1.8439	V	G. Samolyk	0.0005	RZ Com	59357.6720	72436.5	0.0594	V	G. Samolyk	0.0001
CL Aur	59263.3611	21132	0.1885	V	T. Arranz	0.0005	SS Com	59294.7637	83071.5	1.0114	V	G. Samolyk	0.0003
EP Aur	59279.3746	55786	0.0203	V	T. Arranz	0.0002	SS Com	59322.4222	83138.5	1.0128	V	T. Arranz	0.0001
IM Aur	59273.4925	15039	–0.1385	V	T. Arranz	0.0002	SS Com	59380.6301	83279.5	1.0171	V	G. Samolyk	0.0001
TU Boo	59281.7426	80840.5	–0.1685	V	K. Menzies	0.0001	CC Com	59248.8867	89336.5	–0.0362	V	G. Samolyk	0.0002
TU Boo	59295.6875	80883.5	–0.1679	V	K. Menzies	0.0001	CC Com	59312.7736	89626	–0.0379	V	L. Hazel	0.0003
TU Boo	59341.4102	81024.5	–0.1696	V	T. Arranz	0.0001	CC Com	59340.3596	89751	–0.0377	V	T. Arranz	0.0001
TU Boo	59341.5732	81025	–0.1688	V	T. Arranz	0.0001	RW CrB	59345.6868	25597	0.0040	V	G. Samolyk	0.0003
TY Boo	59346.6063	78405.5	0.0573	V	K. Menzies	0.0001	RW CrB	59391.4504	25660	0.0037	V	T. Arranz	0.0001
TY Boo	59348.8270	78412.5	0.0580	V	K. Menzies	0.0002	TW CrB	59371.6945	36465	0.0626	V	G. Samolyk	0.0001
TY Boo	59358.4981	78443	0.0561	V	T. Arranz	0.0001	TW CrB	59376.4054	36473	0.0625	V	T. Arranz	0.0001
TY Boo	59382.6015	78519	0.0562	V	K. Menzies	0.0005	AS CrB	59375.5691	18062	0.0201	V	T. Arranz	0.0001
TY Boo	59414.6328	78620	0.0556	V	G. Samolyk	0.0004	AW CrB	59376.5779	22095	–0.0154	V	T. Arranz	0.0002
TZ Boo	59304.7238	66199	0.0548	V	K. Menzies	0.0004	BD CrB	59373.4566	22849	0.0216	V	T. Arranz	0.0005
TZ Boo	59351.3799	66356	0.0564	V	T. Arranz	0.0002	BD CrB	59393.4464	22905	0.0267	V	T. Arranz	0.0007
TZ Boo	59351.5281	66356.5	0.0560	V	T. Arranz	0.0001	BD CrB	59406.4699	22941.5	0.0244	V	T. Arranz	0.0005
TZ Boo	59370.6948	66421	0.0558	V	G. Samolyk	0.0002	W Crv	59338.4238	50738.5	0.0186	V	T. Arranz	0.0003
UW Boo	59280.8342	16797	–0.0061	V	K. Menzies	0.0003	W Crv	59339.3934	50741	0.0180	V	T. Arranz	0.0001
VW Boo	59293.7476	82146	–0.3125	V	G. Samolyk	0.0002	RV Crv	59334.7049	24497	–0.1138	V	G. Samolyk	0.0003
VW Boo	59293.9183	82146.5	–0.3130	V	G. Samolyk	0.0001	RV Crv	59346.6592	24513	–0.1155	V	G. Samolyk	0.0004
VW Boo	59347.6623	82303.5	–0.3140	V	G. Samolyk	0.0001	RV Crv	59347.4061	24514	–0.1159	V	T. Arranz	0.0002
VW Boo	59371.6241	82373.5	–0.3150	V	G. Samolyk	0.0003	SX Crv	59346.6372	57890	–1.0270	V	G. Samolyk	0.0003
AD Boo	59363.7666	17333	0.0391	V	G. Samolyk	0.0001	V Crt	59321.7189	25532	0.0009	V	G. Samolyk	0.0002
CK Boo	58268.6868	43282.5	0.0267	V	S. Cook	0.0010	V Crt	59322.4195	25533	–0.0005	V	T. Arranz	0.0001
CK Boo	59338.5220	46295	–0.0278	V	L. Corp	0.0005	SW Cyg	59368.8146	3827	–0.3869	V	G. Samolyk	0.0003
V376 Boo	59390.4161	22546	0.0001	V	T. Arranz	0.0009	WW Cyg	59398.8171	5733	0.1614	V	G. Samolyk	0.0001
AL Cam	59303.6903	24762	–0.0231	V	G. Samolyk	0.0001	ZZ Cyg	59383.6474	22881	–0.0823	V	L. Hazel	0.0003
WW Cnc	59294.4061	3002	0.0455	V	T. Arranz	0.0001	BR Cyg	59387.8315	13394	0.0019	V	G. Samolyk	0.0003
WY Cnc	59297.4521	39723	–0.0503	V	T. Arranz	0.0001	BR Cyg	59418.4799	13417	0.0013	V	T. Arranz	0.0001
XZ Cnc	59281.4066	8615	0.0207	V	T. Arranz	0.0001	CG Cyg	59376.8304	31612	0.0804	V	L. Hazel	0.0003
R CMa	59248.6996	13169	0.1382	V	G. Samolyk	0.0004	CG Cyg	59376.8328	31612	0.0814	V	G. Samolyk	0.0001
SX CMa	59248.7118	19180	0.0361	V	G. Samolyk	0.0002	DK Cyg	59380.8424	45425	0.1404	V	G. Samolyk	0.0002
UU CMa	59293.6157	6783	–0.0586	V	G. Samolyk	0.0002	DK Cyg	59405.7890	45478	0.1404	V	G. Samolyk	0.0003
UZ CMi	59286.6789	12308	0.0206	V	L. Hazel	0.0003	DO Cyg	59367.7920	8702	–0.0842	V	L. Hazel	0.0003
XZ CMi	59283.7140	29093	0.0075	V	L. Hazel	0.0003	KV Cyg	59348.8652	10525	0.0685	V	L. Hazel	0.0004
XZ CMi	59286.6097	29098	0.0092	V	L. Hazel	0.0003	MY Cyg	59398.7121	6379.5	0.0127	V	G. Samolyk	0.0002
XZ CMi	59296.4479	29115	0.0076	V	T. Arranz	0.0001	MY Cyg	59400.7027	6380	0.0007	V	G. Samolyk	0.0001
YY CMi	59317.5992	28655	0.0187	V	G. Samolyk	0.0002	V346 Cyg	59380.6844	8637	0.2078	V	G. Samolyk	0.0004
AC CMi	58882.6673	7961	0.0048	C	G. Frey	0.0001	V387 Cyg	59363.8127	48983	0.0182	V	G. Samolyk	0.0001
AK CMi	59278.3914	28586	–0.0265	V	T. Arranz	0.0001	V387 Cyg	59404.8103	49047	0.0177	V	G. Samolyk	0.0001
BH CMi	58888.6649	11423	0.0036	C	G. Frey	0.0003	V388 Cyg	59397.6617	20307	–0.1440	V	G. Samolyk	0.0003
CZ CMi	58872.6389	14945	–0.0160	C	G. Frey	0.0002	V401 Cyg	59349.7558	26624	0.1063	V	L. Hazel	0.0006
TY Cap	59413.8085	10271	0.1036	V	G. Samolyk	0.0003	V445 Cyg	59381.6623	9875	0.3356	V	L. Hazel	0.0005
TW Cas	59426.8228	12195	0.0243	V	K. Menzies	0.0002	V456 Cyg	59414.6687	16274	0.0538	V	G. Samolyk	0.0003

Table continued on following pages

Table 1. Recent times of minima of stars in the AAVSO eclipsing binary program, cont.

Star	JD (min) Hel. 2400000+	Cycle	O–C (day)	F	Observer	Standard Error (day)	Star	JD (min) Hel. 2400000+	Cycle	O–C (day)	F	Observer	Standard Error (day)
V466 Cyg	59361.7205	21980.5	0.0087	V	L. Hazel	0.0004	SW Lac	59420.8266	44105.5	–0.0768	V	G. Samolyk	0.0003
V477 Cyg	59400.7678	6481.5	–0.5157	V	G. Samolyk	0.0002	VX Lac	59382.7835	13145	0.0912	V	L. Hazel	0.0003
V477 Cyg	59404.7603	6483	–0.0437	V	G. Samolyk	0.0003	CM Lac	59426.6409	20191	–0.0032	V	G. Samolyk	0.0004
V704 Cyg	59348.8392	37514	0.0403	V	G. Samolyk	0.0003	CO Lac	59420.7666	20676	0.0115	V	G. Samolyk	0.0001
V1034 Cyg	59400.7465	16851	0.0232	V	G. Samolyk	0.0002	DG Lac	59376.7614	6738	–0.2517	V	L. Hazel	0.0003
TT Del	59366.7939	4923	–0.1439	V	G. Samolyk	0.0002	DG Lac	59398.8273	6748	–0.2511	V	G. Samolyk	0.0003
YY Del	59103.7098	20356	0.0130	C	G. Frey	0.0002	Y Leo	59289.3826	8216	–0.0824	V	T. Arranz	0.0001
YY Del	59417.7772	20752	0.0159	V	L. Hazel	0.0006	Y Leo	59304.5568	8225	–0.0831	V	K. Menzies	0.0001
YY Del	59421.7408	20757	0.0140	V	G. Samolyk	0.0002	UU Leo	59302.5890	8278	0.2378	V	G. Samolyk	0.0001
FZ Del	59426.7501	35881	–0.0302	V	L. Hazel	0.0003	UV Leo	59296.7211	34755	0.0482	V	K. Menzies	0.0001
Z Dra	59301.6524	6967	–0.0031	V	L. Hazel	0.0003	UV Leo	59320.4252	34794.5	0.0490	V	L. Corp	0.0002
RZ Dra	59348.6940	27540	–0.0740	V	G. Samolyk	0.0001	UZ Leo	59320.3537	31583.5	0.0259	V	L. Corp	0.0006
RZ Dra	59416.4515	27663	0.0741	V	T. Arranz	0.0001	VZ Leo	59290.3902	25806	–0.0400	V	T. Arranz	0.0002
TW Dra	59371.7905	5428	–0.0700	V	G. Samolyk	0.0003	WZ Leo	59293.6181	4824	–0.0001	V	G. Samolyk	0.0003
TW Dra	59391.4370	5435	–0.0714	V	T. Arranz	0.0001	XY Leo	59309.3578	50105	0.1920	V	L. Corp	0.0002
UZ Dra	59347.6460	5451	0.0036	V	L. Hazel	0.0002	XY Leo	59309.4993	50105.5	0.1915	V	L. Corp	0.0003
AI Dra	59379.7610	13420	0.0421	V	G. Samolyk	0.0007	XZ Leo	59296.5729	29260	0.0859	V	K. Menzies	0.0001
BH Dra	59380.6536	10654	–0.0043	V	G. Samolyk	0.0002	XZ Leo	59309.4993	29286.5	0.0873	V	L. Corp	0.0006
BH Dra	59420.6338	10676	–0.0034	V	G. Samolyk	0.0004	AM Leo	59304.3534	45957	0.0133	V	L. Corp	0.0004
S Equ	59402.7867	4891	0.0933	V	L. Hazel	0.0003	FS Leo	59309.4432	14901	0.0092	V	L. Corp	0.0006
S Equ	59426.8380	4898	0.0919	V	G. Samolyk	0.0002	SS Lib	59372.4098	12668	0.1873	V	T. Arranz	0.0002
RW Gem	59276.4009	14299	0.0014	V	T. Arranz	0.0001	RY Lyn	59293.6386	11433	–0.0222	V	L. Hazel	0.0003
RX Gem	59308.3556	1536	0.0736	V	T. Arranz	0.0002	RY Lyn	59293.6388	11433	–0.0220	V	G. Samolyk	0.0001
SX Gem	59307.6046	29466	–0.0631	V	L. Hazel	0.0003	TZ Lyr	59419.6887	27675	–0.0051	V	L. Hazel	0.0006
SX Gem	59307.6065	29466	–0.0612	V	G. Samolyk	0.0001	EW Lyr	59345.7375	16855	0.3143	V	L. Hazel	0.0006
TX Gem	59261.3659	14076	–0.0432	V	T. Arranz	0.0001	RU Mon	59290.3836	4895	–0.1575	V	T. Arranz	0.0001
WW Gem	59281.4093	26900	0.0364	V	T. Arranz	0.0003	RW Mon	59296.3540	13439	–0.0930	V	T. Arranz	0.0001
AF Gem	59279.4475	25828	–0.0699	V	T. Arranz	0.0001	BB Mon	59268.3377	44475	–0.0042	V	T. Arranz	0.0002
AF Gem	59285.6678	25833	–0.0671	V	G. Samolyk	0.0003	SX Oph	59374.4880	12589	–0.0005	V	T. Arranz	0.0003
AL Gem	59278.4060	23685	0.1069	V	T. Arranz	0.0001	SX Oph	59382.7403	12593	–0.0015	V	G. Samolyk	0.0002
CX Gem	59297.3607	14380	–0.0438	V	T. Arranz	0.0003	SX Oph	59413.6895	12608	–0.0018	V	G. Samolyk	0.0004
FG Gem	59261.3814	39260	–0.0270	V	T. Arranz	0.0002	V501 Oph	59347.8328	29378	–0.0091	V	G. Samolyk	0.0002
GW Gem	59269.3474	10265	0.0019	V	T. Arranz	0.0001	V501 Oph	59413.6529	29446	–0.0096	V	G. Samolyk	0.0002
V405 Gem	59260.3871	14628.5	–0.0219	V	T. Arranz	0.0004	V502 Oph	59405.4348	24053	–0.0017	V	L. Corp	0.0003
SZ Her	59302.8515	21316	–0.0366	V	G. Samolyk	0.0001	V508 Oph	59352.7708	41388	–0.0288	V	L. Hazel	0.0003
SZ Her	59348.6653	21372	–0.0363	V	L. Hazel	0.0002	V508 Oph	59363.8060	41420	–0.0270	V	G. Samolyk	0.0001
SZ Her	59357.6638	21383	–0.0369	V	G. Samolyk	0.0001	V508 Oph	59371.7362	41443	–0.0270	V	L. Hazel	0.0003
TT Her	59375.7300	21248	0.0426	V	G. Samolyk	0.0008	V508 Oph	59415.5246	41570	–0.0272	V	T. Arranz	0.0001
TT Her	59399.4446	21274	0.0433	V	T. Arranz	0.0001	V839 Oph	59321.8475	46145	0.3456	V	G. Samolyk	0.0001
TU Her	59365.7030	6751	–0.2794	V	L. Hazel	0.0002	V839 Oph	59348.8415	46211	0.3459	V	G. Samolyk	0.0001
TU Her	59399.7089	6766	–0.2785	V	G. Samolyk	0.0003	V839 Oph	59378.6992	46284	0.3469	V	G. Samolyk	0.0001
UX Her	59353.7570	12707	0.1682	V	L. Hazel	0.0004	V839 Oph	59410.3977	46361.5	0.3483	V	T. Arranz	0.0001
UX Her	59370.7984	12718	0.1723	V	G. Samolyk	0.0001	V1010 Oph	59346.8424	30856.5	–0.2220	V	G. Samolyk	0.0004
UX Her	59423.4611	12752	0.1741	V	T. Arranz	0.0001	V1010 Oph	59347.8334	30858	–0.2231	V	G. Samolyk	0.0002
AK Her	59405.4456	40849.5	0.0222	V	L. Corp	0.0002	EQ Ori	59260.3821	15934	–0.0331	V	T. Arranz	0.0001
BO Her	59082.7015	5825	–0.1046	V	L. Hazel	0.0006	ER Ori	59247.5779	41618	0.1558	V	G. Samolyk	0.0001
CC Her	59009.7896	11154	0.3469	C	L. Hazel	0.0002	ER Ori	59259.4355	41646	0.1582	V	T. Arranz	0.0001
CC Her	59075.6789	11192	0.3440	V	L. Hazel	0.0002	ET Ori	59263.3320	34260	–0.0047	V	T. Arranz	0.0001
CC Her	59309.7859	11327	0.3603	V	L. Hazel	0.0003	FL Ori	59295.3865	8993	0.0403	V	T. Arranz	0.0001
CC Her	59349.6710	11350	0.3632	V	L. Hazel	0.0002	FR Ori	58887.6931	35130	0.0471	C	G. Frey	0.0002
CC Her	59370.4802	11362	0.3643	V	T. Arranz	0.0001	FZ Ori	59268.3248	38111	–0.0228	V	T. Arranz	0.0002
CT Her	59366.6710	9429	0.0110	V	G. Samolyk	0.0004	GU Ori	59286.4373	34453.5	–0.0735	V	T. Arranz	0.0002
CT Her	59402.3977	9449	0.0102	V	T. Arranz	0.0001	U Peg	59398.8298	61068.5	–0.1787	V	G. Samolyk	0.0002
LT Her	59381.6704	17182	–0.1639	V	L. Hazel	0.0003	U Peg	59419.8178	61124.5	–0.1785	V	L. Hazel	0.0003
V728 Her	59413.6453	14669	0.0300	V	G. Samolyk	0.0004	BB Peg	59378.8377	43193.5	–0.0367	V	G. Samolyk	0.0001
WY Hya	59294.5894	26150	0.0439	V	G. Samolyk	0.0001	BB Peg	59414.8070	43293	–0.0368	V	G. Samolyk	0.0002
WY Hya	59303.5394	26162.5	0.0438	V	L. Hazel	0.0003	BN Peg	59130.7083	35377	–0.0010	C	G. Frey	0.0001
AV Hya	59292.6334	33098	–0.1210	V	G. Samolyk	0.0003	BO Peg	59124.6721	23280	–0.0636	C	G. Frey	0.0001
AV Hya	59310.4026	33124	–0.1204	V	T. Arranz	0.0001	DI Peg	59151.6737	19605	0.0173	C	G. Frey	0.001
DF Hya	59294.4200	49928.5	0.0207	V	T. Arranz	0.0001	DK Peg	59185.6458	8368	0.1795	C	G. Frey	0.0001
DF Hya	59294.5865	49929	0.0219	V	G. Samolyk	0.0001	GP Peg	59381.8341	18597	–0.0600	V	G. Samolyk	0.0001
DF Hya	59294.7511	49929.5	0.0212	V	G. Samolyk	0.0001	V351 Peg	59143.6507	17940	0.0599	C	G. Frey	0.0001
DF Hya	59299.3792	49943.5	0.0208	V	T. Arranz	0.0001	UV Psc	59157.6555	18293	–0.0217	C	G. Frey	0.0010
DI Hya	59283.4075	45695	–0.0432	V	T. Arranz	0.0001	VZ Psc	59126.6932	58541.5	0.0067	C	G. Frey	0.0001
DK Hya	59306.6613	31543	–0.0013	V	G. Samolyk	0.0001	CP Psc	59134.7438	9699	0.0012	C	G. Frey	0.0001
V470 Hya	59294.6134	16674.5	0.0062	V	G. Samolyk	0.0005	DS Psc	59135.6600	19374	–0.0003	C	G. Frey	0.0001

Table continued on next page

Table 1. Recent times of minima of stars in the AAVSO eclipsing binary program, cont.

Star	JD (min) Hel. 2400000+	Cycle	O–C (day)	F	Observer	Standard Error (day)	Star	JD (min) Hel. 2400000+	Cycle	O–C (day)	F	Observer	Standard Error (day)
DV Psc	59133.6460	21500	0.0139	C	G. Frey	0.0010	XZ UMa	59309.4274	10751	–0.1609	V	T. Arranz	0.0001
DZ Psc	95136.6824	116449	0.2348	C	G. Frey	0.0001	XZ UMa	59320.4281	10760	–0.1611	V	T. Arranz	0.0001
ET Psc	95129.6577	96750	–0.0338	C	G. Frey	0.0001	ZZ UMa	59309.6645	10159	–0.0019	V	L. Hazel	0.0006
GR Psc	59138.6703	15518	–0.0090	C	G. Frey	0.0001	W UMi	59413.8007	15126	–0.2322	V	G. Samolyk	0.0004
HO Psc	59145.6675	5620	0.0055	C	G. Frey	0.0002	RU UMi	59345.6517	33813	–0.0137	V	G. Samolyk	0.0006
UZ Pup	59291.4081	18466	–0.0124	V	T. Arranz	0.0001	VV Vir	59348.6875	63047	–0.0524	V	G. Samolyk	0.0001
UZ Pup	59292.6013	18467.5	–0.0115	V	G. Samolyk	0.0002	VV Vir	59363.4103	63080	–0.0521	V	T. Arranz	0.0001
UZ Pup	59294.5893	18470	–0.0107	V	L. Hazel	0.0003	VV Vir	59369.4373	63093.5	–0.0479	V	T. Arranz	0.0004
AV Pup	59294.6608	50538	0.2745	V	G. Samolyk	0.0002	AG Vir	59244.8867	21493	–0.0205	V	K. Menzies	0.0003
XZ Sgr	59424.6664	5353	0.0004	V	L. Hazel	0.0009	AG Vir	59320.4024	21610.5	–0.0162	V	L. Corp	0.0003
V505 Sgr	59379.8353	12612	–0.1315	V	G. Samolyk	0.0001	AH Vir	59338.4847	33185.5	0.3116	V	L. Corp	0.0003
V1968 Sgr	59376.7879	38271	–0.0180	V	G. Samolyk	0.0004	AH Vir	59346.6355	33205.5	0.3119	V	G. Samolyk	0.0002
DK Sct	59414.4414	5678	0.0173	V	T. Arranz	0.0004	AH Vir	59379.6451	33286.5	0.3123	V	G. Samolyk	0.0002
RS Ser	59397.7753	40920	0.0334	V	G. Samolyk	0.0004	AK Vir	59317.7690	14026	–0.0460	V	G. Samolyk	0.0001
AO Ser	59394.4683	28727	–0.0099	V	T. Arranz	0.0001	AK Vir	59352.3833	14055	–0.0460	V	T. Arranz	0.0001
CC Ser	59302.8534	42286.5	1.1872	V	G. Samolyk	0.0004	AW Vir	59306.8091	40351	0.0332	V	G. Samolyk	0.0001
CC Ser	59361.4249	42400	1.1920	V	T. Arranz	0.0002	AW Vir	59359.3777	40499.5	0.0332	V	T. Arranz	0.0002
CC Ser	59376.6478	42429.5	1.1927	V	G. Samolyk	0.0002	AW Vir	59359.5546	40500	0.0331	V	T. Arranz	0.0001
CC Ser	59395.4813	42466	1.1920	V	T. Arranz	0.0001	AW Vir	59375.6623	40545.5	0.0340	V	G. Samolyk	0.0002
RZ Tau	58880.6460	51011	0.0969	C	G. Frey	0.0002	AX Vir	59357.6759	45247	0.0289	V	G. Samolyk	0.0002
TY Tau	58879.6438	34981	0.2761	C	G. Frey	0.0002	AX Vir	59376.6448	45274	0.0296	V	G. Samolyk	0.0001
WY Tau	59295.5687	31603	0.0670	V	K. Menzies	0.0003	AZ Vir	59294.8255	43808.5	–0.0185	V	G. Samolyk	0.0002
AC Tau	59247.5874	6661	0.2031	V	G. Samolyk	0.0002	AZ Vir	59366.6831	44014	–0.0171	V	G. Samolyk	0.0004
EQ Tau	59181.6678	55569	–0.0509	C	G. Frey	0.0002	BH Vir	59368.7108	19756	–0.0137	V	G. Samolyk	0.0003
V1223 Tau	59186.6263	15487.5	0.0018	C	G. Frey	0.0001	Z Vul	59422.5221	6711	–0.0177	V	T. Arranz	0.0002
W UMa	59321.6365	40631	–0.1269	V	G. Samolyk	0.0005	AW Vul	59375.7443	16232	–0.0400	V	L. Hazel	0.0005
TY UMa	59303.6272	55764.5	0.4629	V	L. Hazel	0.0004	AW Vul	59404.7772	16268	–0.0393	V	G. Samolyk	0.0001
TY UMa	59306.6392	55773	0.4614	V	G. Samolyk	0.0001	AX Vul	59399.7885	7184	–0.0420	V	G. Samolyk	0.0002
TY UMa	59306.8163	55773.5	0.4612	V	G. Samolyk	0.0001	BE Vul	59398.7343	12427	0.1025	V	L. Hazel	0.0003
TY UMa	59321.3539	55814.5	0.4627	V	T. Arranz	0.0001	BE Vul	59426.6696	12445	0.1010	V	G. Samolyk	0.0001
TY UMa	59321.5304	55815	0.4619	V	T. Arranz	0.0001	BE Vul	59426.6699	12445	0.1013	V	L. Hazel	0.0003
UX UMa	59293.8145	111155	–0.0020	V	G. Samolyk	0.0002	BO Vul	59417.6901	12040	–0.0037	V	L. Hazel	0.0003
UX UMa	59317.6119	111276	–0.0019	V	G. Samolyk	0.0001	BO Vul	59425.4718	12044	–0.0054	V	T. Arranz	0.0001
UX UMa	59317.8084	111277	–0.0020	V	G. Samolyk	0.0001	BS Vul	59380.7944	33845	–0.0380	V	G. Samolyk	0.0001
UX UMa	59339.4429	111387	–0.0014	V	T. Arranz	0.0001	BS Vul	59424.5836	33937	–0.0382	V	T. Arranz	0.0001
UX UMa	59352.6196	111454	–0.0016	V	G. Samolyk	0.0002	BU Vul	59405.8063	45470	0.0116	V	G. Samolyk	0.0001
VV UMa	59334.6287	19668	–0.0976	V	G. Samolyk	0.0001							

Daylight Photometry of Bright Stars—Observations of Betelgeuse at Solar Conjunction

Otmar Nickel
Zum Schollberg 11, 55129 Mainz, Germany; otmar.nickel@web.de

Tom Calderwood
1184 NW Mt. Washington Drive, Bend, OR 97703; tjc@cantordust.net

Received October 2, 2021; revised November 26, 2021; accepted November 29, 2021

Abstract Betelgeuse is an important variable star with many observations in the AAVSO International Database, but there is an annual gap of about four months where Betelgeuse is close to the sun and not observable at night. This gap could be filled with daylight observations. The star is bright enough to be imaged with small telescopes during the day, so photometry is possible when the sun is up. We present V-band photometry of α Ori taken with an amateur telescope equipped with an interline-transfer CCD camera and neutral density filter. These data compare favorably with contemporaneous nighttime photometry. The method used is a variation on ensemble photometry (using other bright daytime stars), and involves large stacks of very short exposures. The ensemble method provided V magnitudes of Betelgeuse with calculated errors of (0.020 ± 0.008) mag from February to April 2021. From May to July, at the closest distances to the sun, the photometry of Betelgeuse could be continued with mean errors of (0.040 ± 0.013) mag.

1. Introduction

Betelgeuse has been a subject of great astrophysical interest. Ground-based telescopes can acquire reliable photometry from roughly early September to late April. That leaves four months without data in each year's light curve. Given the extraordinary recent behavior of this star, that break in coverage is most unfortunate. While space-based photometry of Betelgeuse is now practiced during solar conjunctions (Dupree *et al.* 2020), the technique is difficult and cannot be performed in a standard photometric passband.

With care, it is possible to gather daytime aperture photometry of Betelgeuse. Such data were collected from February to July of 2021 with a small telescope located in Central Europe at only 200 m elevation, where the atmospheric clarity is far from exceptional.

2. Observations and data acquisition

The telescope used is of Newtonian configuration with aperture 250 mm and focal length 1250 mm, carried on a computer-controlled German equatorial mount. The optical tube is fitted with a cylindrical sun shade extending 50 cm beyond the aperture. Together with a secondary shade around the focuser aperture, the shielding permits pointings as close as 10 degrees to the sun.

The camera is an ATIK model 460exm CCD, equipped with a wheel of photometric filters. For daylight work an additional neutral density (ND) filter of 1 percent transmission (as proposed by Miles 2007) is mounted ahead of the wheel. The camera is cooled to 0°C for daytime work and has its own sun shade. Pointing of the mount is accurate enough to place a target star directly in the camera field of 20' × 30'.

During photometry only a 687 × 550-pixel region of the CCD is read out. Image frames were typically integrated for 0.1 sec, though sometimes for 0.2 sec. This kept the daytime sky to 50 percent or less of the sensor full-well depth. On each target, consecutive frames were taken to accumulate 10 sec of total exposure time. This process takes approximately 150 sec when using 0.1 sec exposures. Since guiding is not possible in daylight, the pointing is subject to drift. To mitigate this effect, it is helpful to use groups of ten exposures co-added and saved and later re-registered to make the 10-sec stack (this was done for about half of the presented photometry). Flat-field images were acquired by exposing against the daytime sky through a 3 mm-thick white polystyrene foam board. The benefit of the 10-sec stack is illustrated in Figure 1. Depending upon the surrounding sky brightness, stars as dim as V = 6.5 can be successfully sampled this way.

3. Sky background

The brightness of the sky can be measured (in mag/arcsec^2) if a bright star with known magnitude is included in the field. The value is calculated by the formula:

$$m_{sky} = -2.5 \log \left(\frac{N_{sky}}{p^2 N_{star}} \right) + m_{star} \quad (1)$$

where N_{sky} = mean counts/pixel within sky annulus; N_{star} = summed pixel counts within star aperture after subtraction of sky background; and p = pixel scale in arcsec/pixel.

This method was first tested on the nighttime sky and gave values comparable to those of the "Sky Quality Meter" (SQM; Cinzano 2005), which is not usable in daylight. The individual measurements were repeatable within ± 0.1 mag/arcsec^2.

The daylight sky brightness depends on several factors, the prime one being angular distance of the field from the sun. As a test, the sky brightness was measured on 17 cloudless days (July to September 2020 and February to April 2021) with selections from 22 stars at different distances from the sun. The altitude of the sun ranged from 10° to 52°. The results are seen in Figure 2. As expected, the smaller the angular distance from the sun, the higher the sky brightness.

Other factors are the altitude of the sun and the level of water vapor in the atmosphere. The latter is correlated to the extinction coefficient, so there is also a correlation between sky brightness and extinction, which can be seen in Figure 3. This figure shows the measured sky brightness around β Aur versus the measured extinction coefficient during 9 days between May 31 and July 22, 2021, around local solar noon.

Additional factors, e.g. airmass or the strong polarization of light at daytime can influence the sky brightness as well.

4. Photometry methods

4.1. Instrumental and standardized magnitude

Instrumental magnitudes m_{inst} were calculated using rectangular apertures for both star and sky background. The instrumental magnitude is given by:

$$m_{inst} = -2.5 \log \left(\frac{N_{star}}{t} \right) \quad (2)$$

where N_{star} = background-subtracted pixel counts within star aperture; t = exposure time in s.

The processing of the images was done with the Fitsmag software package (Nickel 2021).

The standardized V magnitude of a star can be calculated from the instrumental magnitude by (Da Costa 1992):

$$V = m_{inst} + a_0 + a_1 (B-V) + a_2 X + a_3 X (B-V) + \ldots \quad (3)$$

where $(B-V)$ = color index of star; X = airmass of the observation; a_0 = the zero point magnitude (in the following written as m_0); a_1 = the color-term (in the following called Transformation coefficient T_v; a_2 = primary extinction coefficient (in the following replaced by $-k_v$); a_3 = second order extinction coefficient. The second order extinction coefficient is very small for the V band and can be neglected.

Having in hand values for the transform coefficient and the target star's airmass, color, and instrumental magnitude, a standard magnitude can be calculated if the zero point and primary extinction coefficient can be established. This is done with observations of an ensemble of reference stars at a range of airmasses. For each of these stars, we have:

$$V = m_{inst} + m_0 + T_v (B-V) - k_v X \quad (4)$$

and

$$V - (m_{inst} + T_v (B-V)) = m_0 - k_v X \quad (5)$$

where k_v and m_0 are unknowns. If the differences $V - (m_{inst} + T_v (B-V))$ of several stars are plotted against their airmasses, then k_v and m_0 can be calculated by linear regression (LR) as shown in Figure 4.

With known m_0 and k_v the standard magnitude can be calculated by:

$$V_{var} = m_{inst} + T_v (B-V)_{var} + m_0 - k_v X_{var} \quad (6)$$

It can be shown (Appendix A) that the result of this calculation is mathematically equivalent to differential ensemble photometry, where a mean value of differential magnitudes against an ensemble of comparison stars is calculated.

T_v for the combined V and ND filters was -0.0026 ± 0.0057. For the V filter alone it was $+0.024$, so the ND filter shifts the color significantly. The spectral transmission of the ND filter (Vendor: Antares) was analyzed by the author using a slit spectrograph measuring the fraction of transmitted light of a LED light source: it has a local peak at 550 nm, as shown from the transmission curve in Figure 5.

4.2. Signal to noise ratio

The well known sources of noise in CCD photometry are shot noise, thermal electronic noise, and readout noise. The quantum signal-to-noise-ratio (SNR) of an observation can be calculated using the "CCD equation" (Merline and Howell 1995). In a high-count regime, this equation is approximated by:

$$SNR_q = \frac{N_{star}}{\sqrt{N_{star} + n_{pix} \left(1 + \frac{n_{pix}}{n_{sky}}\right) N_{sky}}} \quad (7)$$

where N_{star} = number of collected photo electrons from star; n_{pix}, n_{sky} = number of pixels in star and sky apertures; N_{sky} = number of collected electrons/pixel from sky background. The inverse of SNR_q may be called "normalized quantum noise" σ_q:

$$\sigma_q = \frac{1}{SNR_q}$$

In short exposures scintillation noise can be significant. The normalized scintillation noise σ_s is the standard deviation of a series of star intensities divided by the mean value of the intensities (if no other noise would be involved). The total noise σ_{total} and the SNR from both CCD noise σ_q and scintillation σ_s are:

$$\sigma_{total} = \sqrt{\sigma_q^2 + \sigma_s^2}, \quad SNR = \sqrt{\frac{1}{\sigma_q^2 + \sigma_s^2}} \quad (8)$$

Scintillation noise can be estimated with the modified "Young formula" (Young 1967 and Osborn et al. 2015):

$$\sigma_s = 0.003953 \times X^{3/2} D^{-2/3} e^{-H/8000} \sqrt{t^{-1}} \quad (9)$$

where D = aperture of telescope in m; H = local height above sea level in m; X = airmass; t = exposure time in s. For the observing station ($D = 0.25$ m, $H = 200$ m) this gives the following values of σ_s:

	$X = 1.0$	$X = 1.5$	$X = 2.0$
$t = 1$ s	0.00976	0.01793	0.0276
$t = 10$ s	0.00309	0.00567	0.00874

Actual scintillation measurements (at daylight) were tested against the formula and demonstrate its reliability in this application.

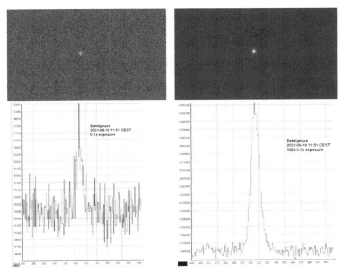

Figure 1. Images and profiles of Betelgeuse near to sun, left: single image, right: stack of 100 images.

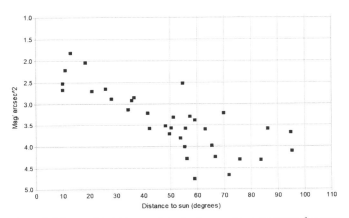

Figure 2. Measured brightness of sky background in mag/arcsec² versus distance from sun.

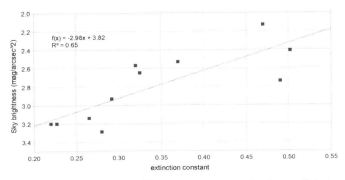

Figure 3. Sky brightness around β Aur at noon versus extinction coefficient.

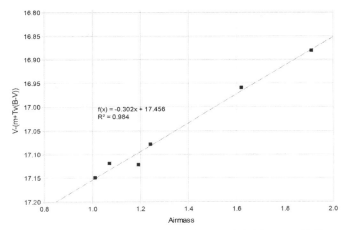

Figure 4. Plot of $V - (m_{inst} + T_v (B-V))$ vs. airmass of 6 bright stars with linear regression line.

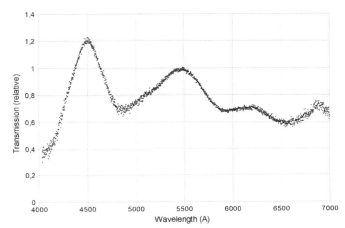

Figure 5. Transmission curve of ND filter (units relative to 550 nm).

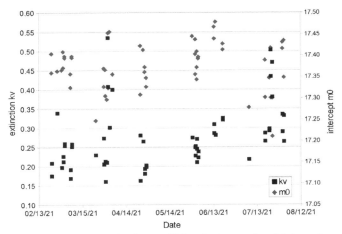

Figure 6. Values for extinction coefficient k_v (squares) and intercept m_0 (diamonds).

4.3. Total photometric error

The statistical error of ordinary differential photometry can be estimated as follows:

$$\text{error [mag]} = \sqrt{\text{error}^2 \text{(variable)} + \text{error}^2 \text{(comparison)}} \quad (10)$$

In the ensemble approach used here, the zero point is the reference magnitude, and its error is a function of the LR:

$$\text{error [mag]} = \sqrt{\text{error}^2 (\text{var}) + (X_{\text{var}} - \bar{X})^2 \sigma^2(k) + \text{SER}^2} \quad (11)$$

where error(star) = $1.0857 \, (1/\text{SNR}(\text{star}))$; X_{var} = airmass of variable; $\bar{X}$ = mean value of the airmass of the reference stars; σ_k = error of extinction coefficient; SER = standard error of the LR (Equation 5). This formula is derived in Appendix A.

4.4. Reference stars for Betelgeuse

The reference stars must be bright enough in comparison to the sky. The SNR should be not much below 50; this is the case if the star magnitude is nearly equal to the sky brightness (in mag/arcsec$_2$). From Figure 2, it follows that down to a distance of 30° to the sun the star magnitude should be at least V=2.8; between 15° and 30° it should be brighter than 2.

Most bright stars have a slight variability, therefore only those with a magnitude range of less than 0.1 mag (from GCVS (Samus *et al.* 2017) were selected as comparison stars. V magnitudes were taken from the Extended Hipparcos catalogue (Anderson and Francis 2012). Table 1 shows the selected stars.

Depending on their distance from the sun, not all of these stars can always be used. Betelgeuse is closest to the sun on June 20, 2021, therefore the distance of the stars to the sun are listed for this date in the table. Around this date β Tau and γ Gem are too close to the sun; from June 30, α and γ Gem cannot be used.

α Gem has two components (A, B), separated by 5.5", which have to be measured together. If it is used in the ensemble, the aperture must be large; that could increase the error. ζ Ori is a double star with 2.2" separation, which may be no problem. β Aur is an EA variable with known period; it should be used only outside of the eclipses.

5. Observation results

5.1. Results of reference stars and extinction

From February 21 to July 31, 2021, daytime observations of Betelgeuse, γ Ori (as check star) and 4 to 7 reference stars were performed during 61 runs on 33 days. The reference stars were selected from among the following: α Gem, α Ari, α Cas, β Ori, α CMi, β Gem, β Tau, ζ Ori, βAur, γ Gem.

The SNR of the star measurements used was typically above or around 100. Results with SNR below 50 (2 of 200) were discarded. The remaining range was between 53 and 290. The LR statistics for Betelgeuse observations are shown in Table 2. The first group represents data from February to April 2021, the second, May to July 2021, where the stars are closer to the sun. The errors of the second group are significantly higher, as can be expected from the brighter sky and from the worse seeing during this time. The extinction was also higher due to many hazy days. All values of k_v and m_0 are shown in Figure 6.

The whole range of extinction values was between 0.16 and 0.53; values above 0.3 corresponded to a very hazy sky. The standard error did not correlate significantly to the extinction value.

For the LR, up to one star was excluded if the deviation of the star magnitude from the regression line was significant (> 2 × std.error) and greater than 0.05 mag. In 4 of 65 regression calculations this rule was used.

5.2. Observations of Betelgeuse (February to July 2021)

The daylight observations of Betelgeuse from February 21 to July 31, 2021, together with nighttime photoelectric photometry (PEP) are shown in Figure 7. The PEP data were collected with

Table 1. Reference stars for Betelgeuse.

Star Name	V mag. (XHip)	B–V mag.	δ mag. (GCVS)	Dist. to α Ori (°)	Sun Dist. (°) 2021-06-20	Comment
β Ori	0.18	–0.03	0.05	19	33	
α CMi	0.40	0.43	0.07	26	32	
β Gem	1.16	0.99	0.07	33	25	
α Gem	1.58	0.03	0	34	24	double (5.5")
γ Ori	1.64	–0.22	0.05	7.5	18	
β Tau	1.65	–0.13	0	22	7	
ζ Ori	1.74	–0.20	0.07	10	25	double (2.2")
β Aur	1.90	0.08	0.09	37	21.5	EA (P=3.96 d)
γ Gem	1.93	0.00	0	14	12	
α Ari	2.01	1.15	0.06	57	51	
α Cas	2.24	1.17	0.07	78	64	

Table 2. Statistics of linear regression from February to July 2021.

	February to April Mean Value (std. dev.)	May to July Mean Value (std. dev.)
Standard error	0.017 (0.008)	0.036 (0.011)
Extinction constant (k_v)	0.247 (0.084)	0.299 (0.073)
Error of (k_v)	0.027 (0.010)	0.045 (0.016)
Intercept (m_0)	17.365 (0.047)	17.381 (0.064)
(R^2)	0.962 (0.036)	0.917 (0.056)

Table 3. The mean values of the differences between day and night magnitudes.

	Difference (day – night) (mag.)	Error (day) (mag.)	Error (night) (mag.)
Mean value (n = 6)	0.006	0.018	0.020
Std. dev.	0.026	0.007	0.012
Error of mean	0.011	0.003	0.005

an Optec SSP-3 photometer mounted on a 235-mm telescope in North America. The PEP V data were taken in concert with B data, and the Δ(B–V) with respect to the comparison star (HD 37160) was established from the instrumental magnitudes and transformation coefficients. During the period 2459255 (Feb 9, 2021) to 2459327 (Apr. 22, 2021) the mean Δ(B–V) was 0.910 with a standard deviation of 0.010. The comparison (B–V) is taken as 0.958, implying that the mean (B–V) of Betelgeuse was 1.868. The color transformation of the daylight measurements were based on a constant (B–V) of Betelgeuse of 1.85 (from Hipparcos).

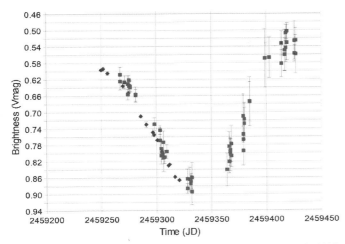

Figure 7. Light curve (V-magnitudes) of Betelgeuse February to July 2021. Blue diamonds: PEP data (night), Red squares: CCD data (daylight). PEP error bars are too small to see.

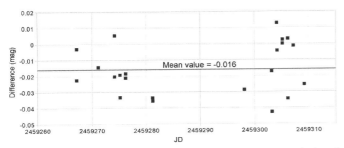

Figure 8. Difference between daylight and interpolated PEP magnitudes of Betelgeuse. Horizontal line: mean value.

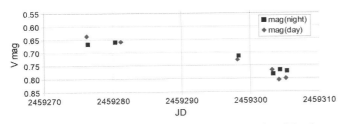

Figure 9. Comparison of daytime and nighttime magnitudes of Betelgeuse resulting from the same method. Blue squares: Nighttime results; Red diamonds: Daytime results.

The PEP magnitudes are in most cases slightly dimmer, and the difference between the daylight results and interpolated PEP magnitudes is shown in Figure 8. The mean value of the differences is -0.016 ± 0.003 mag (n = 22). The angular distance from the sun was between 16° and 114° and the sky brightness around Betelgeuse was between 1.8 and 6.0 mag/arcsec^2.

The mean errors of the daylight results were 0.02 mag (February–April) and 0.04 mag (May–July), which is only slightly higher than the LR standard errors of the linear regression of the reference stars (Table 2); this is a consequence of the instrumental magnitude error of Betelgeuse in the range 0.006–0.008 mag, which was much smaller than the standard error in all cases, as well as the extinction error (airmass term in Equation 11), which was in the range 0.001–0.015 mag.

On 6 nights, from March 02 to March 31, 2021, Betelgeuse was observed also at night with the same CCD equipment and method as in the daytime, and the magnitudes were calculated using the same comparison star ensemble as in the daytime. The results are shown in Figure 9. The mean values of the differences between day and night magnitudes are shown in Table 3.

Day and night magnitudes are not significantly different; the mean values of the calculated errors are also in the same range.

All daylight observations were uploaded to the AAVSO International Database, where the PEP data may also be found; the data may be visualized using the AAVSO Light Curve Generator (www.aavso.org/LCGv2). If more than one daylight observation was available per day, only the mean value of the magnitudes was recorded.

6. Discussion

The accuracy of the magnitude results depends partly on the selection of reference stars. For consistency, the magnitudes of the Betelgeuse ensemble stars were all taken from the Hipparcos catalogue. There are other catalogues, e.g. the *General Catalog of Photometric Data* (GCPD; Mermilliod *et al.* 1997). For some stars used as references, there are differences between Hipparcos and GCPD of up to 0.042 mag. In a sample of three cases, where β Ori, α CMi, and β Gem were used, the V mag result with GCPD values differed by about –0.01 mag and the error was slightly smaller. Therefore, future projects should use the GCPD magnitudes of the reference stars, if available.

The systematic differences of -0.016 (± 0.003) between daylight and PEP magnitudes cannot result from differences in the (B–V) values used for Betelgeuse in transformation: the PEP and daylight (B–V) differ by only about 0.018. This would result in an offset of less than 1 mmag in the daylight photometry. But the difference can be explained by the error of the color transformation constant T_v. Because of the high color index of Betelgeuse of around 1.85 this difference can be induced by a T_v error of -0.0097 (mean value of color indices of the reference stars was around 0.2 mag). The calculated T_v error was only slightly lower (0.0057), therefore this explanation is plausible. If the GCPD magnitudes were used, the systematic error would be on the order of -0.026 and the error of T_v would be on the order of 0.016. Therefore, in future projects the transformation constant should be adjusted in some way.

7. Conclusions

We have shown that CCD photometry in daylight with amateur equipment can be performed with an accuracy on the order of 0.02–0.05 mag. This is possible even within 10° of the sun, at least for stars of brightness on the order of V ≤ 2. Daylight photometry of Betelgeuse was compared to nighttime PEP photometry obtained over a period of eight weeks; systematic differences between the light curves were less than 0.02 mag. Observations of bright variable stars could be collected in this way nearly uninterrupted by conjunction with the sun.

The photometry could be improved in three ways:

(1) A camera with higher framing rates could enable stacking of more than 1000 0.1-sec frames. The additional frames would increase the SNR, reduce the effect of scintillation, and allow the observation of stars with magnitudes in the range up to V = 3.

(2) Observing regions with altitudes above 1,000 m and with low humidity may provide a darker sky and would also enlarge the range of observable stars.

(3) Because the sky brightness is lower at longer wavelengths, it could be better to use redder filters to reduce the background noise, but in this case the transformation to the V magnitude would be more difficult. In any case an additional measurement with photometric R filter would give more information.

References

Anderson, E., and Francis, C. 2012, *Astron. Lett.*, **38**, 331.
Cinzano, P. 2005, *ISTIL Internal Rep. No. 9, v.1.4*, 1.
Da Costa, G. 1992, in *Astronomical CCD Observing and Reduction Techniques*, ed. S. B. Howell, ASP 23, Astronomical Society of the Pacific, San Francisco, 90.
Dupree, A., Guinan, E., and Thompson, W. T. 2020, *Astron. Telegram*, No. 13901, 1.
Merline, W. J., and Howell, S. B. 1995, *Exp. Astron.*, **6**, 163.
Mermilliod, J.-C., Mermilliod, M., and Hauck, B. 1997, *Astron. Astrophys., Suppl. Ser.*, **124**, 349.
Miles, R. 2007, *J. Br. Astron. Assoc.*, **117**, 278.
Nickel, O. 2021, FITSMAG photometry software (https://www.staff.uni-mainz.de/nickel/fitsmag.html).
Osborn, J., Föhring, D., Dhillon, V. S., and Wilson, R. W. 2015, *Mon. Not. Roy. Astron. Soc.*, **452**, 1707.
Samus, N. N., Kazarovets, E. V., Durlevich, O. V., Kireeva, N. N., and Pastukhova, E. N. 2017, *Astron. Rep.*, **61**, 80.
Young, A. T. 1967, *Astron. J.*, **72**, 747.

Appendix A: Ensemble comparison and linear regression

With LR, the values of $V_i - (m_i + T_v Y)$ of a sample of n reference stars with catalogue magnitudes V_i, color index Y_i and instrumental magnitudes m_i are correlated to their airmass X_i via the linear function:

$$V_i - (m_i + T_v Y_i) = m_0 - k_v X_i \quad (12)$$

LR of this formula provides the regression parameters m_0 and k_v. Incorporating the residual ε_i the magnitude of each reference star is given by the equation:

$$V_i = (m_i + T_v Y_i) + \varepsilon_i + m_0 - k_v X_i \quad (13)$$

The differential magnitude V_{var} of a variable with instrumental magnitude m_{var}, with respect to a reference star (with instrumental magnitude m_i) would be:

$$V_{var} = m_{var} - m_i + T_v(Y_{var} - Y_i) - k_v(X_{var} - X_i) + V_i \quad (14)$$

The ensemble method uses the arithmetic mean value of all reference stars, therefore:

$$V_{var} = \frac{1}{n} \sum_{i=1}^{n} (m_{var} - m_i + T_v(Y_{var} - Y_i) - k_v(X_{var} - X_i) + V_i) \quad (15)$$

which simplifies to:

$$V_{var} = m_{var} + T_v Y_{var} - k_v X_{var} - \frac{1}{n} \sum_{i=1}^{n} (m_i + T_v Y_i - k_v X_i - V_i) \quad (16)$$

Now V_i is replaced by the right side of Equation 13:

$$V_{var} = m_{var} + T_v Y_{var} - k_v X_{var}$$
$$- \frac{1}{n} \sum_{i=1}^{n} (m_i + T_v Y_i - k_v X_i - (m_i + T_v Y_i) - \varepsilon_i - m_0 + k_v X_i) \quad (17)$$

which becomes:

$$V_{var} = m_{var} + T_v Y_{var} - k_v X_{var} - \frac{1}{n} \sum_{i=1}^{n} (\varepsilon_i - m_0) \quad (18)$$

Because the sum of the residuals \epsiloni is zero, this yields:

$$V_{var} = m_{var} + T_v Y_{var} - k_v X_{var} + m_0 \quad (19)$$

This is exactly the same as Equation 6. This means that the LR method is equivalent to a differential comparison method with an ensemble of comparison stars. Therefore the error calculation can be based on the mean value of differential magnitudes.

For the evaluation of the resulting error one works with Equation 16. This can be transformed to:

$$V_{var} = m_{var} + T_v(Y_{var} - \frac{1}{n}\sum_{i=1}^{n} Y_i) - k_v(X_{var} - \frac{1}{n}\sum_{i=1}^{n} X_i) - \frac{1}{n}\sum_{i=1}^{n}(m_i - V_i) \quad (20)$$

$\bar{X}$, $\bar{Y}$, and $\bar{V}$ are mean values of X_i, Y_i, and V_i)

$$V_{var} = m_{var} + T_v(Y_{var} - \bar{Y}) - k_v(X_{var} - \bar{X}) - \frac{1}{n}\sum_{i=1}^{n}(m_i + \bar{X}) \quad (21)$$

In this formula $\bar{V}$ is constant, m_{var}, T_v, Y_{var}, k_v, and m_i have statistical errors, which are independent of one another. Therefore $\sigma^2(V_{var})$, the squared error of V_{var}, can now be estimated from a sum of the variances of these errors:

$$\sigma^2(V_{var}) = \sigma^2(m_{var}) + T_v^2 \sigma^2(Y_{var}) + (Y_{var} - \bar{Y})^2 \sigma^2(T_v)$$
$$+ (X_{var} - \bar{X})^2 \sigma^2(k_v) + \frac{1}{n}\sum_{i=1}^{n} \sigma^2(m_i) \quad (22)$$

The variance of $m_i + T_v Y_i$ is obtained from the sum of the residuals of the linear regression divided by $1/(n-2)$:

$$\sigma^2(m_i + T_v Y_i) = \frac{1}{n-2} \sum_{i=1}^{n} \varepsilon_i^2 \quad (23)$$

The right side of the above equation is the square of the standard error of the regression (SER). It follows:

$$\sigma^2(m_i) = SER^2 - \sigma^2(T_v Y_i) \quad (24)$$

and

$$\frac{1}{n}\sum_{i=1}^{n} \sigma^2 m_i = SER^2 - \frac{1}{n}\sum_{i=1}^{n} Y_i^2 \sigma^2(T_v) = SER^2 - \sigma^2(T_v) \frac{1}{n}\sum_{i=1}^{n} Y_i^2 \quad (25)$$

From Equation 22 follows:

$$\sigma^2(V_{var}) = \sigma^2(m_{var}) + T_v^2 \sigma^2(Y_{var}) + (Y_{var} - \bar{Y})^2 \sigma^2(T_v)$$
$$+ (X_{var} - \bar{X})^2 \sigma^2(k_v) + SER^2 - \sigma^2(T_v) \frac{1}{n}\sum_{i=1}^{n} Y_i^2 \quad (26)$$

If the error of T_v is neglected, then the approximate error of V(ensemble) can be calculated by:

$$\sigma(V_{var}) = \sqrt{\sigma^2(m_{var}) + (X_{var} - \bar{X})^2 \sigma^2(k) + SER^2} \quad (27)$$

Erratum: Four New Variable Stars in the Field of KELT-16

Daniel J. Brossard
Department of Physics and Astronomy, Ball State University, 2000 West University Avenue, Muncie, IN 47306; DBrossard25@gmail.com

Ronald H. Kaitchuck
Department of Physics and Astronomy, Ball State University, 2000 West University Avenue, Muncie, IN 47306; rkaitchu@bsu.edu

In the article "Four New Variable Stars in the Field of KELT-16" (*JAAVSO*, 2021, **49**, 24–31), in Table 5, the 2MASS designation for the star designated V7 was given incorrectly. The correct content for Table 5 is given below.

The authors would like to thank Brett Schulz of Minnesota State University Moorhead for bringing the error to our attention.

Table 5. Variable candidate periods and magnitude estimates.

Designation	Star	R-Band Apparent Mag.	ASAS-SN/ZTF Survey Period (days)	Our Reanalysis of ASAS-SN V-band Period (days)	Our R-band Period (days)	Classification
V1	ASASSN-V J205658.12+314215.9	14.8 ± 0.02	0.7543	0.7544 ± 0.0002	0.7643 ± 0.0009	EW
V2	ASASSN-V J205552.88+314615.9	14.1 ± 0.02	0.6002	0.8586 ± 0.0071	0.4388 ± 0.0005	EW
V3	2MASS J20564622+3138394	14.0 ± 0.02	—	—	0.3465 ± 0.0005	EW
V4	2MASS J20560314+3145505	17.1 ± 0.02	—	—	0.3021 ± 0.0006	EW
V5	ZTF J205733.78+314612.6	16.4 ± 0.02	0.4251	—	0.4282 ± 0.0007	EW
V6	ZTF J205627.42+315322.4	16.9 ± 0.02	0.2925	—	0.2981 ± 0.0004	EW
V7	2MASS J20565617+3131253	17.7 ± 0.04	—	—	0.2963 ± 0.0005	Possible EW

Note: As mentioned before, error estimates were not provided for the ASAS-SN and the ZTF survey periods. (Jayasinghe et al. 2018, 2019; Masci et al. 2019; Chen et al. 2020).

Index to Volume 49

Author

Afonina, Marina D., and Richard E. Schmidt, Sergei Yu. Shugarov
 The Photometric Period of V1674 Herculis (Nova Her 2021) 257
Alton, Kevin B., and Edward O. Wiley
 CCD Photometry, Light Curve Modeling, and Period Study of V573 Serpentis, a Totally Eclipsing Overcontact Binary System 170
Alton, Kevin B., and John C. Downing
 CCD Photometry, Light Curve Modeling and Period Study of V1073 Herculis, a Totally Eclipsing Overcontact Binary System 201
 CCD Photometry, Light Curve Modeling, and Period Study of GSC 2624-0941, a Totally Eclipsing Overcontact Binary System 214
Anees, Khola, and Shaukat Naaman Goderya, Fazeel Mahmood Khan
 Differential Photometry of Eclipsing Binary System V798 Her in Globular Cluster NGC 6341 (Abstract) 116
Anon.
 Index to Volume 49 277
Axelsen, Roy A.
 Retraction of and Re-analysis of the Data from "HD 121620: A Previously Unreported Variable Star with Unusual Properties" 197
Axelsen, Roy A., in Tom Richards et al.
 Southern Eclipsing Binary Minima and Light Elements in 2020 251
Bahramian, Arash, in Eric Masington et al.
 An Analysis of X-Ray Hardness Ratios between Asynchronous and Non-Asynchronous Polars 149
Banks, Timothy, in Ross Parker et al.
 Distances for the RR Lyrae Stars UU Ceti, UW Gruis, and W Tucanae 178
Bansal, Avni, and Paul Hamrick, Kalée Tock
 Characterization of NGC 5272, NGC 1904, NGC 3201, and Terzan 3 192
Beckmann, Rev. Kenneth
 Morning Star: The Search for and Discovery of the Stars of Bethlehem According to the Gospel of Matthew (Abstract) 114
Berry, Richard, and Nolan Sottoway, Sol McClain
 New Observations of the SX Phe Star XX Cygni (Abstract) 115
Bewersdorff, Leon, in Ramy Mizrachi et al.
 25 New Light Curves and Updated Ephemeris through Analysis of Exoplanet WASP-50 b with EXOTIC 186
Blackford, Mark, in Tom Richards et al.
 Southern Eclipsing Binary Minima and Light Elements in 2020 251
Boyd, David
 Spectroscopic and Photometric Study of the Mira Stars SU Camelopardalis and RY Cephei 157
Brossard, Daniel J., and Ronald H. Kaitchuck
 Erratum: Four New Variable Stars in the Field of KELT-16 276
 Four New Variable Stars in the Field of KELT-16 24
Brown, Leslie F., in Manny Rosales et al.
 The Blazar BL Lacertae: 2018–2020 V-, R-, and I-Band CCD Photometry (Abstract) 116
Calderwood, Tom, and Otmar Nickel
 Daylight Photometry of Bright Stars—Observations of Betelgeuse at Solar Conjunction 269
Carbonell, Wyatt, in Manny Rosales et al.
 The Blazar BL Lacertae: 2018–2020 V-, R-, and I-Band CCD Photometry (Abstract) 116
Caton, Daniel, in Ronald G. Samec et al.
 Binaries with Mass Ratios Near Unity: The First BVRI Observations, Analysis and Period Studies of TX Canis Minoris and DW Canis Minoris 138
Ciprini, Stefano, in Corrado Spogli et al.
 Photometric Observations of the Dwarf Nova AH Herculis 232
Cristobal, Sandra Moreno, and Demetris Nicolaides, Destiny L. King
 V350 Muscae: RR Lyrae Star Distance Estimate and RRab Reclassification 63
Downing, John C., and Kevin B. Alton
 CCD Photometry, Light Curve Modeling and Period Study of V1073 Herculis, a Totally Eclipsing Overcontact Binary System 201
 CCD Photometry, Light Curve Modeling, and Period Study of GSC 2624-0941, a Totally Eclipsing Overcontact Binary System 214
Duncan, Todd, and Erika Dunning
 Building Connection through Community-Based Astronomy (Abstract) 115
Dunning, Erika, and Todd Duncan
 Building Connection through Community-Based Astronomy (Abstract) 115
Eaton, Joel A., and Gary W. Steffens, Andrew P. Odell
 V963 Persei as a Contact Binary 121
Faulkner, Danny, in Ronald G. Samec et al.
 Binaries with Mass Ratios Near Unity: The First BVRI Observations, Analysis and Period Studies of TX Canis Minoris and DW Canis Minoris 138
Gentry, Davis, in Ronald G. Samec et al.
 Binaries with Mass Ratios Near Unity: The First BVRI Observations, Analysis and Period Studies of TX Canis Minoris and DW Canis Minoris 138
Goderya, Shaukat Naaman, and Khola Anees, Fazeel Mahmood Khan
 Differential Photometry of Eclipsing Binary System V798 Her in Globular Cluster NGC 6341 (Abstract) 116
Gupta, Prateek, and John R. Percy
 Pulsating Red Giants in a Globular Cluster: 47 Tucanae 209
Gurney, Kevin
 Using High Resolution Spectroscopy to Measure Cepheid Pulsation (Abstract) 117

Guzik, Joyce A.
- Search for Variability in 30 Bright Metallic-line A Stars Observed by the TESS Spacecraft (Abstract) 113

Hamilton, Joshua R.
- Light Curve Analysis of 185 YSOs: New Periods Discovered for 9 Stars 49

Hamrick, Paul, and Avni Bansal, Kalée Tock
- Characterization of NGC 5272, NGC 1904, NGC 3201, and Terzan 3 192

Heinke, Craig, in Eric Masington *et al.*
- An Analysis of X-Ray Hardness Ratios between Asynchronous and Non-Asynchronous Polars 149

Jenkins, Robert, in Tom Richards *et al.*
- Southern Eclipsing Binary Minima and Light Elements in 2020 251

Joignant, Rowan, and Jamie Lester, Mariel Meier
- Photometric Determination of the Distance to the RR Lyrae Star YZ Capricorni 247

Kaitchuck, Ronald H., and Daniel J. Brossard
- Erratum: Four New Variable Stars in the Field of KELT-16 276
- Four New Variable Stars in the Field of KELT-16 24

Kaneshiro, Dylan, and Horace A. Smith, Gerard Samolyk
- An Update on the Periods and Period Changes of the Blazhko RR Lyrae Star XZ Cygni 19

Karmakar, Pradip, and Gerard Samolyk
- The Long-term Period Changes of the Cepheid Variable SV Monocerotis 135

Karmakar, Pradip, and Horace A. Smith, Wayne Osborn
- Types of Period Changes of W Virginis Stars (Abstract) 113

Khan, Fazeel Mahmood, and Khola Anees, Shaukat Naaman Goderya
- Differential Photometry of Eclipsing Binary System V798 Her in Globular Cluster NGC 6341 (Abstract) 116

King, Destiny L., and Demetris Nicolaides, Sandra Moreno Cristobal
- V350 Muscae: RR Lyrae Star Distance Estimate and RRab Reclassification 63

Knote, Matthew
- Characterizing the O'Connell Effect in Kepler Eclipsing Binaries (Abstract) 113

Larsen, Kristine
- Identification of Bimodal Period and Long Secondary Period Carbon Red Giants Misclassified as "Miscellaneous" in VSX (Abstract) 117
- Transits, Spots, and Eclipses: The Sun's Unique Role in Outreach 90

Lester, Jamie, and Rowan Joignant, Mariel Meier
- Photometric Determination of the Distance to the RR Lyrae Star YZ Capricorni 247

Liu, Peter, and Neil Thomas, Kyle Ziegler
- High Cadence Millimagnitude Photometric Observation of V1112 Persei (Nova Per 2020) 151

Ludington, E. Whit, and Edward O. Wiley
- Times of Minima for Eclipsing Binaries 2017–2020 from Stellar Skies Observatories and 2004–2009 SuperWasp Data Mining 106

Ly, Dylan, in Ramy Mizrachi *et al.*
- 25 New Light Curves and Updated Ephemeris through Analysis of Exoplanet WASP-50 b with EXOTIC 186

Maccarone, Thomas J.
- A Historical Perspective on the Diversity of Explanations for New Classes of Transient and Variable Stars 83

Maccarone, Thomas J., in Eric Masington *et al.*
- An Analysis of X-Ray Hardness Ratios between Asynchronous and Non-Asynchronous Polars 149

Madsen, Madelyn
- Simultaneous Photometry on VSX Variables and TESS Exoplanet Candidates (Abstract) 113

Martin, John C., and Bill Rea
- Low Resolution Spectroscopy of Miras 2—R Octantis 2

Masington, Eric, and Thomas J. Maccarone, Liliana Rivera Sandoval, Craig Heinke, Arash Bahramian, Aarran W. Shaw
- An Analysis of X-Ray Hardness Ratios between Asynchronous and Non-Asynchronous Polars 149

McClain, Sol, and Richard Berry, Nolan Sottoway
- New Observations of the SX Phe Star XX Cygni (Abstract) 115

Meier, Mariel, and Jamie Lester, Rowan Joignant
- Photometric Determination of the Distance to the RR Lyrae Star YZ Capricorni 247

Menke, John
- Using Bespoke 18-inch Newtonian and R = 3000 Spectrometer for High-Precision Observations (Abstract) 114

Michaels, Edward J.
- A Photometric Study of the Eclipsing Binary LO Ursae Majoris 221

Miller, Mike
- Establishing a New ToM (Time of Minimum) for the Primary Eclipse of the Binary System WZ Ophiuchi (Abstract) 116

Mizrachi, Ramy, and Dylan Ly, Leon Bewersdorff, Kalée Tock
- 25 New Light Curves and Updated Ephemeris through Analysis of Exoplanet WASP-50 b with EXOTIC 186

Moriarty, David J. W., in Tom Richards *et al.*
- Southern Eclipsing Binary Minima and Light Elements in 2020 251

Morrison, Nancy D.
- Conference Proceedings of the AAVSO 1
- The Range of Content in *JAAVSO* 119

Nickel, Otmar, and Tom Calderwood
- Daylight Photometry of Bright Stars—Observations of Betelgeuse at Solar Conjunction 269

Nicolaides, Demetris, and Destiny L. King, Sandra Moreno Cristobal
- V350 Muscae: RR Lyrae Star Distance Estimate and RRab Reclassification 63

Odell, Andrew P., and Joel A. Eaton, Gary W. Steffens
- V963 Persei as a Contact Binary 121

Osborn, Wayne, and Pradip Karmakar, Horace A. Smith
>Types of Period Changes of W Virginis Stars
>(Abstract) 113

Pace, Benedict, and Julian F. West
>A Spectroscopic Study of the Variable Star T Centauri Over a 91-day Cycle, and the Effects of Titanium Oxide on Its Atmosphere 32

Paczkowski, Margaret, and Neil Thomas
>The Confirmation of Three Faint Variable Stars and the Observation of Eleven Others in the Vicinity of Kepler-8b by the Lookout Observatory 12

Parker, Hayden, in Ross Parker et al.
>Distances for the RR Lyrae Stars UU Ceti, UW Gruis, and W Tucanae 178

Parker, Liam, in Ross Parker et al.
>Distances for the RR Lyrae Stars UU Ceti, UW Gruis, and W Tucanae 178

Parker, Ross, and Liam Parker, Hayden Parker, Faraz Uddin, Timothy Banks
>Distances for the RR Lyrae Stars UU Ceti, UW Gruis, and W Tucanae 178

Percy, John R.
>RU Cam: The Reluctant Cepheid Revisited 46

Percy, John R., and Prateek Gupta
>Pulsating Red Giants in a Globular Cluster: 47 Tucanae 209

Pollmann, Ernst
>The Correlation between H? and HeI 6678 Emission Activity in the Be Star ? Cassiopeiae from 1995 to 2021 77

Pyatnytskyy, Maksym
>Refining Ephemeris and Estimating Period Change Rate for V965 Cephei 58

Ray, Jacob, in Ronald G. Samec et al.
>Binaries with Mass Ratios Near Unity: The First BVRI Observations, Analysis and Period Studies of TX Canis Minoris and DW Canis Minoris 138

Rea, Bill, and John C. Martin
>Low Resolution Spectroscopy of Miras 2—R Octantis 2

Reid, Deshawn, and Donald Smith
>Automating a Small Urban College Observatory (Abstract) 114

Richards, Thomas J.
>Researching Eclipsing Binaries "Down Under": Illustrating the Methods and Results of Variable Stars South (Abstract) 115

Richards, Tom, and Roy A. Axelsen, Mark Blackford, Robert Jenkins, David J. W. Moriarty
>Southern Eclipsing Binary Minima and Light Elements in 2020 251

Richie, Helena M.
>Disk Instabilities Caused the 2018 Outburst of AG Draconis (Abstract) 114

Rocchi, Gianni, in Corrado Spogli et al.
>Photometric Observations of the Dwarf Nova AH Herculis 232

Romanov, Filipp Dmitrievich
>Discoveries of Variable Stars by Amateur Astronomers Using Data Mining on the Example of Eclipsing Binary Romanov V20 (Abstract) 115
>Discovery of Romanov V20, an Algol-Type Eclipsing Binary in the Constellation Centaurus, by Means of Data Mining 130

Rosales, Manny, and Christina Singh, Wyatt Carbonell, Leslie F. Brown, Gary Walker
>The Blazar BL Lacertae: 2018-2020 V-, R-, and I-Band CCD Photometry (Abstract) 116

Rosati, Jacopo, in Corrado Spogli et al.
>Photometric Observations of the Dwarf Nova AH Herculis 232

Saini, Anshita, and Nicholas Walker
>Period Determination and Classification Analysis of 25 Pulsating Red Giants 70

Samec, Ronald G., and Daniel Caton, Jacob Ray, Riley Waddell, Davis Gentry, Danny Faulkner
>Binaries with Mass Ratios Near Unity: The First BVRI Observations, Analysis and Period Studies of TX Canis Minoris and DW Canis Minoris 138

Samolyk, Gerard
>Recent Maxima of 79 Short Period Pulsating Stars 103
>Recent Minima of 218 Eclipsing Binary Stars 265
>Recent Minima of 225 Eclipsing Binary Stars 108

Samolyk, Gerard, and Dylan Kaneshiro, Horace A. Smith
>An Update on the Periods and Period Changes of the Blazhko RR Lyrae Star XZ Cygni 19

Samolyk, Gerard, and Pradip Karmakar
>The Long-term Period Changes of the Cepheid Variable SV Monocerotis 135

Sandoval, Liliana Rivera, in Eric Masington et al.
>An Analysis of X-Ray Hardness Ratios between Asynchronous and Non-Asynchronous Polars 149

Schmidt, Richard E.
>The Photometric Period of V1112 Persei (Nova Per 2020) 99
>The Photometric Period of V1391 Cassiopeiae (Nova Cas 2020) 95
>The Photometric Period of V606 Vulpeculae (Nova Vul 2021) 261

Schmidt, Richard E., and Sergei Yu. Shugarov, Marina D. Afonina
>The Photometric Period of V1674 Herculis (Nova Her 2021) 257

Shaw, Aarran
>Measuring the Masses of White Dwarfs with X-rays: A NuSTAR Legacy Survey (Abstract) 115

Shaw, Aarran W., in Eric Masington et al.
>An Analysis of X-Ray Hardness Ratios between Asynchronous and Non-Asynchronous Polars 149

Shugarov, Sergei Yu., and Richard E. Schmidt, Marina D. Afonina
>The Photometric Period of V1674 Herculis (Nova Her 2021) 257

Singh, Christina, in Manny Rosales *et al.*
    The Blazar BL Lacertae: 2018-2020 V-, R-, and
       I-Band CCD Photometry (Abstract)   116
Sivakoff, Gregory
    The Quick and the Deadtime (Abstract)   113
Smith, Donald, and Deshawn Reid
    Automating a Small Urban College Observatory
       (Abstract)   114
Smith, Horace A., and Dylan Kaneshiro, Gerard Samolyk
    An Update on the Periods and Period Changes of
       the Blazhko RR Lyrae Star XZ Cygni   19
Smith, Horace A., and Pradip Karmakar, Wayne Osborn
    Types of Period Changes of W Virginis Stars
       (Abstract)   113
Sottoway, Nolan, and Richard Berry, Sol McClain
    New Observations of the SX Phe Star XX Cygni
       (Abstract)   115
Spogli, Corrado, and Gianni Rocchi, Stefano Ciprini,
  Dario Vergari, Jacopo Rosati
    Photometric Observations of the Dwarf Nova
       AH Herculis   232
Steffens, Gary W., and Joel A. Eaton, Andrew P. Odell
    V963 Persei as a Contact Binary   121
Szkody, Paula
    GW Lib and V386 Ser: CVs Containing Accreting,
       Pulsating White Dwarfs (Abstract)   116
Thomas, Neil, and Kyle Ziegler, Peter Liu
    High Cadence Millimagnitude Photometric
       Observation of V1112 Persei (Nova Per 2020)   151
Thomas, Neil, and Margaret Paczkowski
    The Confirmation of Three Faint Variable Stars and
       the Observation of Eleven Others in the Vicinity of
       Kepler-8b by the Lookout Observatory   12
Tock, Kalée, and Paul Hamrick, Avni Bansal
    Characterization of NGC 5272, NGC 1904,
       NGC 3201, and Terzan 3   192
Tock, Kalée, in Ramy Mizrachi *et al.*
    25 New Light Curves and Updated Ephemeris
       through Analysis of Exoplanet WASP-50 b
       with EXOTIC   186

Uddin, Faraz, in Ross Parker *et al.*
    Distances for the RR Lyrae Stars UU Ceti, UW Gruis,
       and W Tucanae   178
Vergari, Dario, in Corrado Spogli *et al.*
    Photometric Observations of the Dwarf Nova
       AH Herculis   232
Waddell, Riley, in Ronald G. Samec *et al.*
    Binaries with Mass Ratios Near Unity: The First
       BVRI Observations, Analysis and Period Studies of
       TX Canis Minoris and DW Canis Minoris   138
Walker, Gary, in Manny Rosales *et al.*
    The Blazar BL Lacertae: 2018-2020 V-, R-, and
       I-Band CCD Photometry (Abstract)   116
Walker, Nicholas, and Anshita Saini
    Period Determination and Classification Analysis
       of 25 Pulsating Red Giants   70
West, Julian F., and Benedict Pace
    A Spectroscopic Study of the Variable Star T Centauri
       Over a 91-day Cycle, and the Effects of Titanium
       Oxide on Its Atmosphere   32
Wiley, Edward O., and E. Whit Ludington
    Times of Minima for Eclipsing Binaries 2017–2020
       from Stellar Skies Observatories and 2004–2009
       SuperWasp Data Mining   106
Wiley, Edward O., and Kevin B. Alton
    CCD Photometry, Light Curve Modeling, and Period
       Study of V573 Serpentis, a Totally Eclipsing
       Overcontact Binary System   170
Wink, William
    Star "Crawling" with Astronomical Binoculars
       (Abstract)   116
Ziegler, Kyle, and Neil Thomas, Peter Liu
    High Cadence Millimagnitude Photometric
       Observation of V1112 Persei (Nova Per 2020)   151

Subject

AAVSO

Conference Proceedings of the AAVSO
   Nancy D. Morrison   1
Discoveries of Variable Stars by Amateur Astronomers Using Data Mining on the Example of Eclipsing Binary Romanov V20 (Abstract)
   Filipp Romanov   115
The Photometric Period of V1112 Persei (Nova Per 2020)
   Richard E. Schmidt   99
The Range of Content in *JAAVSO*
   Nancy D. Morrison   119
Recent Maxima of 79 Short Period Pulsating Stars
   Gerard Samolyk   103
Recent Minima of 218 Eclipsing Binary Stars
   Gerard Samolyk   265
Recent Minima of 225 Eclipsing Binary Stars
   Gerard Samolyk   108
RU Cam: The Reluctant Cepheid Revisited
   John R. Percy   46
Transits, Spots, and Eclipses: The Sun's Unique Role in Outreach
   Kristine Larsen   90

AAVSO INTERNATIONAL DATABASE

Discovery of Romanov V20, an Algol-Type Eclipsing Binary in the Constellation Centaurus, by Means of Data Mining
   Filipp Dmitrievich Romanov   130
GW Lib and V386 Ser: CVs Containing Accreting, Pulsating White Dwarfs (Abstract)
   Paula Szkody   116
High Cadence Millimagnitude Photometric Observation of V1112 Persei (Nova Per 2020)
   Neil Thomas, Kyle Ziegler, and Peter Liu   151
Identification of Bimodal Period and Long Secondary Period Carbon Red Giants Misclassified as "Miscellaneous" in VSX (Abstract)
   Kristine Larsen   117
Light Curve Analysis of 185 YSOs: New Periods Discovered for 9 Stars
   Joshua R. Hamilton   49
The Long-term Period Changes of the Cepheid Variable SV Monocerotis
   Pradip Karmakar and Gerard Samolyk   135
Low Resolution Spectroscopy of Miras 2—R Octantis
   Bill Rea and John C. Martin   2
Period Determination and Classification Analysis of 25 Pulsating Red Giants
   Anshita Saini and Nicholas Walker   70
Recent Maxima of 79 Short Period Pulsating Stars
   Gerard Samolyk   103
Recent Minima of 218 Eclipsing Binary Stars
   Gerard Samolyk   265
Recent Minima of 225 Eclipsing Binary Stars
   Gerard Samolyk   108

Refining Ephemeris and Estimating Period Change Rate for V965 Cephei
   Maksym Pyatnytskyy   58
RU Cam: The Reluctant Cepheid Revisited
   John R. Percy   46
A Spectroscopic Study of the Variable Star T Centauri Over a 91-day Cycle, and the Effects of Titanium Oxide on Its Atmosphere
   Julian F. West and Benedict Pace   32
Transits, Spots, and Eclipses: The Sun's Unique Role in Outreach
   Kristine Larsen   90
An Update on the Periods and Period Changes of the Blazhko RR Lyrae Star XZ Cygni
   Dylan Kaneshiro, Horace A. Smith, and Gerard Samolyk   19

AAVSO MEETINGS

Conference Proceedings of the AAVSO
   Nancy D. Morrison   1

AAVSO, JOURNAL OF

Conference Proceedings of the AAVSO
   Nancy D. Morrison   1
Index to Volume 49
   Anon.   277
The Range of Content in *JAAVSO*
   Nancy D. Morrison   119

ACTIVE GALACTIC NUCLEI (AGNs)
[See also EXTRAGALACTIC]

The Blazar BL Lacertae: 2018–2020 V-, R-, and I-Band CCD Photometry (Abstract)
   Manny Rosales *et al.*   116

AMPLITUDE ANALYSIS

Pulsating Red Giants in a Globular Cluster: 47 Tucanae
   John R. Percy and Prateek Gupta   209
RU Cam: The Reluctant Cepheid Revisited
   John R. Percy   46

ARCHAEOASTRONOMY
[See also ASTRONOMY, HISTORY OF]

Morning Star: The Search for and Discovery of the Stars of Bethlehem According to the Gospel of Matthew (Abstract)
   Rev. Kenneth Beckmann   114

ASTRONOMERS, AMATEUR; PROFESSIONAL-AMATEUR COLLABORATION

The Correlation between Hα and HeI 6678 Emission Activity in the Be Star γ Cassiopeiae from 1995 to 2021
   Ernst Pollmann   77
GW Lib and V386 Ser: CVs Containing Accreting, Pulsating White Dwarfs (Abstract)
   Paula Szkody   116

ASTRONOMY, HISTORY OF
[See also ARCHAEOASTRONOMY; OBITUARIES]

A Historical Perspective on the Diversity of Explanations for New Classes of Transient and Variable Stars
Thomas J. Maccarone — 83

Morning Star: The Search for and Discovery of the Stars of Bethlehem According to the Gospel of Matthew (Abstract)
Rev. Kenneth Beckmann — 114

Be STARS [See also VARIABLE STARS (GENERAL)]

The Correlation between Hα and HeI 6678 Emission Activity in the Be Star γ Cassiopeiae from 1995 to 2021
Ernst Pollmann — 77

BLACK HOLES

A Historical Perspective on the Diversity of Explanations for New Classes of Transient and Variable Stars
Thomas J. Maccarone — 83

The Quick and the Deadtime (Abstract)
Gregory Sivakoff — 113

CATACLYSMIC VARIABLES
[See also VARIABLE STARS (GENERAL)]

An Analysis of X-Ray Hardness Ratios between Asynchronous and Non-Asynchronous Polars
Eric Masington et al. — 149

GW Lib and V386 Ser: CVs Containing Accreting, Pulsating White Dwarfs (Abstract)
Paula Szkody — 116

A Historical Perspective on the Diversity of Explanations for New Classes of Transient and Variable Stars
Thomas J. Maccarone — 83

Measuring the Masses of White Dwarfs with X-rays: A NuSTAR Legacy Survey (Abstract)
Aarran Shaw — 115

Photometric Observations of the Dwarf Nova AH Herculis
Corrado Spogli et al. — 232

The Photometric Period of V1674 Herculis (Nova Her 2021)
Richard E. Schmidt, Sergei Yu. Shugarov, and Marina D. Afonina — 257

CATALOGUES, DATABASES, SURVEYS

25 New Light Curves and Updated Ephemeris through Analysis of Exoplanet WASP-50 b with EXOTIC
Ramy Mizrachi et al. — 186

An Analysis of X-Ray Hardness Ratios between Asynchronous and Non-Asynchronous Polars
Eric Masington et al. — 149

Binaries with Mass Ratios Near Unity: The First BVRI Observations, Analysis and Period Studies of TX Canis Minoris and DW Canis Minoris
Ronald G. Samec et al. — 138

CCD Photometry, Light Curve Modeling, and Period Study of GSC 2624-0941, a Totally Eclipsing Overcontact Binary System
Kevin B. Alton and John C. Downing — 214

CCD Photometry, Light Curve Modeling, and Period Study of V573 Serpentis, a Totally Eclipsing Overcontact Binary System
Kevin B. Alton and Edward O. Wiley — 170

Characterizing the O'Connell Effect in Kepler Eclipsing Binaries (Abstract)
Matthew Knote — 113

The Confirmation of Three Faint Variable Stars and the Observation of Eleven Others in the Vicinity of Kepler-8b by the Lookout Observatory
Neil Thomas and Margaret Paczkowski — 12

Discoveries of Variable Stars by Amateur Astronomers Using Data Mining on the Example of Eclipsing Binary Romanov V20 (Abstract)
Filipp Romanov — 115

Discovery of Romanov V20, an Algol-Type Eclipsing Binary in the Constellation Centaurus, by Means of Data Mining
Filipp Dmitrievich Romanov — 130

Distances for the RR Lyrae Stars UU Ceti, UW Gruis, and W Tucanae
Ross Parker et al. — 178

Erratum: Four New Variable Stars in the Field of KELT-16
Daniel J. Brossard and Ronald H. Kaitchuck — 276

Establishing a New ToM (Time of Minimum) for the Primary Eclipse of the Binary System WZ Ophiuchi (Abstract)
Mike Miller — 116

Four New Variable Stars in the Field of KELT-16
Daniel J. Brossard and Ronald H. Kaitchuck — 24

High Cadence Millimagnitude Photometric Observation of V1112 Persei (Nova Per 2020)
Neil Thomas, Kyle Ziegler, and Peter Liu — 151

A Historical Perspective on the Diversity of Explanations for New Classes of Transient and Variable Stars
Thomas J. Maccarone — 83

Identification of Bimodal Period and Long Secondary Period Carbon Red Giants Misclassified as "Miscellaneous" in VSX (Abstract)
Kristine Larsen — 117

Light Curve Analysis of 185 YSOs: New Periods Discovered for 9 Stars
Joshua R. Hamilton — 49

The Long-term Period Changes of the Cepheid Variable SV Monocerotis
Pradip Karmakar and Gerard Samolyk — 135

Low Resolution Spectroscopy of Miras 2—R Octantis
Bill Rea and John C. Martin — 2

Measuring the Masses of White Dwarfs with X-rays:
  A NuSTAR Legacy Survey (Abstract)
    Aarran Shaw   115
Period Determination and Classification Analysis of
  25 Pulsating Red Giants
    Anshita Saini and Nicholas Walker   70
Photometric Determination of the Distance to the
  RR Lyrae Star YZ Capricorni
    Jamie Lester, Rowan Joignant, and Mariel Meier   247
A Photometric Study of the Eclipsing Binary
  LO Ursae Majoris
    Edward J. Michaels   221
Pulsating Red Giants in a Globular Cluster:
  47 Tucanae
    John R. Percy and Prateek Gupta   209
Recent Minima of 225 Eclipsing Binary Stars
    Gerard Samolyk   108
Refining Ephemeris and Estimating Period Change
  Rate for V965 Cephei
    Maksym Pyatnytskyy   58
Retraction of and Re-analysis of the Data from
  "HD 121620: A Previously Unreported Variable Star
  with Unusual Properties"
    Roy A. Axelsen   197
RU Cam: The Reluctant Cepheid Revisited
    John R. Percy   46
Southern Eclipsing Binary Minima and Light
  Elements in 2020
    Tom Richards *et al.*   251
Spectroscopic and Photometric Study of the Mira Stars
  SU Camelopardalis and RY Cephei
    David Boyd   157
Times of Minima for Eclipsing Binaries 2017–2020
  from Stellar Skies Observatories and 2004–2009
  SuperWasp Data Mining
    Edward O. Wiley and E. Whit Ludington   106
Transits, Spots, and Eclipses: The Sun's Unique Role
  in Outreach
    Kristine Larsen   90
An Update on the Periods and Period Changes of the
  Blazhko RR Lyrae Star XZ Cygni
    Dylan Kaneshiro, Horace A. Smith, and
    Gerard Samolyk   19
V350 Muscae: RR Lyrae Star Distance Estimate and
  RRab Reclassification
    Demetris Nicolaides, Destiny L. King, and
    Sandra Moreno Cristobal   63
V963 Persei as a Contact Binary
    Joel A. Eaton, Gary W. Steffens, and
    Andrew P. Odell   121

CEPHEID VARIABLES
[See also VARIABLE STARS (GENERAL)]
A Historical Perspective on the Diversity of
  Explanations for New Classes of Transient and
  Variable Stars
    Thomas J. Maccarone   83

The Long-term Period Changes of the Cepheid
  Variable SV Monocerotis
    Pradip Karmakar and Gerard Samolyk   135
RU Cam: The Reluctant Cepheid Revisited
    John R. Percy   46
Types of Period Changes of W Virginis Stars (Abstract)
    Pradip Karmakar, Horace A. Smith, and
    Wayne Osborn   113
Using High Resolution Spectroscopy to Measure
  Cepheid Pulsation (Abstract)
    Kevin Gurney   117

CLUSTERS, GLOBULAR
Characterization of NGC 5272, NGC 1904,
  NGC 3201, and Terzan 3
    Paul Hamrick, Avni Bansal, and Kalée Tock   192
Differential Photometry of Eclipsing Binary System
  V798 Her in Globular Cluster NGC 6341 (Abstract)
    Khola Anees, Shaukat Naaman Goderya, and
    Fazeel Mahmood Khan   116
Pulsating Red Giants in a Globular Cluster:
  47 Tucanae
    John R. Percy and Prateek Gupta   209
Types of Period Changes of W Virginis Stars
  (Abstract)
    Pradip Karmakar, Horace A. Smith, and
    Wayne Osborn   113

COMPUTERS; SOFTWARE;
INTERNET, WORLD WIDE WEB
Identification of Bimodal Period and Long Secondary
  Period Carbon Red Giants Misclassified as
  "Miscellaneous" in VSX (Abstract)
    Kristine Larsen   117
Simultaneous Photometry on VSX Variables and
  TESS Exoplanet Candidates (Abstract)
    Madelyn Madsen   113

COORDINATED OBSERVATIONS
[MULTI-SITE, MULTI-WAVELENGTH OBSERVATIONS]
CCD Photometry, Light Curve Modeling and Period
  Study of V1073 Herculis, a Totally Eclipsing
  Overcontact Binary System
    Kevin B. Alton and John C. Downing   201
Researching Eclipsing Binaries "Down Under":
  Illustrating the Methods and Results of Variable Stars
  South (Abstract)
    Thomas J. Richards   115

DATA MANAGEMENT [See also AAVSO; COMPUTERS]
Simultaneous Photometry on VSX Variables and
  TESS Exoplanet Candidates (Abstract)
    Madelyn Madsen   113

DATA MINING

Discoveries of Variable Stars by Amateur Astronomers Using Data Mining on the Example of Eclipsing Binary Romanov V20 (Abstract)
  Filipp Romanov — 115

Discovery of Romanov V20, an Algol-Type Eclipsing Binary in the Constellation Centaurus, by Means of Data Mining
  Filipp Dmitrievich Romanov — 130

Pulsating Red Giants in a Globular Cluster: 47 Tucanae
  John R. Percy and Prateek Gupta — 209

Times of Minima for Eclipsing Binaries 2017–2020 from Stellar Skies Observatories and 2004–2009 SuperWasp Data Mining
  Edward O. Wiley and E. Whit Ludington — 106

DATA REDUCTION

High Cadence Millimagnitude Photometric Observation of V1112 Persei (Nova Per 2020)
  Neil Thomas, Kyle Ziegler, and Peter Liu — 151

Retraction of and Re-analysis of the Data from "HD 121620: A Previously Unreported Variable Star with Unusual Properties"
  Roy A. Axelsen — 197

DELTA SCUTI STARS
[See also VARIABLE STARS (GENERAL)]

Recent Maxima of 79 Short Period Pulsating Stars
  Gerard Samolyk — 103

Refining Ephemeris and Estimating Period Change Rate for V965 Cephei
  Maksym Pyatnytskyy — 58

Search for Variability in 30 Bright Metallic-line A Stars Observed by the TESS Spacecraft (Abstract)
  Joyce A. Guzik — 113

ECLIPSING BINARIES
[See also VARIABLE STARS (GENERAL)]

Binaries with Mass Ratios Near Unity: The First BVRI Observations, Analysis and Period Studies of TX Canis Minoris and DW Canis Minoris
  Ronald G. Samec et al. — 138

CCD Photometry, Light Curve Modeling, and Period Study of GSC 2624-0941, a Totally Eclipsing Overcontact Binary System
  Kevin B. Alton and John C. Downing — 214

CCD Photometry, Light Curve Modeling and Period Study of V1073 Herculis, a Totally Eclipsing Overcontact Binary System
  Kevin B. Alton and John C. Downing — 201

CCD Photometry, Light Curve Modeling, and Period Study of V573 Serpentis, a Totally Eclipsing Overcontact Binary System
  Kevin B. Alton and Edward O. Wiley — 170

Characterizing the O'Connell Effect in Kepler Eclipsing Binaries (Abstract)
  Matthew Knote — 113

The Confirmation of Three Faint Variable Stars and the Observation of Eleven Others in the Vicinity of Kepler-8b by the Lookout Observatory
  Neil Thomas and Margaret Paczkowski — 12

Differential Photometry of Eclipsing Binary System V798 Her in Globular Cluster NGC 6341 (Abstract)
  Khola Anees, Shaukat Naaman Goderya, and Fazeel Mahmood Khan — 116

Discoveries of Variable Stars by Amateur Astronomers Using Data Mining on the Example of Eclipsing Binary Romanov V20 (Abstract)
  Filipp Romanov — 115

Discovery of Romanov V20, an Algol-Type Eclipsing Binary in the Constellation Centaurus, by Means of Data Mining
  Filipp Dmitrievich Romanov — 130

Erratum: Four New Variable Stars in the Field of KELT-16
  Daniel J. Brossard and Ronald H. Kaitchuck — 276

Establishing a New ToM (Time of Minimum) for the Primary Eclipse of the Binary System WZ Ophiuchi (Abstract)
  Mike Miller — 116

Four New Variable Stars in the Field of KELT-16
  Daniel J. Brossard and Ronald H. Kaitchuck — 24

A Photometric Study of the Eclipsing Binary LO Ursae Majoris
  Edward J. Michaels — 221

Recent Minima of 218 Eclipsing Binary Stars
  Gerard Samolyk — 265

Recent Minima of 225 Eclipsing Binary Stars
  Gerard Samolyk — 108

Refining Ephemeris and Estimating Period Change Rate for V965 Cephei
  Maksym Pyatnytskyy — 58

Search for Variability in 30 Bright Metallic-line A Stars Observed by the TESS Spacecraft (Abstract)
  Joyce A. Guzik — 113

Southern Eclipsing Binary Minima and Light Elements in 2020
  Tom Richards et al. — 251

Times of Minima for Eclipsing Binaries 2017–2020 from Stellar Skies Observatories and 2004–2009 SuperWasp Data Mining
  Edward O. Wiley and E. Whit Ludington — 106

Using Bespoke 18-inch Newtonian and R = 3000 Spectrometer for High-Precision Observations (Abstract)
  John Menke — 114

V963 Persei as a Contact Binary
  Joel A. Eaton, Gary W. Steffens, and Andrew P. Odell — 121

EDITORIAL

Conference Proceedings of the AAVSO
  Nancy D. Morrison — 1

The Range of Content in *JAAVSO*
  Nancy D. Morrison — 119

EDUCATION
Transits, Spots, and Eclipses: The Sun's Unique Role in Outreach
    Kristine Larsen   90

EDUCATION, VARIABLE STARS IN
Automating a Small Urban College Observatory (Abstract)
    Donald Smith and Deshawn Reid   114
New Observations of the SX Phe Star XX Cygni (Abstract)
    Richard Berry, Nolan Sottoway, and Sol McClain   115
Pulsating Red Giants in a Globular Cluster: 47 Tucanae
    John R. Percy and Prateek Gupta   209
The Range of Content in *JAAVSO*
    Nancy D. Morrison   119
Transits, Spots, and Eclipses: The Sun's Unique Role in Outreach
    Kristine Larsen   90

ERRATA
Erratum: Four New Variable Stars in the Field of KELT-16
    Daniel J. Brossard and Ronald H. Kaitchuck   276
Retraction of and Re-analysis of the Data from "HD 121620: A Previously Unreported Variable Star with Unusual Properties"
    Roy A. Axelsen   197

EVOLUTION, STELLAR
Distances for the RR Lyrae Stars UU Ceti, UW Gruis, and W Tucanae
    Ross Parker *et al.*   178
A Historical Perspective on the Diversity of Explanations for New Classes of Transient and Variable Stars
    Thomas J. Maccarone   83
Measuring the Masses of White Dwarfs with X-rays: A NuSTAR Legacy Survey (Abstract)
    Aarran Shaw   115

EXTRAGALACTIC
A Historical Perspective on the Diversity of Explanations for New Classes of Transient and Variable Stars
    Thomas J. Maccarone   83

EXTRAGALACTIC VARIABLE STARS
A Historical Perspective on the Diversity of Explanations for New Classes of Transient and Variable Stars
    Thomas J. Maccarone   83

GALAXIES
A Historical Perspective on the Diversity of Explanations for New Classes of Transient and Variable Stars
    Thomas J. Maccarone   83

GAMMA CASSIOPEIAE VARIABLES
[See also VARIABLE STARS (GENERAL)]
The Correlation between Hα and HeI 6678 Emission Activity in the Be Star γ Cassiopeiae from 1995 to 2021
    Ernst Pollmann   77

GAMMA-RAY BURSTS; GAMMA-RAY EMISSION
The Blazar BL Lacertae: 2018–2020 V-, R-, and I-Band CCD Photometry (Abstract)
    Manny Rosales *et al.*   116
A Historical Perspective on the Diversity of Explanations for New Classes of Transient and Variable Stars
    Thomas J. Maccarone   83

GIANTS, RED
Identification of Bimodal Period and Long Secondary Period Carbon Red Giants Misclassified as "Miscellaneous" in VSX (Abstract)
    Kristine Larsen   117
Period Determination and Classification Analysis of 25 Pulsating Red Giants
    Anshita Saini and Nicholas Walker   70
Pulsating Red Giants in a Globular Cluster: 47 Tucanae
    John R. Percy and Prateek Gupta   209
Spectroscopic and Photometric Study of the Mira Stars SU Camelopardalis and RY Cephei
    David Boyd   157

INDEX, INDICES
Index to Volume 49
    Anon.   277

INSTRUMENTATION
[See also CCD; VARIABLE STAR OBSERVING]
Automating a Small Urban College Observatory (Abstract)
    Donald Smith and Deshawn Reid   114
The Confirmation of Three Faint Variable Stars and the Observation of Eleven Others in the Vicinity of Kepler-8b by the Lookout Observatory
    Neil Thomas and Margaret Paczkowski   12
Daylight Photometry of Bright Stars—Observations of Betelgeuse at Solar Conjunction
    Otmar Nickel and Tom Calderwood   269
The Quick and the Deadtime (Abstract)
    Gregory Sivakoff   113
The Range of Content in *JAAVSO*
    Nancy D. Morrison   119
Simultaneous Photometry on VSX Variables and TESS Exoplanet Candidates (Abstract)
    Madelyn Madsen   113
Star "Crawling" with Astronomical Binoculars (Abstract)
    William Wink   116

Using Bespoke 18-inch Newtonian and R = 3000 Spectrometer for High-Precision Observations (Abstract)
John Menke — 114

Using High Resolution Spectroscopy to Measure Cepheid Pulsation (Abstract)
Kevin Gurney — 117

IRREGULAR VARIABLES
[See also VARIABLE STARS (GENERAL)]
Pulsating Red Giants in a Globular Cluster: 47 Tucanae
John R. Percy and Prateek Gupta — 209

LIGHT POLLUTION
Automating a Small Urban College Observatory (Abstract)
Donald Smith and Deshawn Reid — 114

MIRA VARIABLES
[See also VARIABLE STARS (GENERAL)]
Light Curve Analysis of 185 YSOs: New Periods Discovered for 9 Stars
Joshua R. Hamilton — 49

Period Determination and Classification Analysis of 25 Pulsating Red Giants
Anshita Saini and Nicholas Walker — 70

Spectroscopic and Photometric Study of the Mira Stars SU Camelopardalis and RY Cephei
David Boyd — 157

A Spectroscopic Study of the Variable Star T Centauri Over a 91-day Cycle, and the Effects of Titanium Oxide on Its Atmosphere
Julian F. West and Benedict Pace — 32

MODELS, STELLAR
Binaries with Mass Ratios Near Unity: The First BVRI Observations, Analysis and Period Studies of TX Canis Minoris and DW Canis Minoris
Ronald G. Samec et al. — 138

CCD Photometry, Light Curve Modeling, and Period Study of GSC 2624-0941, a Totally Eclipsing Overcontact Binary System
Kevin B. Alton and John C. Downing — 214

CCD Photometry, Light Curve Modeling and Period Study of V1073 Herculis, a Totally Eclipsing Overcontact Binary System
Kevin B. Alton and John C. Downing — 201

The Correlation between Hα and HeI 6678 Emission Activity in the Be Star γ Cassiopeiae from 1995 to 2021
Ernst Pollmann — 77

GW Lib and V386 Ser: CVs Containing Accreting, Pulsating White Dwarfs (Abstract)
Paula Szkody — 116

A Historical Perspective on the Diversity of Explanations for New Classes of Transient and Variable Stars
Thomas J. Maccarone — 83

Photometric Observations of the Dwarf Nova AH Herculis
Corrado Spogli et al. — 232

A Photometric Study of the Eclipsing Binary LO Ursae Majoris
Edward J. Michaels — 221

Search for Variability in 30 Bright Metallic-line A Stars Observed by the TESS Spacecraft (Abstract)
Joyce A. Guzik — 113

A Spectroscopic Study of the Variable Star T Centauri Over a 91-day Cycle, and the Effects of Titanium Oxide on Its Atmosphere
Julian F. West and Benedict Pace — 32

Types of Period Changes of W Virginis Stars (Abstract)
Pradip Karmakar, Horace A. Smith, and Wayne Osborn — 113

V350 Muscae: RR Lyrae Star Distance Estimate and RRab Reclassification
Demetris Nicolaides, Destiny L. King, and Sandra Moreno Cristobal — 63

V963 Persei as a Contact Binary
Joel A. Eaton, Gary W. Steffens, and Andrew P. Odell — 121

MULTI-SITE OBSERVATIONS
[See COORDINATED OBSERVATIONS]

MULTI-WAVELENGTH OBSERVATIONS
[See also COORDINATED OBSERVATIONS]
The Blazar BL Lacertae: 2018–2020 V-, R-, and I-Band CCD Photometry (Abstract)
Manny Rosales et al. — 116

GW Lib and V386 Ser: CVs Containing Accreting, Pulsating White Dwarfs (Abstract)
Paula Szkody — 116

NOVAE, HISTORICAL
A Historical Perspective on the Diversity of Explanations for New Classes of Transient and Variable Stars
Thomas J. Maccarone — 83

NOVAE; RECURRENT NOVAE; NOVA-LIKE
[See also CATACLYSMIC VARIABLES]
High Cadence Millimagnitude Photometric Observation of V1112 Persei (Nova Per 2020)
Neil Thomas, Kyle Ziegler, and Peter Liu — 151

A Historical Perspective on the Diversity of Explanations for New Classes of Transient and Variable Stars
Thomas J. Maccarone — 83

The Photometric Period of V1112 Persei (Nova Per 2020)
Richard E. Schmidt — 99

The Photometric Period of V1391 Cassiopeiae
(Nova Cas 2020)
  Richard E. Schmidt   95
The Photometric Period of V1674 Herculis
(Nova Her 2021)
  Richard E. Schmidt, Sergei Yu. Shugarov, and
    Marina D. Afonina   257
The Photometric Period of V606 Vulpeculae
(Nova Vul 2021)
  Richard E. Schmidt   261

OBSERVATORIES

25 New Light Curves and Updated Ephemeris
  through Analysis of Exoplanet WASP-50 b with
  EXOTIC
    Ramy Mizrachi et al.   186
Automating a Small Urban College Observatory
  (Abstract)
    Donald Smith and Deshawn Reid   114
The Confirmation of Three Faint Variable Stars and
  the Observation of Eleven Others in the Vicinity of
  Kepler-8b by the Lookout Observatory
    Neil Thomas and Margaret Paczkowski   12

PERIOD ANALYSIS; PERIOD CHANGES

25 New Light Curves and Updated Ephemeris
  through Analysis of Exoplanet WASP-50 b with
  EXOTIC
    Ramy Mizrachi et al.   186
Binaries with Mass Ratios Near Unity: The First BVRI
  Observations, Analysis and Period Studies of
  TX Canis Minoris and DW Canis Minoris
    Ronald G. Samec et al.   138
CCD Photometry, Light Curve Modeling, and Period
  Study of GSC 2624-0941, a Totally Eclipsing
  Overcontact Binary System
    Kevin B. Alton and John C. Downing   214
CCD Photometry, Light Curve Modeling and Period
  Study of V1073 Herculis, a Totally Eclipsing
  Overcontact Binary System
    Kevin B. Alton and John C. Downing   201
CCD Photometry, Light Curve Modeling, and Period
  Study of V573 Serpentis, a Totally Eclipsing
  Overcontact Binary System
    Kevin B. Alton and Edward O. Wiley   170
The Confirmation of Three Faint Variable Stars and
  the Observation of Eleven Others in the Vicinity of
  Kepler-8b by the Lookout Observatory
    Neil Thomas and Margaret Paczkowski   12
Differential Photometry of Eclipsing Binary System
  V798 Her in Globular Cluster NGC 6341 (Abstract)
    Khola Anees, Shaukat Naaman Goderya, and
      Fazeel Mahmood Khan   116
Discovery of Romanov V20, an Algol-Type Eclipsing
  Binary in the Constellation Centaurus, by Means of
  Data Mining
    Filipp Dmitrievich Romanov   130
Disk Instabilities Caused the 2018 Outburst of
  AG Draconis (Abstract)
    Helena M. Richie   114
Erratum: Four New Variable Stars in the Field of
  KELT-16
    Daniel J. Brossard and Ronald H. Kaitchuck   276
Establishing a New ToM (Time of Minimum) for the
  Primary Eclipse of the Binary System
  WZ Ophiuchi (Abstract)
    Mike Miller   116
Four New Variable Stars in the Field of KELT-16
  Daniel J. Brossard and Ronald H. Kaitchuck   24
High Cadence Millimagnitude Photometric
  Observation of V1112 Persei (Nova Per 2020)
    Neil Thomas, Kyle Ziegler, and Peter Liu   151
Identification of Bimodal Period and Long Secondary
  Period Carbon Red Giants Misclassified as
  "Miscellaneous" in VSX (Abstract)
    Kristine Larsen   117
Light Curve Analysis of 185 YSOs: New Periods
  Discovered for 9 Stars
    Joshua R. Hamilton   49
The Long-term Period Changes of the Cepheid Variable
  SV Monocerotis
    Pradip Karmakar and Gerard Samolyk   135
New Observations of the SX Phe Star XX Cygni
  (Abstract)
    Richard Berry, Nolan Sottoway, and Sol McClain   115
Period Determination and Classification Analysis of
  25 Pulsating Red Giants
    Anshita Saini and Nicholas Walker   70
Photometric Determination of the Distance to the
  RR Lyrae Star YZ Capricorni
    Jamie Lester, Rowan Joignant, and Mariel Meier   247
The Photometric Period of V1112 Persei
  (Nova Per 2020)
    Richard E. Schmidt   99
The Photometric Period of V1391 Cassiopeiae
  (Nova Cas 2020)
    Richard E. Schmidt   95
The Photometric Period of V1674 Herculis
  (Nova Her 2021)
    Richard E. Schmidt, Sergei Yu. Shugarov, and
      Marina D. Afonina   257
The Photometric Period of V606 Vulpeculae
  (Nova Vul 2021)
    Richard E. Schmidt   261
A Photometric Study of the Eclipsing Binary
  LO Ursae Majoris
    Edward J. Michaels   221
Pulsating Red Giants in a Globular Cluster:
  47 Tucanae
    John R. Percy and Prateek Gupta   209
Recent Maxima of 79 Short Period Pulsating Stars
  Gerard Samolyk   103
Recent Minima of 218 Eclipsing Binary Stars
  Gerard Samolyk   265

Recent Minima of 225 Eclipsing Binary Stars
  Gerard Samolyk   108
Refining Ephemeris and Estimating Period Change
  Rate for V965 Cephei
  Maksym Pyatnytskyy   58
Retraction of and Re-analysis of the Data from
  "HD 121620: A Previously Unreported Variable Star
  with Unusual Properties"
  Roy A. Axelsen   197
RU Cam: The Reluctant Cepheid Revisited
  John R. Percy   46
Search for Variability in 30 Bright Metallic-line
  A Stars Observed by the TESS Spacecraft (Abstract)
  Joyce A. Guzik   113
Southern Eclipsing Binary Minima and Light Elements
  in 2020
  Tom Richards et al.   251
Spectroscopic and Photometric Study of the Mira Stars
  SU Camelopardalis and RY Cephei
  David Boyd   157
Times of Minima for Eclipsing Binaries 2017–2020
  from Stellar Skies Observatories and 2004–2009
  SuperWasp Data Mining
  Edward O. Wiley and E. Whit Ludington   106
Types of Period Changes of W Virginis Stars
  (Abstract)
  Pradip Karmakar, Horace A. Smith, and
    Wayne Osborn   113
An Update on the Periods and Period Changes of the
  Blazhko RR Lyrae Star XZ Cygni
  Dylan Kaneshiro, Horace A. Smith, and
    Gerard Samolyk   19
V350 Muscae: RR Lyrae Star Distance Estimate and
  RRab Reclassification
  Demetris Nicolaides, Destiny L. King, and
    Sandra Moreno Cristobal   63
V963 Persei as a Contact Binary
  Joel A. Eaton, Gary W. Steffens, and
    Andrew P. Odell   121

PHOTOMETRY, CCD

Binaries with Mass Ratios Near Unity: The First BVRI
  Observations, Analysis and Period Studies of
  TX Canis Minoris and DW Canis Minoris
  Ronald G. Samec et al.   138
The Blazar BL Lacertae: 2018–2020 V-, R-, and
  I-Band CCD Photometry (Abstract)
  Manny Rosales et al.   116
CCD Photometry, Light Curve Modeling, and Period
  Study of GSC 2624-0941, a Totally Eclipsing
  Overcontact Binary System
  Kevin B. Alton and John C. Downing   214
CCD Photometry, Light Curve Modeling and Period
  Study of V1073 Herculis, a Totally Eclipsing
  Overcontact Binary System
  Kevin B. Alton and John C. Downing   201
CCD Photometry, Light Curve Modeling, and Period
  Study of V573 Serpentis, a Totally Eclipsing
  Overcontact Binary System
  Kevin B. Alton and Edward O. Wiley   170
Characterization of NGC 5272, NGC 1904,
  NGC 3201, and Terzan 3
  Paul Hamrick, Avni Bansal, and Kalée Tock   192
The Confirmation of Three Faint Variable Stars and
  the Observation of Eleven Others in the Vicinity of
  Kepler-8b by the Lookout Observatory
  Neil Thomas and Margaret Paczkowski   12
Daylight Photometry of Bright Stars—Observations
  of Betelgeuse at Solar Conjunction
  Otmar Nickel and Tom Calderwood   269
Differential Photometry of Eclipsing Binary System
  V798 Her in Globular Cluster NGC 6341 (Abstract)
  Khola Anees, Shaukat Naaman Goderya, and
    Fazeel Mahmood Khan   116
Discovery of Romanov V20, an Algol-Type Eclipsing
  Binary in the Constellation Centaurus, by Means of
  Data Mining
  Filipp Dmitrievich Romanov   130
Disk Instabilities Caused the 2018 Outburst of
  AG Draconis (Abstract)
  Helena M. Richie   114
Distances for the RR Lyrae Stars UU Ceti, UW Gruis,
  and W Tucanae
  Ross Parker et al.   178
Erratum: Four New Variable Stars in the Field of
  KELT-16
  Daniel J. Brossard and Ronald H. Kaitchuck   276
Establishing a New ToM (Time of Minimum) for the
  Primary Eclipse of the Binary System
  WZ Ophiuchi (Abstract)
  Mike Miller   116
Four New Variable Stars in the Field of KELT-16
  Daniel J. Brossard and Ronald H. Kaitchuck   24
High Cadence Millimagnitude Photometric
  Observation of V1112 Persei (Nova Per 2020)
  Neil Thomas, Kyle Ziegler, and Peter Liu   151
Light Curve Analysis of 185 YSOs: New Periods
  Discovered for 9 Stars
  Joshua R. Hamilton   49
The Long-term Period Changes of the Cepheid
  Variable SV Monocerotis
  Pradip Karmakar and Gerard Samolyk   135
Low Resolution Spectroscopy of Miras 2—R Octantis
  Bill Rea and John C. Martin   2
New Observations of the SX Phe Star XX Cygni
  (Abstract)
  Richard Berry, Nolan Sottoway, and Sol McClain   115
Period Determination and Classification Analysis of
  25 Pulsating Red Giants
  Anshita Saini and Nicholas Walker   70
Photometric Determination of the Distance to the
  RR Lyrae Star YZ Capricorni
  Jamie Lester, Rowan Joignant, and Mariel Meier   247

Photometric Observations of the Dwarf Nova
AH Herculis
   Corrado Spogli et al. 232
The Photometric Period of V1112 Persei
(Nova Per 2020)
   Richard E. Schmidt 99
The Photometric Period of V1391 Cassiopeiae
(Nova Cas 2020)
   Richard E. Schmidt 95
The Photometric Period of V1674 Herculis
(Nova Her 2021)
   Richard E. Schmidt, Sergei Yu. Shugarov, and
   Marina D. Afonina 257
The Photometric Period of V606 Vulpeculae
(Nova Vul 2021)
   Richard E. Schmidt 261
A Photometric Study of the Eclipsing Binary
LO Ursae Majoris
   Edward J. Michaels 221
Pulsating Red Giants in a Globular Cluster:
47 Tucanae
   John R. Percy and Prateek Gupta 209
The Quick and the Deadtime (Abstract)
   Gregory Sivakoff 113
Recent Maxima of 79 Short Period Pulsating Stars
   Gerard Samolyk 103
Recent Minima of 218 Eclipsing Binary Stars
   Gerard Samolyk 265
Recent Minima of 225 Eclipsing Binary Stars
   Gerard Samolyk 108
RU Cam: The Reluctant Cepheid Revisited
   John R. Percy 46
Southern Eclipsing Binary Minima and Light Elements
in 2020
   Tom Richards et al. 251
Spectroscopic and Photometric Study of the Mira Stars
SU Camelopardalis and RY Cephei
   David Boyd 157
A Spectroscopic Study of the Variable Star T Centauri
Over a 91-day Cycle, and the Effects of Titanium
Oxide on Its Atmosphere
   Julian F. West and Benedict Pace 32
Times of Minima for Eclipsing Binaries 2017–2020
from Stellar Skies Observatories and 2004–2009
SuperWasp Data Mining
   Edward O. Wiley and E. Whit Ludington 106
An Update on the Periods and Period Changes of the
Blazhko RR Lyrae Star XZ Cygni
   Dylan Kaneshiro, Horace A. Smith, and
   Gerard Samolyk 19
Using High Resolution Spectroscopy to Measure
Cepheid Pulsation (Abstract)
   Kevin Gurney 117
V350 Muscae: RR Lyrae Star Distance Estimate and
RRab Reclassification
   Demetris Nicolaides, Destiny L. King, and
   Sandra Moreno Cristobal 63

V963 Persei as a Contact Binary
   Joel A. Eaton, Gary W. Steffens, and
   Andrew P. Odell 121

PHOTOMETRY, CMOS

High Cadence Millimagnitude Photometric
Observation of V1112 Persei (Nova Per 2020)
   Neil Thomas, Kyle Ziegler, and Peter Liu 151
The Quick and the Deadtime (Abstract)
   Gregory Sivakoff 113

PHOTOMETRY, DSLR

Recent Maxima of 79 Short Period Pulsating Stars
   Gerard Samolyk 103
Retraction of and Re-analysis of the Data from
"HD 121620: A Previously Unreported Variable Star
with Unusual Properties"
   Roy A. Axelsen 197
Star "Crawling" with Astronomical Binoculars
(Abstract)
   William Wink 116

PHOTOMETRY, PHOTOELECTRIC

Daylight Photometry of Bright Stars—Observations
of Betelgeuse at Solar Conjunction
   Otmar Nickel and Tom Calderwood 269
Refining Ephemeris and Estimating Period Change
Rate for V965 Cephei
   Maksym Pyatnytskyy 58
RU Cam: The Reluctant Cepheid Revisited
   John R. Percy 46

PHOTOMETRY, PHOTOGRAPHIC

A Photometric Study of the Eclipsing Binary
LO Ursae Majoris
   Edward J. Michaels 221

PHOTOMETRY, VISUAL

The Long-term Period Changes of the Cepheid
Variable SV Monocerotis
   Pradip Karmakar and Gerard Samolyk 135
Low Resolution Spectroscopy of Miras 2—R Octantis
   Bill Rea and John C. Martin 2
Period Determination and Classification Analysis of
25 Pulsating Red Giants
   Anshita Saini and Nicholas Walker 70
A Photometric Study of the Eclipsing Binary
LO Ursae Majoris
   Edward J. Michaels 258
RU Cam: The Reluctant Cepheid Revisited
   John R. Percy 46
A Spectroscopic Study of the Variable Star T Centauri
Over a 91-day Cycle, and the Effects of Titanium
Oxide on Its Atmosphere
   Julian F. West and Benedict Pace 32

PLANETS, EXTRASOLAR (EXOPLANETS)

25 New Light Curves and Updated Ephemeris through Analysis of Exoplanet WASP-50 b with EXOTIC
  Ramy Mizrachi *et al.* — 186

The Confirmation of Three Faint Variable Stars and the Observation of Eleven Others in the Vicinity of Kepler-8b by the Lookout Observatory
  Neil Thomas and Margaret Paczkowski — 12

Erratum: Four New Variable Stars in the Field of KELT-16
  Daniel J. Brossard and Ronald H. Kaitchuck — 276

Four New Variable Stars in the Field of KELT-16
  Daniel J. Brossard and Ronald H. Kaitchuck — 24

The Range of Content in *JAAVSO*
  Nancy D. Morrison — 119

POETRY, THEATER, DANCE, SOCIETY

Building Connection through Community-Based Astronomy (Abstract)
  Todd Duncan and Erika Dunning — 115

Morning Star: The Search for and Discovery of the Stars of Bethlehem According to the Gospel of Matthew (Abstract)
  Rev. Kenneth Beckmann — 114

PULSATING VARIABLES

Low Resolution Spectroscopy of Miras 2—R Octantis
  Bill Rea and John C. Martin — 2

Period Determination and Classification Analysis of 25 Pulsating Red Giants
  Anshita Saini and Nicholas Walker — 70

Pulsating Red Giants in a Globular Cluster: 47 Tucanae
  John R. Percy and Prateek Gupta — 209

A Spectroscopic Study of the Variable Star T Centauri Over a 91-day Cycle, and the Effects of Titanium Oxide on Its Atmosphere
  Julian F. West and Benedict Pace — 32

RADIAL VELOCITY

Using High Resolution Spectroscopy to Measure Cepheid Pulsation (Abstract)
  Kevin Gurney — 117

V963 Persei as a Contact Binary
  Joel A. Eaton, Gary W. Steffens, and Andrew P. Odell — 121

RADIO ASTRONOMY; RADIO OBSERVATIONS

A Historical Perspective on the Diversity of Explanations for New Classes of Transient and Variable Stars
  Thomas J. Maccarone — 83

REMOTE OBSERVING

The Blazar BL Lacertae: 2018–2020 V-, R-, and I-Band CCD Photometry (Abstract)
  Manny Rosales *et al.* — 116

Discoveries of Variable Stars by Amateur Astronomers Using Data Mining on the Example of Eclipsing Binary Romanov V20 (Abstract)
  Filipp Romanov — 115

RR LYRAE STARS
[See also VARIABLE STARS (GENERAL)]

Characterization of NGC 5272, NGC 1904, NGC 3201, and Terzan 3
  Paul Hamrick, Avni Bansal, and Kalée Tock — 192

Distances for the RR Lyrae Stars UU Ceti, UW Gruis, and W Tucanae
  Ross Parker *et al.* — 178

Photometric Determination of the Distance to the RR Lyrae Star YZ Capricorni
  Jamie Lester, Rowan Joignant, and Mariel Meier — 247

The Range of Content in *JAAVSO*
  Nancy D. Morrison — 119

Recent Maxima of 79 Short Period Pulsating Stars
  Gerard Samolyk — 103

An Update on the Periods and Period Changes of the Blazhko RR Lyrae Star XZ Cygni
  Dylan Kaneshiro, Horace A. Smith, and Gerard Samolyk — 19

V350 Muscae: RR Lyrae Star Distance Estimate and RRab Reclassification
  Demetris Nicolaides, Destiny L. King, and Sandra Moreno Cristobal — 63

RV TAURI STARS
[See also VARIABLE STARS (GENERAL)]

An Analysis of X-Ray Hardness Ratios between Asynchronous and Non-Asynchronous Polars
  Eric Masington *et al.* — 149

A Historical Perspective on the Diversity of Explanations for New Classes of Transient and Variable Stars
  Thomas J. Maccarone — 83

Measuring the Masses of White Dwarfs with X-rays: A NuSTAR Legacy Survey (Abstract)
  Aarran Shaw — 115

SATELLITE OBSERVATIONS

An Analysis of X-Ray Hardness Ratios between Asynchronous and Non-Asynchronous Polars
  Eric Masington *et al.* — 149

The Blazar BL Lacertae: 2018–2020 V-, R-, and I-Band CCD Photometry (Abstract)
  Manny Rosales *et al.* — 116

Characterization of NGC 5272, NGC 1904, NGC 3201, and Terzan 3
  Paul Hamrick, Avni Bansal, anéd Kalée Tock — 192

Characterizing the O'Connell Effect in Kepler Eclipsing Binaries (Abstract)
  Matthew Knote — 113

Index, JAAVSO Volume 48, 2020

Discovery of Romanov V20, an Algol-Type Eclipsing
  Binary in the Constellation Centaurus, by Means of
  Data Mining
    Filipp Dmitrievich Romanov   130
Distances for the RR Lyrae Stars UU Ceti, UW Gruis,
  and W Tucanae
    Ross Parker et al.   178
RU Cam: The Reluctant Cepheid Revisited
    John R. Percy   46
Search for Variability in 30 Bright Metallic-line
  A Stars Observed by the TESS Spacecraft (Abstract)
    Joyce A. Guzik   113
Simultaneous Photometry on VSX Variables and
  TESS Exoplanet Candidates (Abstract)
    Madelyn Madsen   113

SCIENTIFIC WRITING, PUBLICATION OF DATA
Conference Proceedings of the AAVSO
    Nancy D. Morrison   1

SELF-CORRELATION ANALYSIS
Period Determination and Classification Analysis of
  25 Pulsating Red Giants
    Anshita Saini and Nicholas Walker   70

SEMIREGULAR VARIABLES
[See also VARIABLE STARS (GENERAL)]
Daylight Photometry of Bright Stars—Observations
  of Betelgeuse at Solar Conjunction
    Otmar Nickel and Tom Calderwood   269
Light Curve Analysis of 185 YSOs: New Periods
  Discovered for 9 Stars
    Joshua R. Hamilton   49
Low Resolution Spectroscopy of Miras 2—R Octantis
    Bill Rea and John C. Martin   2
Period Determination and Classification Analysis of
  25 Pulsating Red Giants
    Anshita Saini and Nicholas Walker   70
Pulsating Red Giants in a Globular Cluster:
  47 Tucanae
    John R. Percy and Prateek Gupta   209
A Spectroscopic Study of the Variable Star T Centauri
  Over a 91-day Cycle, and the Effects of Titanium
  Oxide on Its Atmosphere
    Julian F. West and Benedict Pace   32

SOLAR
Transits, Spots, and Eclipses: The Sun's Unique Role
  in Outreach
    Kristine Larsen   90

SPECTRA, SPECTROSCOPY
An Analysis of X-Ray Hardness Ratios between
  Asynchronous and Non-Asynchronous Polars
    Eric Masington et al.   149
The Correlation between Hα and HeI 6678
  Emission Activity in the Be Star γ Cassiopeiae from
  1995 to 2021
    Ernst Pollmann   77
Low Resolution Spectroscopy of Miras 2—R Octantis
    Bill Rea and John C. Martin   2
Spectroscopic and Photometric Study of the Mira Stars
  SU Camelopardalis and RY Cephei
    David Boyd   157
A Spectroscopic Study of the Variable Star T Centauri
  Over a 91-day Cycle, and the Effects of Titanium
  Oxide on Its Atmosphere
    Julian F. West and Benedict Pace   32
Using Bespoke 18-inch Newtonian and R = 3000
  Spectrometer for High-Precision Observations
  (Abstract)
    John Menke   114
Using High Resolution Spectroscopy to Measure
  Cepheid Pulsation (Abstract)
    Kevin Gurney   117

SPECTROSCOPIC ANALYSIS
An Analysis of X-Ray Hardness Ratios between
  Asynchronous and Non-Asynchronous Polars
    Eric Masington et al.   149
The Correlation between Hα and HeI 6678 Emission
  Activity in the Be Star γ Cassiopeiae from 1995
  to 2021
    Ernst Pollmann   77
Low Resolution Spectroscopy of Miras 2—R Octantis
    Bill Rea and John C. Martin   2
Spectroscopic and Photometric Study of the Mira Stars
  SU Camelopardalis and RY Cephei
    David Boyd   157
A Spectroscopic Study of the Variable Star T Centauri
  Over a 91-day Cycle, and the Effects of Titanium
  Oxide on Its Atmosphere
    Julian F. West and Benedict Pace   32

STATISTICAL ANALYSIS
25 New Light Curves and Updated Ephemeris
  through Analysis of Exoplanet WASP-50 b with
  EXOTIC
    Ramy Mizrachi et al.   186
Binaries with Mass Ratios Near Unity: The First BVRI
  Observations, Analysis and Period Studies of
  TX Canis Minoris and DW Canis Minoris
    Ronald G. Samec et al.   138
CCD Photometry, Light Curve Modeling, and Period
  Study of GSC 2624-0941, a Totally Eclipsing
  Overcontact Binary System
    Kevin B. Alton and John C. Downing   214
CCD Photometry, Light Curve Modeling and Period
  Study of V1073 Herculis, a Totally Eclipsing
  Overcontact Binary System
    Kevin B. Alton and John C. Downing   201

CCD Photometry, Light Curve Modeling, and Period
Study of V573 Serpentis, a Totally Eclipsing
Overcontact Binary System
    Kevin B. Alton and Edward O. Wiley   170
Characterization of NGC 5272, NGC 1904,
NGC 3201, and Terzan 3
    Paul Hamrick, Avni Bansal, and Kalée Tock   192
Characterizing the O'Connell Effect in Kepler
Eclipsing Binaries (Abstract)
    Matthew Knote   113
Daylight Photometry of Bright Stars—Observations
of Betelgeuse at Solar Conjunction
    Otmar Nickel and Tom Calderwood   269
Distances for the RR Lyrae Stars UU Ceti, UW Gruis,
and W Tucanae
    Ross Parker *et al.*   178
Identification of Bimodal Period and Long Secondary
Period Carbon Red Giants Misclassified as
"Miscellaneous" in VSX (Abstract)
    Kristine Larsen   117
Measuring the Masses of White Dwarfs with X-rays:
A NuSTAR Legacy Survey (Abstract)
    Aarran Shaw   115
Photometric Determination of the Distance to the
RR Lyrae Star YZ Capricorni
    Jamie Lester, Rowan Joignant, and Mariel Meier   247
Photometric Observations of the Dwarf Nova
AH Herculis
    Corrado Spogli *et al.*   232
A Photometric Study of the Eclipsing Binary
LO Ursae Majoris
    Edward J. Michaels   221
V963 Persei as a Contact Binary
    Joel A. Eaton, Gary W. Steffens, and
    Andrew P. Odell   121

SUNSPOTS, SUNSPOT COUNTS
Transits, Spots, and Eclipses: The Sun's Unique Role
in Outreach
    Kristine Larsen   90

SUPERNOVAE
[See also VARIABLE STARS (GENERAL)]
A Historical Perspective on the Diversity of
Explanations for New Classes of Transient and
Variable Stars
    Thomas J. Maccarone   83

SUPERNOVAE, HISTORICAL
A Historical Perspective on the Diversity of
Explanations for New Classes of Transient and
Variable Stars
    Thomas J. Maccarone   83

SX PHOENICIS VARIABLES
[See also VARIABLE STARS (GENERAL)]
New Observations of the SX Phe Star XX Cygni
(Abstract)
    Richard Berry, Nolan Sottoway, and Sol McClain   115
Recent Maxima of 79 Short Period Pulsating Stars
    Gerard Samolyk   103

SYMBIOTIC STARS
[See also VARIABLE STARS (GENERAL)]
Disk Instabilities Caused the 2018 Outburst of
AG Draconis (Abstract)
    Helena M. Richie   114

TRANSITS (EXOPLANET)
25 New Light Curves and Updated Ephemeris
through Analysis of Exoplanet WASP-50 b with
EXOTIC
    Ramy Mizrachi *et al.*   186
The Range of Content in *JAAVSO*
    Nancy D. Morrison   119

UNKNOWN; UNSTUDIED VARIABLES
A Historical Perspective on the Diversity of
Explanations for New Classes of Transient and
Variable Stars
    Thomas J. Maccarone   83
Identification of Bimodal Period and Long Secondary
Period Carbon Red Giants Misclassified as
"Miscellaneous" in VSX (Abstract)
    Kristine Larsen   117
Retraction of and Re-analysis of the Data from
"HD 121620: A Previously Unreported Variable Star
with Unusual Properties"
    Roy A. Axelsen   197
Search for Variability in 30 Bright Metallic-line
A Stars Observed by the TESS Spacecraft (Abstract)
    Joyce A. Guzik   113

VARIABLE STAR OBSERVING ORGANIZATIONS
Researching Eclipsing Binaries "Down Under":
Illustrating the Methods and Results of Variable Stars
South (Abstract)
    Thomas J. Richards   115
Southern Eclipsing Binary Minima and Light Elements
in 2020
    Tom Richards *et al.*   251
Transits, Spots, and Eclipses: The Sun's Unique Role
in Outreach
    Kristine Larsen   90

VARIABLE STAR OBSERVING
[See also INSTRUMENTATION]
Automating a Small Urban College Observatory
(Abstract)
    Donald Smith and Deshawn Reid   114

Building Connection through Community-Based
  Astronomy (Abstract)
    Todd Duncan and Erika Dunning 115
Daylight Photometry of Bright Stars—Observations
  of Betelgeuse at Solar Conjunction
    Otmar Nickel and Tom Calderwood 269
Discoveries of Variable Stars by Amateur Astronomers
  Using Data Mining on the Example of Eclipsing
  Binary Romanov V20 (Abstract)
    Filipp Romanov 115
Researching Eclipsing Binaries "Down Under":
  Illustrating the Methods and Results of Variable Stars
  South (Abstract)
    Thomas J. Richards 115
Simultaneous Photometry on VSX Variables and
  TESS Exoplanet Candidates (Abstract)
    Madelyn Madsen 113
Star "Crawling" with Astronomical Binoculars
  (Abstract)
    William Wink 116
Transits, Spots, and Eclipses: The Sun's Unique Role
  in Outreach
    Kristine Larsen 90
Using Bespoke 18-inch Newtonian and R = 3000
  Spectrometer for High-Precision Observations
  (Abstract)
    John Menke 114
Using High Resolution Spectroscopy to Measure
  Cepheid Pulsation (Abstract)
    Kevin Gurney 117

VARIABLE STARS (GENERAL)

An Analysis of X-Ray Hardness Ratios between
  Asynchronous and Non-Asynchronous Polars
    Eric Masington *et al.* 149
A Historical Perspective on the Diversity of
  Explanations for New Classes of Transient and
  Variable Stars
    Thomas J. Maccarone 83
The Quick and the Deadtime (Abstract)
    Gregory Sivakoff 113
The Range of Content in *JAAVSO*
    Nancy D. Morrison 119
Search for Variability in 30 Bright Metallic-line
  A Stars Observed by the TESS Spacecraft (Abstract)
    Joyce A. Guzik 113
Transits, Spots, and Eclipses: The Sun's Unique Role
  in Outreach
    Kristine Larsen 90

VARIABLE STARS (INDIVIDUAL); OBSERVING TARGETS

[R And] Period Determination and Classification
  Analysis of 25 Pulsating Red Giants
    Anshita Saini and Nicholas Walker 70
[V And] Period Determination and Classification
  Analysis of 25 Pulsating Red Giants
    Anshita Saini and Nicholas Walker 70
[RW And] Period Determination and Classification
  Analysis of 25 Pulsating Red Giants
    Anshita Saini and Nicholas Walker 70
[V1432 Aql] An Analysis of X-Ray Hardness Ratios
  between Asynchronous and Non-Asynchronous Polars
    Eric Masington *et al.* 149
[R Cam] Period Determination and Classification
  Analysis of 25 Pulsating Red Giants
    Anshita Saini and Nicholas Walker 70
[T Cam] Period Determination and Classification
  Analysis of 25 Pulsating Red Giants
    Anshita Saini and Nicholas Walker 70
[RU Cam] RU Cam: The Reluctant Cepheid Revisited
    John R. Percy 46
[SU Cam] Spectroscopic and Photometric Study of the
  Mira Stars SU Camelopardalis and RY Cephei
    David Boyd 157
[TX Cam] Period Determination and Classification
  Analysis of 25 Pulsating Red Giants
    Anshita Saini and Nicholas Walker 70
[AU Cam] Period Determination and Classification
  Analysis of 25 Pulsating Red Giants
    Anshita Saini and Nicholas Walker 70
[BY Cam] An Analysis of X-Ray Hardness Ratios
  between Asynchronous and Non-Asynchronous Polars
    Eric Masington *et al.* 149
[TX CMi] Binaries with Mass Ratios Near Unity:
  The First BVRI Observations, Analysis and Period
  Studies of TX Canis Minoris and DW Canis Minoris
    Ronald G. Samec *et al.* 138
[DW CMi] Binaries with Mass Ratios Near Unity:
  The First BVRI Observations, Analysis and Period
  Studies of TX Canis Minoris and DW Canis Minoris
    Ronald G. Samec *et al.* 138
[YZ Cap] Photometric Determination of the Distance
  to the RR Lyrae Star YZ Capricorni
    Jamie Lester, Rowan Joignant, and Mariel Meier 247
[R Cas] Period Determination and Classification
  Analysis of 25 Pulsating Red Giants
    Anshita Saini and Nicholas Walker 70
[S Cas] Period Determination and Classification
  Analysis of 25 Pulsating Red Giants
    Anshita Saini and Nicholas Walker 70
[T Cas] Period Determination and Classification
  Analysis of 25 Pulsating Red Giants
    Anshita Saini and Nicholas Walker 70
[W Cas] Period Determination and Classification
  Analysis of 25 Pulsating Red Giants
    Anshita Saini and Nicholas Walker 70
[AZ Cas] Using Bespoke 18-inch Newtonian and
  R = 3000 Spectrometer for High-Precision
  Observations (Abstract)
    John Menke 114
[V1391 Cas] The Photometric Period of
  V1391 Cassiopeiae (Nova Cas 2020)
    Richard E. Schmidt 95

[γ Cas] The Correlation between Hα and HeI 6678
    Emission Activity in the Be Star γ Cassiopeiae from
    1995 to 2021
        Ernst Pollmann   77
[T Cen] A Spectroscopic Study of the Variable Star
    T Centauri Over a 91-day Cycle, and the Effects of
    Titanium Oxide on Its Atmosphere
        Julian F. West and Benedict Pace   32
[V834 Cen] An Analysis of X-Ray Hardness Ratios
    between Asynchronous and Non-Asynchronous Polars
        Eric Masington et al.   149
[S Cep] Period Determination and Classification
    Analysis of 25 Pulsating Red Giants
        Anshita Saini and Nicholas Walker   70
[T Cep] Period Determination and Classification
    Analysis of 25 Pulsating Red Giants
        Anshita Saini and Nicholas Walker   70
[X Cep] Period Determination and Classification
    Analysis of 25 Pulsating Red Giants
        Anshita Saini and Nicholas Walker   70
[Y Cep] Period Determination and Classification
    Analysis of 25 Pulsating Red Giants
        Anshita Saini and Nicholas Walker   70
[RY Cep] Spectroscopic and Photometric Study of the
    Mira Stars SU Camelopardalis and RY Cephei
        David Boyd   157
[SV Cep] Light Curve Analysis of 185 YSOs:
    New Periods Discovered for 9 Stars
        Joshua R. Hamilton   49
[V965 Cep] Refining Ephemeris and Estimating
    Period Change Rate for V965 Cephei
        Maksym Pyatnytskyy   58
[UU Cet] Distances for the RR Lyrae Stars UU Ceti,
    UW Gruis, and W Tucanae
        Ross Parker et al.   178
[XX Cyg] New Observations of the SX Phe Star
    XX Cygni (Abstract)
        Richard Berry, Nolan Sottoway, and Sol McClain   115
[XZ Cyg] An Update on the Periods and Period
    Changes of the Blazhko RR Lyrae Star XZ Cygni
        Dylan Kaneshiro, Horace A. Smith, and
        Gerard Samolyk   19
[V561 Cyg] Light Curve Analysis of 185 YSOs:
    New Periods Discovered for 9 Stars
        Joshua R. Hamilton   49
[V1331 Cyg] Light Curve Analysis of 185 YSOs:
    New Periods Discovered for 9 Stars
        Joshua R. Hamilton   49
[V1334 Cyg] Using High Resolution Spectroscopy
    to Measure Cepheid Pulsation (Abstract)
        Kevin Gurney   117
[V1500 Cyg] An Analysis of X-Ray Hardness Ratios
    between Asynchronous and Non-Asynchronous Polars
        Eric Masington et al.   149
[V1515 Cyg] Light Curve Analysis of 185 YSOs:
    New Periods Discovered for 9 Stars
        Joshua R. Hamilton   49

[R Dra] Period Determination and Classification
    Analysis of 25 Pulsating Red Giants
        Anshita Saini and Nicholas Walker   70
[T Dra] Period Determination and Classification
    Analysis of 25 Pulsating Red Giants
        Anshita Saini and Nicholas Walker   70
[U Dra] Period Determination and Classification
    Analysis of 25 Pulsating Red Giants
        Anshita Saini and Nicholas Walker   70
[W Dra] Period Determination and Classification
    Analysis of 25 Pulsating Red Giants
        Anshita Saini and Nicholas Walker   70
[RY Dra] Period Determination and Classification
    Analysis of 25 Pulsating Red Giants
        Anshita Saini and Nicholas Walker   70
[AG Dra] Disk Instabilities Caused the 2018 Outburst
    of AG Draconis (Abstract)
        Helena M. Richie   114
[IW Eri] An Analysis of X-Ray Hardness Ratios
    between Asynchronous and Non-Asynchronous Polars
        Eric Masington et al.   149
[UW Gru] Distances for the RR Lyrae Stars UU Ceti,
    UW Gruis, and W Tucanae
        Ross Parker et al.   178
[AH Her] Photometric Observations of the Dwarf Nova
    AH Herculis
        Corrado Spogli et al.   232
[AM Her] An Analysis of X-Ray Hardness Ratios
    between Asynchronous and Non-Asynchronous Polars
        Eric Masington et al.   149
[V1073 Her] CCD Photometry, Light Curve
    Modeling and Period Study of V1073 Herculis,
    a Totally Eclipsing Overcontact Binary System
        Kevin B. Alton and John C. Downing   201
[V1674 Her] The Photometric Period of
    V1674 Herculis (Nova Her 2021)
        Richard E. Schmidt, Sergei Yu. Shugarov, and
        Marina D. Afonina   257
[V798 Her] Differential Photometry of Eclipsing
    Binary System V798 Her in Globular Cluster
    NGC 6341 (Abstract)
        Khola Anees, Shaukat Naaman Goderya, and
        Fazeel Mahmood Khan   116
[CD Ind] An Analysis of X-Ray Hardness Ratios
    between Asynchronous and Non-Asynchronous Polars
        Eric Masington et al.   149
[BL Lac] The Blazar BL Lacertae: 2018–2020 V-, R-,
    and I-Band CCD Photometry (Abstract)
        Manny Rosales et al.   116
[FY Lac] Light Curve Analysis of 185 YSOs:
    New Periods Discovered for 9 Stars
        Joshua R. Hamilton   49
[GW Lib] GW Lib and V386 Ser: CVs Containing
    Accreting, Pulsating White Dwarfs (Abstract)
        Paula Szkody   116
[MY Lup] Light Curve Analysis of 185 YSOs:
    New Periods Discovered for 9 Stars
        Joshua R. Hamilton   49

[V351 Lyr] The Confirmation of Three Faint Variable
Stars and the Observation of Eleven Others in the
Vicinity of Kepler-8b by the Lookout Observatory
  Neil Thomas and Margaret Paczkowski 12
[SV Mon] The Long-term Period Changes of the
Cepheid Variable SV Monocerotis
  Pradip Karmakar and Gerard Samolyk 135
[V350 Mus] V350 Muscae: RR Lyrae Star Distance
Estimate and RRab Reclassification
  Demetris Nicolaides, Destiny L. King, and
  Sandra Moreno Cristobal 63
[R Oct] Low Resolution Spectroscopy of Miras 2
—R Octantis
  Bill Rea and John C. Martin 2
[WZ Oph] Establishing a New ToM
(Time of Minimum) for the Primary Eclipse of the
Binary System WZ Ophiuchi (Abstract)
  Mike Miller 116
[V2301 Oph] An Analysis of X-Ray Hardness Ratios
between Asynchronous and Non-Asynchronous Polars
  Eric Masington et al. 149
[α Ori] Daylight Photometry of Bright Stars—
Observations of Betelgeuse at Solar Conjunction
  Otmar Nickel and Tom Calderwood 269
[V369 Per] Light Curve Analysis of 185 YSOs:
New Periods Discovered for 9 Stars
  Joshua R. Hamilton 49
[V963 Per] V963 Persei as a Contact Binary
  Joel A. Eaton, Gary W. Steffens, and
  Andrew P. Odell 121
[V1112 Per] High Cadence Millimagnitude
Photometric Observation of V1112 Persei
(Nova Per 2020)
  Neil Thomas, Kyle Ziegler, and Peter Liu 151
[V1112 Per] The Photometric Period of V1112 Persei
(Nova Per 2020)
  Richard E. Schmidt 99
[V386 Ser] GW Lib and V386 Ser: CVs Containing
Accreting, Pulsating White Dwarfs (Abstract)
  Paula Szkody 116
[V573 Ser] CCD Photometry, Light Curve Modeling,
and Period Study of V573 Serpentis, a Totally
Eclipsing Overcontact Binary System
  Kevin B. Alton and Edward O. Wiley 170
[W Tuc] Distances for the RR Lyrae Stars UU Ceti,
UW Gruis, and W Tucanae
  Ross Parker et al. 178
[LO UMa] A Photometric Study of the Eclipsing
Binary LO Ursae Majoris
  Edward J. Michaels 221
[R UMi] Period Determination and Classification
Analysis of 25 Pulsating Red Giants
  Anshita Saini and Nicholas Walker 70
[S UMi] Period Determination and Classification
Analysis of 25 Pulsating Red Giants
  Anshita Saini and Nicholas Walker 70
[T UMi] Period Determination and Classification
Analysis of 25 Pulsating Red Giants
  Anshita Saini and Nicholas Walker 70
[U UMi] Period Determination and Classification
Analysis of 25 Pulsating Red Giants
  Anshita Saini and Nicholas Walker 70
[X UMi] Period Determination and Classification
Analysis of 25 Pulsating Red Giants
  Anshita Saini and Nicholas Walker 70
[WW Vul] Light Curve Analysis of 185 YSOs:
New Periods Discovered for 9 Stars
  Joshua R. Hamilton 49
[CT Vul] Light Curve Analysis of 185 YSOs:
New Periods Discovered for 9 Stars
  Joshua R. Hamilton 49
[V606 Vul] The Photometric Period of
V606 Vulpeculae (Nova Vul 2021)
  Richard E. Schmidt 261
[1RXS 145341.1-552146] An Analysis of X-Ray
Hardness Ratios between Asynchronous and
Non-Asynchronous Polars
  Eric Masington et al. 149
[68 eclipsing binary stars] Times of Minima for
Eclipsing Binaries 2017–2020 from Stellar
Skies Observatories and 2004–2009 SuperWasp
Data Mining
  Edward O. Wiley and E. Whit Ludington 106
[77 eclipsing binary stars] Southern Eclipsing Binary
Minima and Light Elements in 2020
  Tom Richards et al. 251
[218 eclipsing binary stars] Recent Minima of
218 Eclipsing Binary Stars
  Gerard Samolyk 265
[225 eclipsing binary stars] Recent Minima of
225 Eclipsing Binary Stars
  Gerard Samolyk 108
[79 short period pulsator stars] Recent Maxima of
79 Short Period Pulsating Stars
  Gerard Samolyk 103
[2MASS 18441165+4201591] The Confirmation of
Three Faint Variable Stars and the Observation
of Eleven Others in the Vicinity of Kepler-8b by the
Lookout Observatory
  Neil Thomas and Margaret Paczkowski 12
[2MASS 18452610+4231055] The Confirmation of
Three Faint Variable Stars and the Observation
of Eleven Others in the Vicinity of Kepler-8b by the
Lookout Observatory
  Neil Thomas and Margaret Paczkowski 12
[2MASS 18465788+4156020] The Confirmation of
Three Faint Variable Stars and the Observation
of Eleven Others in the Vicinity of Kepler-8b by the
Lookout Observatory
  Neil Thomas and Margaret Paczkowski 12
[2MASS J20560314+3145505] Erratum: Four New
Variable Stars in the Field of KELT-16
  Daniel J. Brossard and Ronald H. Kaitchuck 276

[2MASS J20560314+3145505] Four New Variable
Stars in the Field of KELT-16
    Daniel J. Brossard and Ronald H. Kaitchuck   24

[2MASS J20564622+3138394] Erratum: Four New
Variable Stars in the Field of KELT-16
    Daniel J. Brossard and Ronald H. Kaitchuck   276

[2MASS J20564622+3138394] Four New Variable
Stars in the Field of KELT-16
    Daniel J. Brossard and Ronald H. Kaitchuck   24

[2MASS J20565617+3131253] Erratum: Four New
Variable Stars in the Field of KELT-16
    Daniel J. Brossard and Ronald H. Kaitchuck   276

[2MASS J20565617+3131253] Four New Variable
Stars in the Field of KELT-16
    Daniel J. Brossard and Ronald H. Kaitchuck   24

[ASASSN-V J002217.84-720612.7] Pulsating Red
Giants in a Globular Cluster: 47 Tucanae
    John R. Percy and Prateek Gupta   209

[ASASSN-V J002235.85-721110.9] Pulsating Red
Giants in a Globular Cluster: 47 Tucanae
    John R. Percy and Prateek Gupta   209

[ASASSN-V J002258.50-720656.3] Pulsating
Red Giants in a Globular Cluster: 47 Tucanae
    John R. Percy and Prateek Gupta   209

[ASASSN-V J002307.35-720029.8] Pulsating
Red Giants in a Globular Cluster: 47 Tucanae
    John R. Percy and Prateek Gupta   209

[ASASSN-V J002330.09-722236.3] Pulsating
Red Giants in a Globular Cluster: 47 Tucanae
    John R. Percy and Prateek Gupta   209

[ASASSN-V J002355.01-715729.7] Pulsating
Red Giants in a Globular Cluster: 47 Tucanae
    John R. Percy and Prateek Gupta   209

[ASASSN-V J002422.81-715329.0] Pulsating
Red Giants in a Globular Cluster: 47 Tucanae
    John R. Percy and Prateek Gupta   209

[ASASSN-V J002452.03-715611.0] Pulsating
Red Giants in a Globular Cluster: 47 Tucanae
    John R. Percy and Prateek Gupta   209

[ASASSN-V J002503.68-720931.8] Pulsating
Red Giants in a Globular Cluster: 47 Tucanae
    John R. Percy and Prateek Gupta   209

[ASASSN-V J002509.10-720215.3] Pulsating
Red Giants in a Globular Cluster: 47 Tucanae
    John R. Percy and Prateek Gupta   209

[ASASSN-V J002516.00-720355.0] Pulsating
Red Giants in a Globular Cluster: 47 Tucanae
    John R. Percy and Prateek Gupta   209

[ASASSN-V J002522.94-721105.1] Pulsating
Red Giants in a Globular Cluster: 47 Tucanae
    John R. Percy and Prateek Gupta   209

[ASASSN-V J112124.71-522143.6] Discovery of
Romanov V20, an Algol-Type Eclipsing Binary in the
Constellation Centaurus, by Means of Data Mining
    Filipp Dmitrievich Romanov   130

[ASASSN-V J184116.40+421342] The Confirmation
of Three Faint Variable Stars and the Observation
of Eleven Others in the Vicinity of Kepler-8b by the
Lookout Observatory
    Neil Thomas and Margaret Paczkowski   12

[ASASSN-V J205552.88+314615.9] Erratum: Four
New Variable Stars in the Field of KELT-16
    Daniel J. Brossard and Ronald H. Kaitchuck   276

[ASASSN-V J205552.88+314615.9] Four New
Variable Stars in the Field of KELT-16
    Daniel J. Brossard and Ronald H. Kaitchuck   24

[ASASSN-V J205658.12+314215.9] Erratum: Four
New Variable Stars in the Field of KELT-16
    Daniel J. Brossard and Ronald H. Kaitchuck   276

[ASASSN-V J205658.12+314215.9] Four New
Variable Stars in the Field of KELT-16
    Daniel J. Brossard and Ronald H. Kaitchuck   24

[Betelgeuse] Daylight Photometry of Bright Stars
—Observations of Betelgeuse at Solar Conjunction
    Otmar Nickel and Tom Calderwood   269

[CSS_J184816.3+414748] The Confirmation of Three
Faint Variable Stars and the Observation of Eleven
Others in the Vicinity of Kepler-8b by the
Lookout Observatory
    Neil Thomas and Margaret Paczkowski   12

[Gaia DR2 1864883699097368448] Erratum: Four
New Variable Stars in the Field of KELT-16
    Daniel J. Brossard and Ronald H. Kaitchuck   276

[Gaia DR2 1864883699097368448] Four New
Variable Stars in the Field of KELT-16
    Daniel J. Brossard and Ronald H. Kaitchuck   24

[GSC 2624-0941] CCD Photometry, Light Curve
Modeling, and Period Study of GSC 2624-0941,
a Totally Eclipsing Overcontact Binary System
    Kevin B. Alton and John C. Downing   214

[HD 121620] Retraction of and Re-analysis of the
Data from "HD 121620: A Previously Unreported
Variable Star with Unusual Properties"
    Roy A. Axelsen   197

[IGR J19552+0044] An Analysis of X-Ray Hardness
Ratios between Asynchronous and Non-
Asynchronous Polars
    Eric Masington *et al.*   149

[KIC 6836140] The Confirmation of Three Faint
Variable Stars and the Observation of Eleven Others in
the Vicinity of Kepler-8b by the Lookout Observatory
    Neil Thomas and Margaret Paczkowski   12

[KIC 6836820] The Confirmation of Three Faint
Variable Stars and the Observation of Eleven Others in
the Vicinity of Kepler-8b by the Lookout Observatory
    Neil Thomas and Margaret Paczkowski   12

[KIC 7173910] The Confirmation of Three Faint
Variable Stars and the Observation of Eleven Others in
the Vicinity of Kepler-8b by the Lookout Observatory
    Neil Thomas and Margaret Paczkowski   12

[KIC 7176440] The Confirmation of Three Faint
 Variable Stars and the Observation of Eleven Others in
 the Vicinity of Kepler-8b by the Lookout Observatory
    Neil Thomas and Margaret Paczkowski   12
[MarSEC_V13] The Confirmation of Three Faint
 Variable Stars and the Observation of Eleven Others in
 the Vicinity of Kepler-8b by the Lookout Observatory
    Neil Thomas and Margaret Paczkowski   12
[N Cas 2020] The Photometric Period of
 V1391 Cassiopeiae (Nova Cas 2020)
    Richard E. Schmidt   95
[N Her 2021] The Photometric Period of
 V1674 Herculis (Nova Her 2021)
    Richard E. Schmidt, Sergei Yu. Shugarov, and
    Marina D. Afonina   257
[N Per 2020] The Photometric Period of V1112 Persei
 (Nova Per 2020)
    Richard E. Schmidt   99
[N Vul 2021] The Photometric Period of
 V606 Vulpeculae (Nova Vul 2021)
    Richard E. Schmidt   261
[NGC 1904] Characterization of NGC 5272,
 NGC 1904, NGC 3201, and Terzan 3
    Paul Hamrick, Avni Bansal, and Kalée Tock   192
[NGC 3201] Characterization of NGC 5272,
 NGC 1904, NGC 3201, and Terzan 3
    Paul Hamrick, Avni Bansal, and Kalée Tock   192
[NGC 5272] Characterization of NGC 5272,
 NGC 1904, NGC 3201, and Terzan 3
    Paul Hamrick, Avni Bansal, and Kalée Tock   192
[NGC 6341] Differential Photometry of Eclipsing
 Binary System V798 Her in Globular Cluster
 NGC 6341 (Abstract)
    Khola Anees, Shaukat Naaman Goderya, and
    Fazeel Mahmood Khan   116
[NSVS 8092487] CCD Photometry, Light Curve
 Modeling and Period Study of V1073 Herculis,
 a Totally Eclipsing Overcontact Binary System
    Kevin B. Alton and John C. Downing   201
[Romanov V20] Discoveries of Variable Stars by
 Amateur Astronomers Using Data Mining on the
 Example of Eclipsing Binary Romanov V20 (Abstract)
    Filipp Romanov   115
[Romanov V20] Discovery of Romanov V20, an
 Algol-Type Eclipsing Binary in the Constellation
 Centaurus, by Means of Data Mining
    Filipp Dmitrievich Romanov   130
[ROTSE1 J184234.00+420947.9] The Confirmation of
 Three Faint Variable Stars and the Observation
 of Eleven Others in the Vicinity of Kepler-8b by the
 Lookout Observatory
    Neil Thomas and Margaret Paczkowski   12
[ROTSE1 J184517.00+424010.4] The Confirmation of
 Three Faint Variable Stars and the Observation
 of Eleven Others in the Vicinity of Kepler-8b by the
 Lookout Observatory
    Neil Thomas and Margaret Paczkowski   12
[RX J0838.7-2827] An Analysis of X-Ray Hardness
 Ratios between Asynchronous and Non-
 Asynchronous Polars
    Eric Masington et al.   149
[Sun] Transits, Spots, and Eclipses: The Sun's Unique
 Role in Outreach
    Kristine Larsen   90
[Swift J231930.4+261517] An Analysis of X-Ray
 Hardness Ratios between Asynchronous and
 Non-Asynchronous Polars
    Eric Masington et al.   149
[TCP J18573095+1653396] The Photometric
 Period of V1674 Herculis (Nova Her 2021)
    Richard E. Schmidt, Sergei Yu. Shugarov, and
    Marina D. Afonina   257
[Terzan 3] Characterization of NGC 5272, NGC 1904,
 NGC 3201, and Terzan 3
    Paul Hamrick, Avni Bansal, and Kalée Tock   192
[TYC 2688-139-1] Four New Variable Stars in the
 Field of KELT-16
    Daniel J. Brossard and Ronald H. Kaitchuck   24
[WASP-50 b] 25 New Light Curves and Updated
 Ephemeris through Analysis of Exoplanet WASP-50 b
 with EXOTIC
    Ramy Mizrachi et al.   186
[WISE J184227.5+422724] The Confirmation of Three
 Faint Variable Stars and the Observation of Eleven
 Others in the Vicinity of Kepler-8b by the
 Lookout Observatory
    Neil Thomas and Margaret Paczkowski   12
[ZTF J205627.42+315322.4] Erratum: Four New
 Variable Stars in the Field of KELT-16
    Daniel J. Brossard and Ronald H. Kaitchuck   276
[ZTF J205627.42+315322.4] Four New Variable Stars
 in the Field of KELT-16
    Daniel J. Brossard and Ronald H. Kaitchuck   24
[ZTF J205733.78+314612.6] Erratum: Four New
 Variable Stars in the Field of KELT-16
    Daniel J. Brossard and Ronald H. Kaitchuck   276
[ZTF J205733.78+314612.6] Four New Variable Stars
 in the Field of KELT-16
    Daniel J. Brossard and Ronald H. Kaitchuck   24

WHITE DWARFS
CCD Photometry, Light Curve Modeling, and Period
 Study of GSC 2624-0941, a Totally Eclipsing
 Overcontact Binary System
    Kevin B. Alton and John C. Downing   214
Daylight Photometry of Bright Stars—
 Observations of Betelgeuse at Solar Conjunction
    Otmar Nickel and Tom Calderwood   269
A Photometric Study of the Eclipsing Binary
 LO Ursae Majoris
    Edward J. Michaels   221

YSO—YOUNG STELLAR OBJECTS
Light Curve Analysis of 185 YSOs: New Periods
 Discovered for 9 Stars
    Joshua R. Hamilton   49

NOTES

Made in the USA
Middletown, DE
10 January 2022

58376874R00104